W9-CHP-442

Robert M. Poole was executive editor of *National Geographic*. He has written for *The New York Times, The Washington Post,* and *Preservation,* and is a contrtibuting editor at *Smithsonian Magazine*. He lives in Virginia.

◆

Praise for *Explorers House*

"At last, someone has ripped away the veneer of spun sugar and told the truth about the National Geographic Society. And who better for the job than Bob Poole—twenty-year veteran insider, superb writer and diligent researcher? *Explorers House* is fascinating and important, an honest account of the rise and fall of an American icon." —Peter Benchley

"Robert M. Poole's wonderful history of *National Geographic* teems with delights—Whistlerian portraits of the eccentric amateurs who made the magazine, stirring tales from the Brigadoon that is its headquarters, even moments of darkness resulting from power struggles, broken friendships, and the recent intrusion into the sunny kingdom of realists with MBAs who do not understand that the *Geographic*'s stock-in-trade is its incurable enthusiasm for the world and all that is in it. Poole's deft pen, insider's knowledge, and keen eye for the telling detail make more than a century of the magazine's history fly by. Readers will want to keep this book just as the *Geographic*'s forty million monthly readers save every issue of the magazine, as something to go back to, something to trust, something for their children." —Charles McCarry

"Poole, recently retired as *National Geographic*'s executive editor, maintains objectivity without sacrificing scope and detail; the book has been written with all the painstaking care you'd expect." —*Publishers Weekly*

"Double doses of exposé . . . an insightful read in what is an enormously entertaining book." —*Los Angeles Times*

"Poole is not . . . blinkered by his association with the publication and writes candidly. . . . Combined with his insight into what has editorially made the yellow-bordered icon one of the most successful in the magazine industry, that frankness ensures the work against becoming a dreary institutional history." —*Booklist*

"Poole guides readers with special acumen through the mazelike backroom politics (in) a natty tour of the society's house: closets, skeletons, and all." —*Kirkus Reviews*

"A richly textured and delightful history of the National Geographic Society, its magazine, and the Grosvenor family . . . [with] candid descriptions [that] render people, places, and events in vivid colors. . . . A first-class account." —*The Toledo Blade*

"Robert M. Poole's *Explorers House* . . . seizes a topic you might take for granted and provides round factual analysis, rich in anecdotes, thick with surprising detail. It conveys it all through a smooth narrative tone that, while clearly loving, doesn't shy away from the darker side of its subject. Above all, it leaves you with wonderful stories and unforgettable images. . . . Poole, himself an editor at the *Geographic* for decades, has not only the insider's understanding of layers of history—and intrigue—but the *Geographic*'s flair for making a complex topic clear, readable and memorable." —*The News & Observer*

EXPLORERS HOUSE

NATIONAL GEOGRAPHIC AND THE WORLD IT MADE

ROBERT M. POOLE

PENGUIN BOOKS

PENGUIN BOOKS

Published by the Penguin Group

Penguin Group (USA) Inc., 375 Hudson Street, New York, New York 10014, U.S.A.

Penguin Group (Canada), 90 Eglinton Avenue East, Suite 700, Toronto,
Ontario, Canada M4P 2Y3 (a division of Pearson Penguin Canada Inc.)

Penguin Books Ltd, 80 Strand, London WC2R 0RL, England

Penguin Ireland, 25 St Stephen's Green, Dublin 2, Ireland (a division of Penguin Books Ltd)

Penguin Group (Australia), 250 Camberwell Road, Camberwell,
Victoria 3124, Australia (a division of Pearson Australia Group Pty Ltd)

Penguin Books India Pvt Ltd, 11 Community Centre, Panchsheel Park, New Delhi – 110 017, India

Penguin Group (NZ), cnr Airborne and Rosedale Roads, Albany,
Auckland 1310, New Zealand (a division of Pearson New Zealand Ltd)

Penguin Books (South Africa) (Pty) Ltd, 24 Sturdee Avenue,
Rosebank, Johannesburg 2196, South Africa

Penguin Books Ltd, Registered Offices:
80 Strand, London WC2R 0RL, England

First published in the United States of America by The Penguin Press,
a member of Penguin Group (USA) Inc. 2004
Published in Penguin Books 2006

3   5   7   9   10   8   6   4   2

Frontispiece illustrations (*top to bottom*): Photos by E. C. Erdis, Yale Peabody Museum; Horace M.
Albright, Grosvenor Collection, Library of Congress; © Hulton-Deutsche Collection/Corbis;
Grosvenor Collection, Library of Congress; Leet Brothers, © National Geographic Society

Additional illustration credits appear on page 343.

THE LIBRARY OF CONGRESS HAS CATALOGED THE HARDCOVER EDITION AS FOLLOWS:
Poole, Robert M.
Explorers house : National Geographic and the world it made / Robert Poole.
p.   cm.
Includes index.
ISBN 1-59420-032-7 (hc.)
ISBN 0 14 30.3593 2 (pbk.)
1. National Geographic Society (U.S.)   I. Title.
G3.P66 2004
910'.6'073—dc22   2004050536

Printed in the United States of America
Designed by Marysarah Quinn

THE WORLD AND ALL THAT IS IN IT is our theme, and if we can't find anything to interest ordinary people in that subject we better shut up shop. . . .

—Alexander Graham Bell to
Gilbert H. Grosvenor, 1900

CONTENTS

Hubbard

Gardiner Greene Hubbard —— m. —— Gertrude McCurdy
(1822–97)
PRESIDENT,
NATIONAL GEOGRAPHIC SOCIETY
(1827–1909)

Robert Hubbard
(1847–49)

Gertrude Hubbard
(1849–86)
m. Maurice Grossman
(1843–84)
1 CHILD

Roberta Hubbard
(1859–85)
m. Charles J. Bell
(C. 1860–1929)
2 CHILDREN

Grace Hubbard
(1861–1948)
m. (1887) Charles J. Bell
(C. 1860–1929)
3 CHILDREN

Mabel Gardiner Hubbard
(1857–1923)

Edward Bell
(B. & D. 1881)

Robert Bell
(B. & D. 1883)

Marian Hubbard "Daisy" Bell
(1880–1962)
m. David Grandison Fairchild
(1869–1954)
3 CHILDREN

Elsie May Bell
(1878–1964)

Gertrude Hubbard Grosvenor
(1903–66)
m. 1ST Paxton Blair
(1892–1974)
3 CHILDREN
m. 2ND Samuel A. Gayley
(1899–1959)

Mabel "Dr. Mabel"
Harlakenden Grosvenor
(1905–)

Lilian Waters Grosvenor
(1907–85)
m. 1ST Cabot Coville
(1902–87)
2 CHILDREN
m. 2ND Joseph M. Jones
(C. 1908–90)

Melville Bell Grosvenor —— m. 1ST —— Helen North Rowland
(1901–82)
EDITOR, NATIONAL GEOGRAPHIC MAGAZINE
PRESIDENT, NATIONAL GEOGRAPHIC SOCIETY
(1903–85)

Alexander Graham Bell Grosvenor
(1927–78)
m. Marcia Braman
(C. 1930–2002)
2 CHILDREN

Helen Rowland
"Teeny" Grosvenor
(1925–88)
m. 1ST Robert C. Watson
(1924–91)
2 CHILDREN
m. 2ND Richard Lemmerman
(UNK.)

Gilbert Melville "Gil" —— m. 1S
Grosvenor
(1931–)
EDITOR,
NATIONAL GEOGRAPHIC MAGAZINE
PRESIDENT,
NATIONAL GEOGRAPHIC SOCIETY

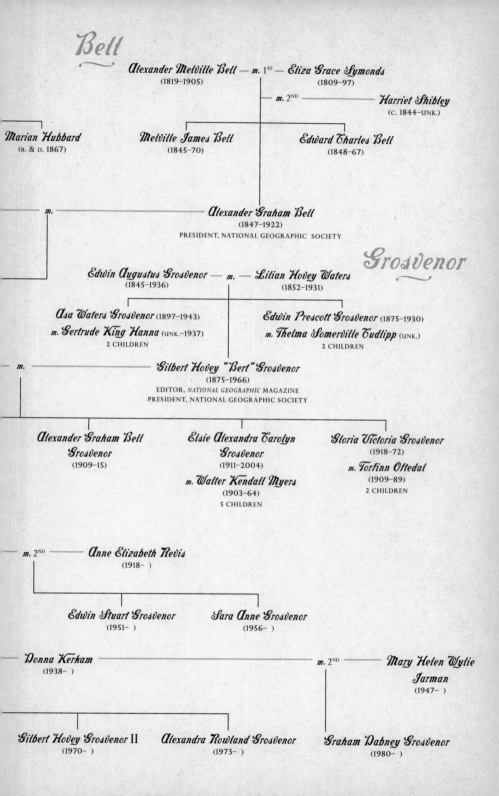

Bell

Alexander Melville Bell — m. 1st — Eliza Grace Symonds
(1819–1905)   (1809–97)

m. 2nd ——— Harriet Shibley
(c. 1844–unk.)

Marian Hubbard
(b. & d. 1867)

Melville James Bell
(1845–70)

Edward Charles Bell
(1848–67)

m. ——— Alexander Graham Bell
(1847–1922)
PRESIDENT, NATIONAL GEOGRAPHIC SOCIETY

Grosvenor

Edwin Augustus Grosvenor — m. — Lilian Hovey Waters
(1845–1936)   (1852–1931)

Asa Waters Grosvenor (1897–1943)
m. Gertrude King Hanna (unk.–1937)
2 CHILDREN

Edwin Prescott Grosvenor (1875–1930)
m. Thelma Somerville Cudlipp (unk.)
2 CHILDREN

m. ——— Gilbert Hovey "Bert" Grosvenor
(1875–1966)
EDITOR, NATIONAL GEOGRAPHIC MAGAZINE
PRESIDENT, NATIONAL GEOGRAPHIC SOCIETY

Alexander Graham Bell
Grosvenor
(1909–15)

Elsie Alexandra Carolyn
Grosvenor
(1911–2004)
m. Walter Kendall Myers
(1903–64)
5 CHILDREN

Gloria Victoria Grosvenor
(1918–72)
m. Torfinn Oftedal
(1909–89)
2 CHILDREN

m. 2nd ——— Anne Elizabeth Revis
(1918–)

Edwin Stuart Grosvenor
(1951–)

Sara Anne Grosvenor
(1956–)

Donna Kerkam
(1938–)

m. 2nd ——— Mary Helen Wylie
Jarman
(1947–)

Gilbert Hovey Grosvenor II
(1970–)

Alexandra Rowland Grosvenor
(1973–)

Graham Dabney Grosvenor
(1980–)

PROLOGUE

JACQUES COUSTEAU AND GILBERT M. GROSVENOR
AT CENTENNIAL DINNER, 1988.

Jacques-Yves Cousteau charmed his way through the crowd of VIPs in the
Maryland Room, past Sir Edmund Hillary and Jane Goodall, past John Glenn
and Richard Leakey, past white-gloved waiters navigating a packed ocean of
black ties and sequined gowns. Down the hall of the Sheraton Washington Hotel, 1,400 more guests trickled through Secret Service checkpoints, fidgeting in
line before metal detectors. Just beyond them, through doors opening onto a
softly lit ballroom, the Marine Corps Band, resplendent in a show of red wool
and gleaming brass, made its final checks for the performance ahead. A sense of
expectancy charged the air.

It was November 17, 1988, and the crowds were gathering in the nation's
capital to celebrate the National Geographic Society's centennial. The organization, in turn, was honoring fifteen noted explorers and researchers to whom

it had given grants or recognition at the start of their careers. This centennial dinner was the last in a year full of parties, and it would be the most lavish, with a salon orchestra, awards ceremonies, and a toast from the president-elect, George H. W. Bush.

From its origins as an obscure learned organization in 1888, the National Geographic Society had evolved into the world's largest nonprofit educational and scientific institution. Its official journal, *National Geographic* magazine, was read by more than forty million people, and its Emmy award–winning documentaries had become a popular fixture on public television. The organization's broad reach and ecumenical appeal took in movie stars, inmates, farmers, fishermen, and more than a few world leaders. Emperor Akihito of Japan read the magazine, as did Danny Glover, Martha Stewart, the Mir of Hunza, and Oliver Sacks. The late Walt Disney and Rudolf Hess had both subscribed. Many admired the pictures, some the maps, and a few the text, but the writer Flannery O'Connor particularly liked the magazine's smell, which she found "a distinct unforgettable transcendent apotheotic . . . and very grave odor. Like no other mere magazine. If *Time* smelled like the *Nat'l. Geo.* there would be some excuse for its being printed."[1]

Cousteau made his way from the VIP reception to the ballroom, tossing his head to laugh at some private joke with Luis Marden, a dapper man whose gravelly voice rumbled with echoes of his native Boston. They were old friends, dating from 1955, when *National Geographic* magazine sent Marden, then a young staff writer and photographer, to cover the *Calypso* voyage that launched Cousteau's fame. Marden was something of a legend around the office, the quintessential *National Geographic* contributor. He lived in a house designed by Frank Lloyd Wright, spoke five languages, flew airplanes, and had his safari shirts tailored in Hong Kong. He recorded his spending in an official buff-colored booklet, which included a line for "Gifts to Natives." He had twice crossed the Atlantic in his own ketch, and posited a controversial new landing site for Christopher Columbus. He had unearthed the remains of the HMS *Bounty* off Pitcairn Island. He had discovered a species of sand flea (*Dolobrotus mardeni*) and a new orchid (*Epistephium mardeni*), both named for him.

Marden was one golden child in an extended family of men and women who roamed the world, climbing mountains, photographing obscure tribes, visiting shamans, collecting beetles, and interviewing dictators for the National Geographic Society and its members. Other contributors included generals and admirals, a convicted felon, a former Baptist missionary, a submarine captain, several Supreme Court justices, two astronauts, several presidents, and a number of social misfits. One of the latter, a botanist who packed a Colt .45, wept when he talked about certain plants.

This eccentric—and often maddening—parade of characters might have marched off into oblivion if not for the influence of one family—the staid and steady Grosvenors—who set high standards, recruited new talent, paid the bills, and kept things moving forward. A proud and private New England clan, the Grosvenor family (pronounced "Grov-uh-nor") was in its fifth generation of guiding the National Geographic Society at the time of the centennial. Although Grosvenors had been at the top of *National Geographic* magazine's masthead since William McKinley was president, they had been content to bask in the glow of the heroes and heroines they created. Unlike other prominent media dynasties—say the Sulzbergers, Luces, or Binghams—the Grosvenors thrived under protective coloration like the moths or leafhoppers celebrated in their television shows and magazine layouts.

At the head table, Sir Edmund Hillary, looking ruddy and sturdy, chatted with Barry Bishop, a barrel-shaped man with a swooping handlebar mustache and a slight limp he earned in 1963, when the first Americans topped Everest, following Hillary by a decade. Then a young photographer for *National Geographic* magazine, Bishop was now vice chairman of the National Geographic Society's prestigious Research and Exploration Committee, which handed out grants for people like Jane Goodall, who sat just down the table from Hillary.

Goodall, whose research on chimps changed ideas about animal behavior and conservation, was placed at a discreet distance from Mary Leakey, the matriarch of the remarkable family of Kenyan paleontologists. Mrs. Leakey had strong feelings about the women her late husband, Louis, recruited. Two generations of Leakeys had been sponsored by the National Geographic Society, which brought their discoveries to the world's attention. Near the center of the head table, Senator John Glenn crinkled his freckled face into the familiar A-OK grin, acknowledging a well-wisher in the audience below; he had carried a National Geographic flag on his historic space flight. Bob Ballard, the reserve navy commander who had discovered the *Titanic,* was there, too, giving credit to Cousteau for his own interest in ocean exploration. With the help of National Geographic grants, Ballard had extended what Cousteau began, taking remote diving vessels to depths that unprotected humans could never endure.

The table fairly sagged under the weight of all the egos, which only served to emphasize the seeming mildness of Gilbert M. Grosvenor, the president and chairman of the National Geographic Society. A reserved, unassuming man in his late fifties, he sat at center stage wearing oversized glasses and a lopsided grin. He endured evenings like this one, but he preferred solitude or a quiet supper with friends, with whom he could be warm and witty. At the office he was known for studying his shoes during long, silent elevator rides with colleagues.

Like Bush, who sat beside him, Gil Grosvenor had been raised in privileged surroundings. He had attended Deerfield Academy and Yale and sailed through the long vacations at Baddeck, Nova Scotia, where his family had summered for years. Gil worked hard at staying unpretentious, favoring inexpensive suits, traveling cheaply, and bragging about boarding a plane with a peanut-butter-and-jelly sandwich in his pocket. Although this dinner was an affirmation of National Geographic's staying power, Gil worried about the legacy he had inherited from his father, Melville Bell Grosvenor, who had inherited it from *his* father, Gilbert H. "Bert" Grosvenor, who had inherited it from *his* father-in-law, Alexander Graham Bell, who had inherited it from *his* father-in-law, Gardiner Greene Hubbard. Now it was Gil's turn and he was determined to protect this heritage, which had its origins in another Washington gathering a century before.

The inaugural dinner had been organized by Gil's great-great-grandfather, Gardiner Greene Hubbard, a lawyer and entrepreneur. Although not a scientist himself, Hubbard was a shrewd, interested amateur who helped bankroll the publication of *Science* magazine, served as a regent of the Smithsonian Institution, and financed surveying expeditions to Alaska. Now he and some friends had a new idea, which he proposed to discuss with thirty-two other Washingtonians.

It was just before eight o'clock on a damp January night when the carriages began to draw up to the Cosmos Club, on Lafayette Square near the White House. Hubbard's guests were prominent men of action and learning who saw scientific endeavor as a way to build a young nation. The crowd included people like John Wesley Powell, the indestructible Civil War veteran who lost an arm at Shiloh only to risk his life again leading a crew on the first survey of the Colorado River, a harrowing nine-hundred-mile voyage through brown rapids and vertiginous canyons. He was joined by Edward Everett Hayden, a naval officer and meteorologist who had lost a leg in an avalanche while exploring the West for the U.S. Geological Survey.

They sat around a gleaming mahogany table with William Healey Dall, the naturalist for whom the Dall sheep was named; Arthur Powell Davis, the chief hydrographer for the Panama Canal; John Russell Bartlett, an authority on the Gulf Stream; and Brigadier General Adolphus W. Greely, chief signal officer of the army. Greely carried himself with the dignity one would expect of a man who not only was a Civil War hero but who had also led a government expedition farther north than any previous Arctic explorer. Stranded, Greely and a few starving survivors had been rescued by Winfield Scott Schley, a naval officer who would later achieve distinction in the Spanish-American War. Schley, too, joined the others at the table.[2]

The proposal Hubbard put before the distinguished group that night was deceptively simple and straightforward: to establish a society of kindred souls who would meet regularly in Washington to share ideas about geography, which by Hubbard's broad definition took in all of life. This idea found a receptive audience in his guests at the Cosmos Club, which had itself come into existence just ten years before, with a wave of scientific and self-improvement associations that flourished in the capital after the war.

Before they scraped back their chairs and went home that night, the group had agreed to launch the National Geographic Society "for the increase and diffusion of geographic knowledge." Their guiding spirit, Gardiner Greene Hubbard, would become the organization's first president. In his first address to the newly formed National Geographic Society, Hubbard said, "I am not a scientific man. I possess only the same general interest in the subject of geography that should be felt by every educated man. By my election you notify the public that the membership of our Society will not be confined to professional geographers, but will include that large number who, like myself, desire to promote special researches by others, and to diffuse the knowledge so gained among men, so that we may all know more of the world upon which we live."[3] The salient features of the National Geographic Society's character were already there in Hubbard's words—the appeal to nonspecialists, the urge for self-improvement, the desire to promote useful research, the sense of wonder for the world. All would contribute to the broad appeal on which the organization grew.

When Gardiner Hubbard died in 1897, family members prevailed upon his son-in-law Alexander Graham Bell to step in as the new president of the National Geographic Society. The magazine soon reflected the inventor's catholic interests. He saw prospective stories everywhere and bombarded *National Geographic* magazine with suggestions for pieces on waltzing mice, China's influence in the Philippines, the meteorites of Canyon Diablo, breeding frogs, polar exploration, Spanish earthquakes, yellow fever, waves, and auroras. He wanted maps to help readers follow events in the news. And he wanted photographs "of the dynamical order" to draw readers into the magazine.[4]

"THE WORLD AND ALL THAT IS IN IT is our theme," Bell wrote in 1900 to Bert Grosvenor, the young editor he had installed at *National Geographic* magazine, "and if we can't find anything to interest ordinary people in that subject we better shut up shop and become a strict, technical, scientific journal for high class geographers and geological experts."[5] For that first Grosvenor, fielding proposals from this erratic genius must have felt like facing a tsunami with a teaspoon. But Bert Grosvenor had a capacity for hard work, a reverence for Bell, and an unshakable confidence that made him willing to try things others had not. He was the action officer for Bell the dreamer, and together they made the National Geographic Society into a unique institution.

To all outward appearances, their descendant Gil Grosvenor had proven himself an able heir—this despite a chorus of doubters who had wondered if the family tradition had run too long when he took over the editor's job in 1970. If such doubts persisted, they did not show at the centennial dinner. Gil was the picture of confidence, enjoying his time on stage with Ted Koppel and the magazine's honorees, taking his place in the spotlight like a man who belonged there. He looked at peace with himself, comfortable at the top of an admired institution cruising toward its second century with few signs of trouble.

The organization had recently paid $34 million in cash for a new headquarters in Washington. The building, shaped like a Mayan temple, was a monument to the National Geographic Society's future, with room to produce new television programs and several new magazines. But the past was never far away; in the domed ceiling of the pink granite lobby a river of stars swirled in a painted night sky, just as they had on January 13, 1888, the night of National Geographic's birth.

In all the years since—through two world wars and the Great Depression, through business setbacks and family travails, through the shifting tastes and challenges to the clan's dominance—those stars had shone good fortune on the National Geographic Society and its founding family. Their fates had been joined since 1873, when a girl went out for a walk in Boston and met a genius named Alexander Graham Bell.

CHAPTER ONE
ALEC AND MABEL

ALEXANDER GRAHAM BELL AND MABEL HUBBARD BELL
WITH DAUGHTERS ELSIE (*LEFT*) AND DAISY, 1885.

Y ou would not have known that Mabel Hubbard was deaf. On a crisp
October morning in 1873, she chatted with Mary True all the way up
Beacon Hill with hardly a pause, their conversation flowing back and forth as
they climbed through the smell of crunching leaves, up past the stone walls of
the old Boston reservoir, up toward the Old Granary Burying Ground. A tall,
thin girl just shy of sixteen, Mabel had lost her hearing at age five in a struggle
with scarlet fever. She kept her eyes fixed on her friend's lips, reading them
expertly.[1]

Mary was saying something about a young teacher named Alexander Graham Bell who had an office just up the street. When he was not teaching at Boston University's School of Oratory he gave private lessons in speech. Mary thought that perhaps Professor Bell could help Mabel improve her articulation, which was perfectly understandable but delivered in the slightly nasal monotone characteristic of those who had been deaf when they learned to talk.[2]

Mabel knew of this Professor Bell. She thought he was a crank. But since they were in the neighborhood and she wanted to be polite, she agreed to stop at 18 Beacon Street. The two young friends climbed the stairs, entered a dark green reception room, and settled in to wait for the professor. When he appeared, Mabel was frankly unimpressed. To her young eye, this twenty-six-year-old Scotsman looked impossibly old and unfashionable, with a thatch of unruly hair that stood up like a pileated woodpecker's crest; Bell nervously tried to calm his cowlick by combing his fingers through it, which did little to improve Mabel's opinion of him.[3]

"I did not like him," Mabel concluded. "To one accustomed to the dainty neatness of Harvard students, he seemed hardly a gentleman. He was tall and dark, with jet black hair and eyes, but dressed badly and carelessly in an old-fashioned suit of black broadcloth, making his hair look shiny. . . . I could never marry such a man!"[4] Even so, Mabel thought that she might learn something from him, so she agreed to a few lessons.

Over the next few weeks, she went daily to Mr. Bell's classroom. Almost against her will, she found herself looking forward to the sessions, where Bell chalked sketches of lips and mouths on the blackboard to show how she could speak more clearly. He opened his mouth and let her look inside to see how he shaped each word. The professor for his part was impressed by Mabel's total concentration, the way her eyes stayed locked on his face. She was a creative listener who quickly grasped not only the content but the spirit of a conversation.[5]

Mabel's view of Bell began to soften. He was "so quick, so enthusiastic, so compelling, I had whether I would or no to follow all he said and tax my brains to respond as he desired," Mabel wrote her mother that autumn. Infinitely patient, enthusiastic, and solicitous of his young student, Bell flattered Mabel.[6] "Mr. Bell said today my voice was naturally sweet," Mabel wrote to her mother. "Think of that! If I can only learn to use it properly, perhaps I will yet rival you in sweetness of voice. He continues pleased with me. . . . I enjoy my lessons very much and am glad you want me to stay. Everyone says it would be a pity to go away just as I am really trying to improve."[7]

For most of her young life Mabel Hubbard had been trying to improve, fighting against the barriers imposed upon her by deafness. She was able to do so because she was smart, adaptable, and full of confidence, but she was also lucky. Her family, from a prominent New England line, had refused to send Mabel away to an asylum, where she would have been confined with other deaf students and taught only sign language, the standard treatment in her day. Such an education was fine for communicating within the realm of the deaf but it did very little to integrate its graduates into the world at large, and Mabel's family had determined that their daughter would live fully in society. They spoke to Mabel as if she could hear so that she soon learned to read lips, and they hired private teachers for her. They traveled with her to Austria, where a leading instructor helped to perfect her lip reading. By the time she returned to the United States and began lessons with Bell, Mabel Hubbard was functioning smoothly in the world of the hearing. When it came to their daughter's interests, Gertrude and Gardiner Greene Hubbard never hesitated to use their prominence, which was considerable.

Hubbard traced his lineage to the 1600s and the European settlement of New England. John Haynes, the first governor of Connecticut, was an ancestor, as was Sir John Leverett, governor of the Massachusetts Bay Colony in 1673. Another ancestor had been an early president of Harvard College, and Gardiner's father, Samuel Hubbard, served as a judge on the Supreme Judicial Court of Massachusetts. Gardiner himself, after studying at Dartmouth and Harvard Law School, followed his father into the law, opening a lucrative practice of his own in 1848.[8]

Known for his frank manner, which often strayed into bluntness, Hubbard made for an imposing figure, standing just under six feet tall, with a high forehead, a Roman nose, a massive beard, and eyes variously described as hazel or black.[9] Perhaps out of boredom or an excess of energy, he found himself unable to focus on a conventional career. He dived into a range of business interests and public causes that might have drowned one less driven. He beat competitors to start the first rail trolley linking Cambridge and Boston in 1856; he brought the first waterworks to Cambridge; he founded the Cambridge Gas Light Company; he helped start the Boston Rubber Shoe Company and served as an officer there;[10] he invested in Nova Scotia coal mines and a Washington State ranch. He also explored a scheme to corner the Boston market on pimento liquor, which came to naught,[11] as did his brash plot to nationalize the country's telegraphic system, then under the monopolistic control of Western Union. Failure did little to discourage Gardiner Greene Hubbard. Even when he did not get his way, this stiff, stubborn aristocrat always behaved as if he would—and should—prevail.

Restless confidence was a trait Hubbard shared with Alexander Graham Bell, whom Hubbard had met through his contacts in New England's thriving educational community. Hubbard had been impressed by this bright, earnest young man, whose pursuits seemed as varied as his own. Bell hurried through his days, teaching students, giving lectures, and sorting through the many subjects crowding his mind, which flitted from vocal physiology to music to the challenges of flight, a subject he often considered when he glimpsed seagulls coasting above the Boston Common. "I cannot manage all that I lay myself out to do," Bell said. "I rush from one thing to another and before I know it the day is gone."[12] By night Bell retreated to the small laboratory he kept in Salem, where he was feverishly taking notes and experimenting with designs for a multiple telegraph, so named because it would allow multiple messages to be carried on a single telegraphic wire; if such a device could be made to work, it would greatly enhance the carrying capacity of the country's telegraphic system and bring great wealth to its inventor. After a long night in the lab, Bell would lock away his notes, sleep a few hours, and renew the cycle.

The professor stood a lanky six feet tall,[13] with a penetrating gaze that unnerved some people. With his thick black muttonchops and dark features, he might have been a buccaneer or an actor. His flair for the dramatic had been honed by intensive training in London, where his grandfather and namesake, Alexander Bell, had coached the youth in public speaking. The elder Bell, a prominent elocutionist and teacher, took young Alec in hand at age fourteen, taught him Latin, immersed him in Shakespeare and other essential literature, and scrubbed away all traces of the boy's native Edinburgh. When his grandfather was done, Alexander Graham Bell went out into the world with a polish and dignity remarkable in one so young. He could be compelling company. People were disarmed by his childlike enthusiasm, which might launch him into a Mohawk war dance in moments of happiness, or keep him banging away on the piano half the night. Just beneath that surface, though, was another Alec Bell, one who fought all his life against a preference for solitude, a tendency that threatened to cut him off from society.[14] Since his youth, spent in Scotland and London, he had moved with his parents to Ontario in 1870; from there he emigrated to Boston, where he quickly established himself as a gifted teacher of the deaf.

Sympathetic and sensitive, Alec knew how to draw deaf children into the world of the hearing. He invented games for them, helped them draw pictures to illustrate their thoughts, and taught them to spell and to read. He encouraged roomfuls of them to shout at the top of their lungs so that, even if they could not hear their own words, they could feel the power of their voices vibrating through the furniture.[15] These often difficult and misunderstood children had few allies as firm as Bell.

In Mabel Hubbard's case, Bell's devotion was developing into something more—a romantic interest in his student. After resisting these feelings for weeks, Bell, now twenty-seven, had to admit that he was in love with a pupil ten years his junior. He confessed this only to himself. He could not reveal this affection to one so young, nor could he continue teaching her, nor could he bear to tell her parents about his feelings. So, after less than a dozen lessons with Mabel, Bell passed her off to an assistant who continued the girl's instruction.[16] Surprised by this sudden shift in teaching assignments, Gertrude Hubbard asked for an explanation. Bell told her that there was nothing wrong with Mabel's work. She was a good student. "I have no fault whatever to find in her, and I will gladly supervise and be present at her lessons whenever possible," Bell said. "But it seems unwise for me to teach her personally at this time." This was the extent of his explanation, which seemed to satisfy Mrs. Hubbard; perhaps she sensed what was afoot.[17]

In any event, both she and her husband were pleased with the progress Mabel had made in Bell's classroom. They began to incorporate the bachelor professor into their busy family life at Cambridge, where he became a frequent dinner guest and enjoyed discussions with the professors who dropped by from Harvard and MIT. Bell spoke often with Gardiner Hubbard, with whom he shared interests in science and technology. Both had a particular obsession with telegraphic matters, which came up for discussion after tea on an October afternoon in 1874.

That evening, after the dishes and cups had been taken away, the family gathered around the piano. Bell, who had youthful aspirations as a concert pianist, settled onto the stool and began to play, probably a homey Scottish favorite such as "The Laird of the Cockpen" or "Loch Lomond."[18]

As the notes of the last tune faded, Bell suddenly turned and addressed Gardiner Hubbard, who had been trying to read in the corner. "Mr. Hubbard, sir, do you know that if I depress the forte pedal and sing 'do' into the piano, the proper note will answer me? Like this?" Alec bent forward and sang the note. The piano answered. Hubbard stopped reading.

Was Bell's gesture just a parlor trick? An off-duty professor's need to continue teaching? A bid to impress Mabel's father? We cannot know, but what happened next, whether contrived or spontaneous, would put Bell and Hubbard onto a common trajectory for the rest of their lives.

"And here's more," said Bell. "If two pianos in two different places were connected by a wire, and a note struck on one, the same note would respond on the other."

"And what value is there in that fact?" Hubbard asked.

This principle of sympathetic vibration, Bell explained, might be applied to the telegraphic system. If the pitch of tuning forks at either end of a wire could be set to exactly the same frequency, a person could send multiple messages over a single wire. Such a device, which Bell called the multiple or harmonic telegraph, would be of inordinate value to Western Union or other carriers. Bell described how he was experimenting with the principle in his workshop. Hubbard's face lit up and he made a decision on the spot.

"Mr. Bell," he said, "I believe you have a very sound idea there, and I am willing and ready to finance it and assist you in taking out the necessary patents."[19]

Within days, Bell and Hubbard shook hands as partners, joining a third named Thomas Sanders, a prominent Boston leather merchant. The two businessmen would support Bell's experiments and all three would share the profits. Bell hoped the money came soon. Mabel Hubbard could not know it yet, but he had determined to marry her.[20]

He kept his feelings secret and tried to focus on the multiple telegraph. Now that he had some financial backing, Bell moved his experiments from Salem to the cluttered offices of Charles Williams, Jr., at 109 Court Street in Boston, which provided electrical equipment for the city's flourishing community of technical wizards. In a corner of the attic where Thomas Edison had once worked, Bell could often be seen conferring with Thomas Watson, the laboratory assistant with whom he collaborated through the spring of 1875. It was here among the steam-powered lathes and squeaking pulleys that Bell offhandedly dropped one of his greatest ideas.[21]

"Watson," he said, "if I can get a mechanism which will make a current of electricity vary in its intensity, as the air varies in density when a sound is passing through it, I can telegraph any sound, even the sound of speech."[22]

In less than fifty words, the inventor had crisply explained the theory behind the telephone, a device that would revolutionize business and communications, shrink the world to a fraction of its former immensity, and secure Bell's name in history. On this night, however, the idea just hung there, floating with the dust motes in the stale air, and Bell went on to more pressing work, testing his multiple telegraph. That, in the view of Gardiner Greene Hubbard, promised more than Bell's farfetched notions about "speaking" telegraphs. The problem with Bell, Hubbard concluded, was that he never focused on any particular thing, but jumped from problem to problem without ever settling on one for long enough to solve it.[23]

Bell was a big-concept man often bored by the tedium of the laboratory, where he strung and restrung wires, changed batteries, and shaped new steel reeds for transmitters in his telegraphic work. Just as he was about to lose patience, a lucky accident led Bell away from his telegraphic experiments and on to the telephone. The decisive moment came on the night of June 2, 1875, as Bell and Watson worked in Williams's attic, divided by a partition with a transmitter on one side and a receiver on the other, the two instruments connected by wires. When a transmitter's steel reed stuck to an electromagnet on Watson's side, he idly picked at the reed to free it. On the other side, an alert Bell heard a second reed answer the first one, just as the piano wire had quivered in sympathetic response to his own voice months before. By plucking the first reed, Watson had inadvertently induced a weak current, which caused the receiving reed to vibrate at the same frequency. As so often happens with scientific discovery, this revelation came by chance. From this moment, Bell knew that he could make the wires speak.[24]

Teaching the wires to speak was one thing—giving voice to the confused emotions of the heart was quite another, even for a professor of oratory. After a year of admiring Mabel Hubbard from afar, Bell was becoming obsessive about the girl, now seventeen. "I do not know how—or why it is that Mabel has so won my heart," he wrote that June. "Had my *mind* chosen—or had others chosen for me all would have been different. . . . However, my *heart* has chosen for me—and I cannot but think it is for the best."[25]

Overworked and deeply repressed, Bell almost cracked under the strain that summer. At the bursting point, he finally admitted to the Hubbards that he loved their daughter. They took the news in surprisingly good stride, but neither parent wanted their daughter to know yet, given her youth and inexperience. They urged Bell to keep his secret a year or two longer and he agreed to do so. In the meantime, they made plans for Mabel to spend some weeks on

Nantucket, where she would be among friends her age, away from Bell, and under the care of a trusted cousin.

After a few weeks without her, Bell could stand the separation no longer. He packed a valise, informed the protesting Hubbards of his intentions, and caught a steamer for Nantucket on August 7. There he checked into a hotel and stayed up all night pouring out his emotions in a letter to Mabel. He wrote seventeen pages. He explained why he had behaved strangely in her presence. He expressed his admiration for her parents. What did he want from her? He did not propose marriage—she was too young for that—nor did he ask that she reveal her feelings for him. He merely sought permission to approach her openly, and to spend time with her, so that she could know him better.

"You of course can see no more of me than the mere outside—and I can well understand how little there is there that can prove attractive to you," Bell wrote. "I want you to look within—I want you to know me better before you dislike me. There is much to admire—there is much to lament and deplore. . . . But there is a wealth of love for you. There is a heart that sympathizes with you and for you a thousand times more than you do for yourself." He closed with a pledge to stay away if that was Mabel's desire.[26]

Satisfied that he had done his best, Bell arose early the next morning, pocketed the thick letter, crossed the island to the Ocean View House, and pressed the envelope on Mabel's guardian. "Tell Mabel not to be distressed on my account," Bell said. Then he left Nantucket.[27]

Instead of frightening Mabel away, the energy and honesty of Bell's approach touched her. She agreed to see him upon her return to the mainland that summer. They met at the Hubbard home on August 26 and went to the greenhouse alone. There, among the roses and lilies, they spoke about all that had happened. Mabel thrilled Bell by revealing that she did not dislike him. She did not love him, but she was willing to continue seeing him. That was enough for Bell, who declared that Thursday as the happiest day of his life.[28]

In what remained of the summer, Bell visited his parents in Ontario and rested from his recent nervous strain. He was pleased to find that his father, with whom he had a rather distant relationship, now made efforts to close the gap. Bell also gave some thought to pending telegraphy projects, which had languished during his romantic crisis, and he began to think more optimistically about his prospects for the telephone, to the point of lining up sponsors to help him with foreign rights.[29]

Back in Boston, Gardiner Hubbard's patience was wearing thin. Bell returned that autumn with very little progress to show in his telegraphic work. Then he launched into his old routine of teaching at Boston University, tutoring the deaf, and lecturing all over the region. His patron snapped. "I have been sorry to see how little interest you seem to have taken in telegraph matters, & to hear that you are lecturing upon various subjects in different places," Hubbard wrote. "I hear that you were to lecture on astronomy," he charged, as if Bell's interest in the heavens was a betrayal to those on Earth. "I trust this is not true for it will be a great injury to you, & confirm the tendency of your mind to undertake every new thing that interests you & accomplish nothing of any value to any one. . . . I will close by saying that your whole conduct since you returned has been a very great disappointment to me, & a sore trial. I am very truly Your Friend, Gardiner G. Hubbard."[30]

Hubbard followed this note with a heavy-handed suggestion that Bell give up his lectures and forgo his plans for training instructors for the deaf. If Bell focused entirely on scientific work, Hubbard would cover all of his living expenses. However, if Bell continued teaching, Hubbard portentously suggested that this might jeopardize his future with Mabel. Hubbard's timing could not have been worse, for Bell had been under countervailing pressure from his father to spend more time teaching and less time on scientific research, which the elder Bell considered impractical and speculative. Alec erupted at Hubbard, first in a meeting, then more moderately but no less firmly in a letter.[31]

"I do not intend to make any alteration in my profession," Bell wrote, "merely on account of any feeling or prejudice against it that may be in your mind or those of others. . . . Should Mabel learn to love me as devotedly and truly as I love her—she will not object to any work in which I may be engaged as long as it is honorable and profitable. If she does not come to love me well enough to accept me whatever my profession or business may be—I do not want her at all." Bell rejected Hubbard's offer of payment, citing a conflict of interest. "You are Mabel's father—and I will not urge you to give—nor will I accept if offered—any pecuniary assistance whatever otherwise than what was agreed upon before my affection for Mabel was known."[32]

The incident, which drew upon all of Bell's self-confidence and resolve, would set the tone for the often wary Hubbard-Bell relationship, which would be severely tested over the years, despite the reservoir of affection and respect the two men felt for each other. The Hubbard women, caught between the impulsive Bell and the algid Hubbard, often soothed such rough passages, as they did this time. Mabel helped to convince her father that Bell meant no disrespect by rejecting his overture. And Gertrude, seeing Mabel's exasperation at this tug-of-war between father and suitor, suggested that the time had arrived for a decision—jettison Bell or marry him.[33]

Mabel took the initiative. When Bell came to visit on Thanksgiving in 1875—which also happened to be Mabel's eighteenth birthday—she surprised him by saying she had grown to love him as well as anyone except for her mother. They could be engaged if he liked. A flabbergasted Bell felt bound to remind Mabel of her youth, and to ask if she really meant it. She did. They agreed to be married, following an engagement that would last eighteen months.[34]

That night, Bell wondered if the events of the day had actually occurred: "I am afraid to go to sleep," Bell wrote to Mabel, "lest I should find it all a dream—so I shall lie awake and think of you. It is so cold and selfish living all for oneself! A man is only half a man who has no one to love and cherish."[35]

For her part, Mabel seemed as amazed as Alec, as she was now entitled to call Mr. Bell. "When Alec had gone," she wrote to an old friend just after their engagement, "I was so frightened at what I had done, I was perfectly miserable and hardly knew if I really cared for him. But as soon as I saw him again I was quite satisfied and happy and have been more so every day."[36]

By the autumn of 1875, Alec had managed, despite his romantic distractions, to draft a description of his new device for "transmitting vocal utterance telegraphically" by means of a unique principle he termed "undulatory current." Finally seeing the promise in Bell's telephone, Hubbard now urged him to perfect and patent it as quickly as possible. Bell stalled, tinkering with the patent specification through December, finally grinding out a finished version of the description for Hubbard in mid-January.[37] After another week of delay on Bell's part, Hubbard took matters into his own hands and, without checking with the inventor, filed the patent for him on February 14, 1876, in Washington. A few days later Bell was awarded United States Patent No. 174,465 for the telephone. It would be one of the most valuable ever granted, and therefore one of the most contested. More than six hundred challenges to it were raised, including one that reached the Supreme Court. Bell won every case. The challenges would continue long after his death, but his claim to the invention remained secure.[38]

In Bell's day, this was due largely to the vigilance of Gardiner Greene Hubbard, who had finally discovered his great role in life, to orchestrate Bell's defense in patent litigation and to organize the Bell Telephone Company, of which Hubbard would serve as the founding president. Had it not been for the meticulous and insistent Hubbard, Bell certainly would have been the poorer, his name lost to history in a fog of false starts and unfinished dreams. The Hubbard-Bell partnership, while far from the perfect union, ultimately proved a fruitful one. Hubbard found absorbing, important work to do, drawing upon his well-placed contacts in the legal and political worlds. This freed his gifted associate from the procedural details that drove Bell to distraction. The inventor saved himself for the big ideas. It is doubtful that either man would have accomplished as much without the other.

The same might be said of Bell's partnership with Mabel Hubbard, whose love for him deepened through their engagement, despite his impulsiveness, stubborn nature, and tendency to procrastinate, qualities that came to annoy Mabel as much as they had her father. Even worse was his habit of working through the night, which by now had become essential to him. Mabel never liked it, but she accepted it, just as she learned to appreciate the other attributes of this odd, tender man. "Every day I see something new in him to love and admire," she wrote. "It is wonderful that he should be so clever . . . so utterly without conceit of any kind, so very true, and as thoughtful for others as a woman, far more so than I."[39]

They were married in July of 1877 in the house where Alec sat at the piano and demonstrated the principle of sympathetic vibration almost two years before. As a wedding gift, the agnostic Alec presented Mabel with a pearl crucifix. Even though she could not hear, she bought him a piano and asked that he play for her each day.[40] Alec also signed over most of his stock in the Bell Telephone Company to Mabel, giving her ownership of one-third of the company. The Hubbard and Bell families were joined not only in holy matrimony but also in business.[41]

It would become a lucrative enterprise, of course, but the first years of the telephone company were tense ones, as Hubbard fended off competitors at home and abroad, defended Alec's invention, fell into debt, and fought with his son-in-law over their increasingly complicated financial affairs. Both families struggled through the anxious two years after Alec and Mabel were married. By 1879 Hubbard, fifty-seven, could finally see brighter days ahead. "Eighteen months ago," he wrote to Gertrude, "I knew not where to beg or borrow one hundred dollars. . . . It does seem as though the good time might really come at last. I wished several times last night that we were ten years younger and might have that much more time to enjoy it."[42]

To keep an eye on the continuing legal challenges to Bell's patent—and to escape winters in Boston—he prevailed upon the rest of the family to relocate to Washington. There he proceeded to make the most of the time remaining to him. He settled into semiretirement, thinking big, spending big, and living big. Both families moved into the leafy neighborhood near Dupont Circle, where Hubbard indulged his love of books, assembled a French wine cellar, and displayed his growing collection of etchings and engravings, with a particular emphasis on Napoleonic works. The family also bought an eighteen-acre estate in the hills outside of Washington, where the elevation provided views of the distant Blue Ridge Mountains and relief from the summer heat. They called the place Twin Oaks, for the stately giants guarding their porch.[43] Gertrude Hubbard oversaw the garden, which sprawled around the orchid house they had disassembled in Cambridge and rebuilt at Twin Oaks, brick for brick.[44]

As was expected of an educated man of this era, Hubbard immersed himself in scientific study and conversation. He fell into easy friendship with the community of scientists and explorers who flocked to Washington after the Civil War to fill new government bureaus. Many of these people became regular guests at the Hubbards' dinner table, among them Brig. Gen. Adolphus W. Greely, who established the first international weather station on Ellesmere Island in 1881; Simon Newcomb, an astronomer and economist who might drop by for an animated talk about the promise of manned flight; Andrew White, diplomat and educator, who consulted Hubbard about his plans for a new college called Cornell University; and Spencer Baird, zoologist and secretary of the Smithsonian Institution. He invited Hubbard to join the Smithsonian's board of regents.

Many of the scientists around Hubbard's table brought colorful histories with them. More than mere academics, they had been tempered by experience in the harsh world. None of this breed was more notable than Hubbard's friend John Wesley Powell, the scrawny, one-armed army major with a bulbous nose, a tobacco-stained beard, and a face creased by years of hard seasoning. By the time Powell was fifty, he had survived the Civil War, explored the Colorado River Basin, and founded the U.S. Bureau of Ethnology. Like other scientists of this era, Powell put great emphasis on fieldwork and careful observation. Through Hubbard, Powell established a firm friendship with Alexander Graham Bell and the other newcomers who were transforming Washington from a sleepy prewar village into the capital of a nation fixated on science.

Bell continued inventing. With the help of his cousin Chichester Bell, he spent several years trying to perfect the photophone, a wireless device that transmitted sound on a concentrated beam of light and anticipated laser technology by many decades.[45] He went on to other experiments, such as improving upon Thomas Edison's phonograph.[46] He also perfected an audiometer, which measured hearing ability in standard increments of sound intensity; these measurements became known as decibels, a word that still honors the inventor's name.[47] He even devised a hastily built induction balance to search for the assassin's bullet buried in President James A. Garfield, whom Bell could not save.[48]

Bell had invented the telephone at age twenty-nine. Now in his mid-thirties, he wondered whether that would be his last great achievement. This kept him striving in the laboratory, making public appearances, writing articles for technical journals, and acting as a patron for other scientific enterprises. He helped finance the physics experiments of Albert A. Michelson, who became America's first Nobel laureate for science in 1907.[49] And Bell joined Hubbard to support *Science,* a struggling young weekly that ultimately gave both men an expensive lesson in the uncertain economics of magazine publishing.

Science magazine had been born in 1880, wobbled through its infancy, and almost died in 1882. Bell agreed to buy the journal for $5,000, but he had no intention of publishing it without changes.[50] He suspended publication for a year, conferred with Hubbard, and the two men decided to relaunch *Science* as a lively weekly equivalent of *Nature,* the venerable British magazine. They hired a new editor named Samuel H. Scudder, a prominent entomologist and assistant librarian at Harvard. He not only had scientific credentials but his middle name was Hubbard—he was Gardiner's cousin. Bell agreed to hire him and put up $25,000 to carry the reincarnated *Science* through its first year.[51]

Despite its quality and high standing in the scientific community, the magazine sputtered along, starved for advertising and readers, never crossing the threshold of the six thousand subscribers it needed to climb out of debt.[52] After several years of losses, Bell and Hubbard sold the magazine for $25 in 1891. Bell had lost $60,000 on the enterprise, Hubbard, $20,000. Taken over in 1894 by the American Association for the Advancement of Science (AAAS), the magazine finally began to thrive. "I think it looks as if *Science* may now go & be good for something," Hubbard wrote Bell in 1892. This time the magazine's godfather was right. *Science,* still the flagship publication of the AAAS, remains the premier science weekly in the U.S.[53]

The most notable aspect of Bell and Hubbard's joint publishing venture, given their stormy history, was the pacific quality of the experience. Despite *Science*'s significant cost and disappointing result, they remained friendly throughout their work on it. By this time each man had mellowed enough to enjoy the other's trust and affection.

CHAPTER TWO
A QUIET BIRTH

GARDINER GREENE HUBBARD, 1876.

As he aged, Gardiner Greene Hubbard became a fixture in Washington's civic life, an angular, upright, and magnificently bearded personage delivering papers at the Columbia Historical Society, overseeing management of the Smithsonian Institution, promoting new programs for deaf students, and directing the growth of the national college that would become George Washington University. With the money he had made from the telephone, he might have settled down to a quiet retirement, but even at sixty-six Hubbard needed to stay busy and involved. His old friends Major Powell and General Greely also remained engaged in public affairs and scientific inquiry, and the three met often at the Cosmos Club on Lafayette Square for speeches, cigars, and gossip.[1]

In the clubhouse, the capital's intellectual community gathered to relive Civil War battles, to retrace the sequence of the 1886 Charleston earthquake, or to talk their way off to Siberia, where one of their number had recently helped lay a new telegraph line. They had experience of the world and they yearned to share it with others. They honored the spirit of the late Rear Adm. Charles Wilkes, who had owned the old mansion where the Cosmologists now met. Wilkes, an American James Cook, had led the U.S. Exploring Expedition from 1838 to 1842, returning from his long Pacific voyage with artifacts, natural-history specimens, new charts of Antarctica, and tales of an exotic planet Americans were just beginning to discover.

It was in the old explorer's house that the National Geographic Society was born on January 13, 1888. The event launched one of the most successful and influential enterprises of modern times, one that would send explorers and adventurers scattering across the globe in the years ahead braving ocean depths, living among foreign tribes, elucidating vanished cultures, and reaching the edge of space—all for the sake of advancing human knowledge. But like other births that turn out well, this one began quietly. Just three days before it took place, a terse note went out to a handful of prominent Washingtonians requesting their presence at an evening gathering. "Dear Sir," it began,

> You are invited to be present at a meeting to be held in the Assembly Hall of the Cosmos Club, on Friday evening, January 13, at 8 o'clock P.M. for the purpose of considering the advisability of organizing a society for the increase and diffusion of geographical knowledge.

The stiff, lawyerly language suggests that the invitation was written by Gardiner Hubbard, whose name led the list of hosts for that evening.[2] The other signatories were Adolphus Greely, Comdr. J. R. Bartlett of the navy, Henry Mitchell of the U.S. Coast and Geodetic Survey, and, from the U.S. Geological Survey, Henry Gannett and A. H. Thompson. Their names were well known to the capital's tight community of scientists, explorers, and leading business figures. Despite the short notice and cold weather enveloping Washington in a damp mist, thirty-three people converged on the old Wilkes mansion at the appointed hour. Hubbard and Greely were there, of course, as was John Wesley Powell, who arrived with a retinue of enthusiasts in tow—his brother, his brother-in-law, his nephew, and ten other employees from the Geological Survey and the Bureau of Ethnology.[3] The Cosmos gathering comprised a talented lot.

They settled into wide leather chairs to hear what Hubbard and his collaborators had in mind for this new geographical society, as yet unnamed. The basic concept, which borrowed the model of Britain's Royal Geographical Society, was to create a geographic organization in which members would have an opportunity to meet regularly, to share their knowledge about the world with one another, and to broadcast what they had learned by publishing their proceedings. The organization would exist for the self-improvement of its members, for the improvement of the scientific community in Washington, and for that of the nation at large. This altruistic notion struck a chord that night, and the group unanimously adopted the following resolution:

1. As the sense of this meeting that it is both advisable and practicable to organize at the present time a geographic society in Washington.
2. That this society should be organized on as broad and liberal a basis in regard to qualification for membership as is consistent with its own well-being and the dignity of the science it represents.
3. That a committee of nine be appointed by the chairman to prepare a draft of a constitution and plan of organization, to be presented at an adjourned meeting to be held in this hall on Friday evening, January 20, 1888.[4]

Within two weeks, the committee had met twice and chosen a name for the new group—it would be called the National Geographic Society. Its officers were elected and its bylaws in place. Its dues were fixed at $5 per year.[5] It was organized for "the increase and diffusion of geographic knowledge." Any person "interested in geographic science" could be a member, provided that he or she had been nominated by another member and approved by the Society's board of managers. Its members would meet once a year to elect officers and, if necessary, to change the bylaws. Because many members were government scientists who spent the spring and summer in the field, the early life of the National Geographic Society was seasonal; the group met every other Friday during the winter to hear lectures and discuss the organization's business. Gardiner Hubbard was elected the first president.[6]

From the start, this new National Geographic Society gave the appearance of being a confederacy of amateurs. It paid nothing for articles or speeches. It had no employees. It had no office or auditorium, but lived a nomadic existence, moving from one borrowed hall to another for meetings—to the Cosmos Club, the Smithsonian Institution, Lincoln Hall, Columbian University, the Builders' Exchange Hall, the National Rifles' Armory Hall, back to the Cosmos again. Geography, as defined by Gardiner Hubbard, was so broadly construed that it could include anything animal, vegetable, or mineral—on the land, in the sea, or in the clouds, as he explained in his inaugural address:

"The Greek origin of the word," he said, combined γη, meaning the earth, with γραφη, meaning description, forming "description of the earth. But the 'earth' known to the Greeks was a very different thing from the earth with which we are acquainted.

To the ancient Greek it meant land—not all land, but only a limited territory, in the centre of which he lived. His earth comprised simply the Persian Empire, Italy, Egypt, and the borders of the Black and Mediterranean seas, besides his own country. Beyond these limits, the land extended indefinitely to an unknown distance—till it reached the borders of the great ocean which completely surrounded it.

To the members of this Society the word "earth" suggests a very different idea. The term arouses in our minds the conception of an enormous globe suspended in empty space, one side in shadow and the other bathed in the rays of the sun. The outer surface of this globe consists of a uniform, unbroken ocean of air, enclosing a more solid surface . . . which teems with countless forms of animal and vegetable life. This is the earth of which geography gives *us* a description.[7]

This flexible view of geography manifested itself in the roster of vice presidents serving under President Hubbard. General Greely was designated vice president for the geography of the air, which took in everything from flight to meteorology. Another vice president was assigned the land; another, the oceans; another, life. Two other vice presidents were added a few years later—one for the geography of art and one for commercial geography. Each of these gentlemen was expected to find speakers from his area of expertise to give lectures before the Society. These lectures, in turn, were often converted into articles for the *National Geographic* magazine, which began its life as quietly as its parent had done a few months before.

Every scientific organization needed a journal. But the National Geographic Society's official publication—which would later become the organization's main reason for being—was very much an afterthought. Hubbard and company made no mention of a magazine or a journal of any sort in the bylaws they adopted within weeks of their organizing meeting; they referred only to a "publications committee" composed of seven members, duties unspecified. The first official mention of the *National Geographic* magazine does not occur until October 1888 when the journal makes its appearance ten months into the life of its parent. That first issue, like others that followed, was a slim folio of sixteen pages, bound in muddy brown in a tall octavo format, with a sprinkling of illustrations inside, plenty of maps and charts, and miles of close-set gray text. It contained lectures and monographs from Hubbard and other leaders of the Society, as well as a listing of all members, which numbered 209 in that first October. The next issue did not appear for another six months, establishing the irregular publication pattern that was to characterize the magazine's first seven years. The seven vice presidents and secretaries who made up the publications committee pulled together an issue when enough lectures and other material accumulated to make the printing job worthwhile.[8] It was not until 1896 that the *National Geographic* magazine began to be published on a regular monthly schedule, when John Hyde, the chief statistician of the Agriculture Department, took over as the journal's honorary, unpaid editor.[9]

Throughout these years the magazine was often tedious and sometimes somnolent. It was occasionally impenetrable, as in William Morris Davis's 1889 description of Pennsylvania's hydrology: "Let AB, fig. 9, be a stream whose initial consequent course led it down the gently sloping axial trough of a syncline," he wrote.[10]

Charles Willard Hayes and Marius R. Campbell did little to lighten the tone a few years later, when they wrote: "The long period of quiescence, during which the Cretaceous peneplain was produced, was terminated by a general elevation of the larger part of the province. Like most of the oscillations that have occurred since, it was compound in character, combining epeirogenic and orogenic movements . . ."[11]

The magazine could be unabashedly racist on occasion, placing Europeans at the pinnacle of civilization while portraying other cultures as mere curiosities, a bias that would show itself persistently, if not consistently, into the twentieth century. In 1897, Henry Gannett, a prominent geographer who would later become president of the National Geographic Society, wrote without apology about the dangers of annexing Hawaii: "While the governing race is largely made up of our own kin," he warned in an article, "the vast body of the population is Kanaka, Chinese, Japanese, and Portuguese—not by any means a desirable addition to our numbers."[12] In the same vein, he worried about the alarming influx of immigrants. "This flood of immigration," he wrote in 1893, "has lowered the average intelligence and morality of the community. The illiterate of the northern states are mainly foreign born, the proportion of illiterates among them being four times as great as among the native born. Again, the criminals of foreign birth in the northern states are double their due proportion as compared with the native born."[13] By "native born" Gannett did not, of course, mean true natives, only those born between the arrival of the first English colonists on Roanoke Island in 1585 and the European flood of the late 1800s.

Even the generous Gardiner Hubbard embraced this hierarchical view of cultures, which amounted to bigotry politely dressed up as geographical determinism. Those who lived within the temperate latitudes were civilized, Hubbard explained in an 1894 article, while those who lived beyond that zone were, with few exceptions, barbarians. "If parallels of latitude were drawn around the earth about fifteen degrees north and fifteen degrees south of Washington, the land within these parallels would include all the countries of the world that have been highly civilized and distinguished for art and science." No great men have ever lived, no great poems have ever been written, no literary or scientific work ever produced, in other parts of the globe, he wrote.[14]

Africa was in a class of its own, in Hubbard's view. "It is the only continent in which the largest part of its territory lies within the tropics. As the earth here spontaneously furnishes food for the substance of man, and as only scanty clothing is required, all inducements to either mental or manual labor are wanting." The effect of this salubrious environment was to keep Africans in a "savage" state, Hubbard said. These attitudes were not unusual for a gentleman of Hubbard's time and status, but it makes them no less unsettling today.[15]

Alongside these instances of the racial bias of its era, the early National Geographic magazine also published open-minded and respectful work from pioneering anthropologists such as J. B. Hatcher, who traveled far and wide to study aboriginal cultures. His unsentimental article on the nomadic Indians of Patagonia remains an admirable model.[16]

The magazine under Hubbard also began a steady coverage of polar exploration, keeping readers abreast of Fridtjof Nansen's northern probes over several seasons, as well as those of the balloonist Salomon August Andrèe, who vanished as he searched for the Pole in 1897.[17] When interest in Alaskan subjects peaked with the Yukon Gold Rush, Hubbard and other members privately contributed money to support the first National Geographic Society expedition, a survey of the region around Mount St. Elias, the highest peak on the newly established Alaska-Canada border. As a result, Hubbard had an Alaskan glacier and a mountain named for him.[18]

Hubbard's magazine also showed an early interest in the weather and natural disasters—no doubt because the organization's founders included influential meteorologists. One of these, Edward Everett Hayden, the one-legged chief of the Naval Marine Meteorology Department, produced a riveting account of 1888's Perfect Storm, a northwester that raked the East Coast from Chesapeake Bay to New England, sinking or damaging 185 vessels, altering the course of the Gulf Stream, and testing the nerve of mariners like those of the *Charles H. Marshall,* which, though she was covered with ice, thrown on her beam ends, and driven a hundred miles before the storm, returned to port "without the loss of a spar or a sail," according to Hayden.[19] His piece made such natural disasters a staple for the magazine, which continues to chase volcanoes and earthquakes around the globe.

The best of this genre was also one of the earliest, an account of a Japanese tsunami that devastated the northeast coast of Honshu in 1896, killing almost 27,000. The storm was covered by an exceptional contributor named Eliza Ruhamah Scidmore, a sometime newspaper reporter who had befriended the Hubbards while covering Washington's social scene. Her tsunami piece, which led the September 1896 issue, described the development and aftermath of the great wave with the clear-eyed precision, sympathy, and authority that became her trademark style.

"One loyal schoolmaster carried the emperor's portrait to a place of safety before seeking out his own family," she wrote. "A half demented soldier, retired since the late war and continually brooding on a possible attack by the enemy, became convinced that the first cannonading sound was from a hostile fleet, and, seizing his sword, ran down to the beach to meet the foe. . . . The good Père Raspail had just reached Kamaishi from his all-day walk of fifty miles over the mountains and entered his inn, when his assistant called to him from the street. The priest came to the veranda, but in an instant the water was upon him. He was seen later, swimming, but evidently was struck by timbers or swept out to sea, as his body has not been recovered."[20] Scidmore, a fearless and inveterate traveler, would contribute many other articles over a seventeen-year alliance with *National Geographic.* In that time, she became an associate editor and the first woman elected to the Society's board of managers.

An equally exceptional individual who would become a famous contributor stepped out of obscurity and onto the stage of the National Geographic Society about this time. Robert E. Peary, a skinny navy lieutenant with a walrus mustache, looked more like a dandified actor than an explorer. In a photograph from the period, the young officer slouches under a soft felt hat, his legs sheathed in knee-length gaiters, a compass hanging from his neck, a revolver holstered on his hip, a riding crop in one hand, a rifle in the other. Peary's gaze, dark and imperious, aims just over the viewer's shoulder, as if contemplating some distant object. He has just emerged from the wilds of Nicaragua, where he led a surveying party through swamps and thickets to map a course for a proposed isthmus canal (which eventually went to Panama).[21]

Aloof and almost pathologically driven, Peary was a superb organizer with a gift for public speaking and an unquenchable thirst for fame.[22] "I shall not be satisfied that I have done my best until my name is known from one end of the world to the other," he wrote to his mother in 1880. "I will make powerful friends with whom I can shape my future."[23] He found those friends in Adolphus Greely and Gardiner Hubbard, who saw something they liked in this brash young man. Greely invited him to lecture about his Nicaraguan exploits for the National Geographic Society, which also published the lecture as an article, "Across Nicaragua with Transit and Machete," in the magazine.[24] Peary returned a few years later, standing onstage with Hubbard at Lincoln Hall, where a thousand members and guests heard Peary describe his plans for exploring northern Greenland in the summer of 1891.[25] Hubbard crossed the stage at the conclusion of Peary's lecture, unfurled an American flag, and presented it to the young explorer.

"Now take this flag and place it as far north on this planet as you possibly can!"[26]

Peary, visibly moved, took the flag, bowed his thanks, and departed for Greenland.[27] There he began almost two decades of nonstop polar travel, building his own reputation and that of the National Geographic Society as he went.

Over almost a decade, Hubbard and other founders had watched the membership rise from 209 in the year of the National Geographic Society's birth to almost 1,500 by 1897.[28] The numbers stalled there. Despite the best efforts of Hubbard, Greely, and others, the Society remained essentially a local club, with most of its members residing in Washington. Without some attention, the young organization would meet the fate of other learned societies that had recently blossomed and faded on the Potomac. It needed a boost.

At this crucial point in its development, the National Geographic Society lost its chief advocate and founding father. Gardiner Greene Hubbard, seventy-five, died on December 11, 1897, while working on preparations for the tenth anniversary of the organization. If the National Geographic Society was to survive, it needed a strong leader to carry it into the next century. Gertrude Hubbard knew how important the organization had been to her husband, and she knew that there was only one person who could breathe fresh life into it—Alexander Graham Bell.

Her son-in-law was not interested. Obsessed with flight experiments, he felt that he could spare little time for the National Geographic Society. But Gertrude Hubbard and Mabel Bell prevailed upon the inventor, who grudgingly yielded to them. Vowing to abandon the job as soon as possible, Alexander Graham Bell stepped in as president of the National Geographic Society in 1898.[29]

CHAPTER THREE
THE FIRST GROSVENOR

GILBERT H. GROSVENOR AND ALEXANDER GRAHAM BELL.

By the end of Alexander Graham Bell's first year as president of the National Geographic Society, it looked as if the organization might not survive the waning months of the nineteenth century. Membership had sunk from 1,500 to 1,000 and the organization was $2,000 in debt.[1] Bell had been distracted by flight experiments in Canada, but now he snapped awake, suddenly aware of the danger.

If the National Geographic Society was to avoid extinction, Bell knew that it must evolve beyond its clubby origins in Washington, beyond Gardiner Hubbard's loose collection of friends, their occasional lectures and meetings, and the erratically published *National Geographic* magazine. The magazine was, in

Bell's words, "a valuable technical journal that every one put upon his library shelf and few people read. It was valuable, it was important, but it did not contribute anything to the financial support of the Society."[2]

He had in mind a new invention, a National Geographic Society that would exist primarily to publish a more ambitious magazine for members outside of Washington. This new National Geographic Society would be not a local club but a truly national enterprise, its scattered members bound together by a popular magazine that infected ordinary, intelligent Americans with the sense of wonder Bell felt for science, for foreign cultures, and for the natural world. This new arrangement would serve the idealistic purpose of the original National Geographic Society, "the increase and diffusion of geographic knowledge," and, with luck, it might also generate some much needed revenues.[3]

If this new journal was to stand for the whole National Geographic Society, it needed an energetic, full-time editor, not one of the busy scientists or government officials who had been haphazardly pulling together each month's *National Geographic* magazine when their day jobs were done. With a few notable exceptions, they had filled the magazine with begged essays and abstruse technical articles and delivered the results later with each passing month. Just recently, the first half of the magazine had gone to press while the editor scrambled to find material for the second half.[4] Bell had seen enough of that. His new magazine would be on time, so that it could be reviewed by other publications and so that it supplemented the news stories readers followed in daily papers.

Bell announced to the Society's board of managers that he was going to find a full-time editor for the magazine—and that he was going to pay him the grand salary of $1,200 per year out of his own pocket. He began to search for a candidate. He wanted someone young and energetic, bright and well educated, with some diplomatic skills, along with the confidence to stand up to such seasoned board members as Adolphus Greely, Henry Gannett, and WJ McGee.*

---

*McGee was known as "No Points" McGee because he insisted on no periods after his initials. He greatly influenced President Theodore Roosevelt's progressive views on conservation.

The new editor needed to work independently because Bell, still wedded to his experiments, could sacrifice little time toward running a magazine. Finally, Bell wanted an editor who was somewhat worldly, but without too much experience, without preconceived notions about publishing that would interfere with Bell's own clear vision of the magazine.

He foresaw a popular journal with broad appeal, lively, timely, and accessible, but with nothing vulgar about it. It would be a smart, high-quality magazine like the *Century* and *McClure's,* two of Bell's favorite periodicals. Like them, the revitalized *National Geographic* would be well written, solidly reported, and flawlessly printed. Unlike other magazines, it would be profusely illustrated with maps and photographs.

All he needed now was a master builder to put his architectural scheme in place. Bell found his ideal collaborator on the first try—actually, he found two. They were the identical twins Gilbert H. Grosvenor and Edwin P. Grosvenor, the twenty-three-year-old sons of Lilian Waters and Edwin Augustus Grosvenor. The twins, both recent magna cum laude graduates of Amherst, taught at the Englewood School for Boys in New Jersey. Except for an interlude when the Russo-Turkish war separated the family, the boys had grown up in Constantinople, where the elder Grosvenor had taught for twenty years at Robert College. There on the seam of Asia and Europe, surrounded by antiquities, a lively Turkish culture, and a close-knit enclave of diplomats and academics, the Grosvenor boys received a cosmopolitan education that would serve them well in later years. Family letters were spiced with *Mashallahs* and other reminders of that exotic upbringing, but the clan was firmly rooted in New England soil, from a respected ancestry that included English ironworkers, gun makers, an author of children's books, two Salem witches, several doctors, numerous clergymen, and a prominent Byzantine scholar.[5]

The Byzantine scholar, Edwin Augustus Grosvenor, now taught European history at Amherst. At the invitation of Gardiner Greene Hubbard, Professor Grosvenor had recently lectured on Constantinople at the National Geographic Society and had thereby fallen into an easy companionship with the Hubbards and Bells. When their families exchanged visits, Bell had been impressed by the bright, well-mannered Grosvenor twins, who displayed a confidence and earnestness that might have reminded the inventor of himself as a young man. He remembered them when he needed an editor for *National Geographic,* and wrote to Professor Grosvenor in February of 1899, offering the position to whichever twin expressed interest.

That Bell turned to the professor is indicative of the faith he placed in Edwin A. Grosvenor, one of few people admitted into the inventor's minute— and very close—circle of friends. If Professor Grosvenor thought that either of his sons was suited for the editor's job, that endorsement was good enough for Bell, who often made important judgments on the basis of instinct. It may have also helped that the Grosvenor twins had demonstrated some interest in Bell's daughters, Elsie and Daisy. Bell, who had a strong interest in genetic studies, did not discourage the romance. The Bell daughters were approaching marrying age, and one could hardly find a more advantageous match than the intelligent, hardworking, solidly middle-class Grosvenors. If one of the professor's sons took the job, who knew where that would lead? The new position would carry the title of assistant editor, in deference to John Hyde, a former editor of *Banker's Monthly,* whose grandiose title at *National Geographic* magazine was editor in chief.[6]

After a year of teaching chemistry, algebra, and French to the youths of New Jersey, Gilbert H. Grosvenor was eager for the change. "It is too good a plum to throw away for some one else to pick up," Bert wrote.[7] "I think it is very good of Prof. Bell to remember us in this way, for none of us have ever asked him for a lift, tho he could give a pretty good one if he chose. Of course it is out of friendship and belief in you," Bert told his father. "I think we had better write to him saying that one of us would be glad to accept the position but we needn't decide just yet which one."[8] It was clear who wanted the job. With his brother Ed's blessing, Bert went to Washington to nail it down.

He arrived in the capital that March and visited the Bells for an interview. The inventor saw something in Bert that went beyond the good feelings he had for the young man's father. Bert was eager, well organized, and meticulous. And unlike Hyde, who had already threatened to resign once and seemed to drag his feet whenever Bell suggested innovations, Bert Grosvenor was open-minded toward change. Above all, Bell sensed that Bert would be loyal—just the fellow to implement his grand plan for *National Geographic*. While Mabel sat by, Alec offered Bert the job on the spot. Then he sketched out his goals for the magazine, assured the young man that he would have the support of the board, and repeated his pledge of $1,200 in salary. Beyond that, Bell offered operating funds if the magazine needed them, just as he had done in his ill-fated *Science* venture.

Bert interrupted. He would "be glad to assume the undertaking" but he objected to further subsidy for the magazine, which he believed should pay its own way. "I caught a smile on Mrs. Bell's face," he recalled years later. With the *Science* losses a recent memory, "it seemed to please her very much that her husband wasn't going to be asked to put up any more money."[9] Bert had not only landed the job, he had also ingratiated himself with Mabel Bell, who would be an important ally in the years to come.

Bert returned to Englewood to do what any sensible young man would do for an important career move—he bought new underwear and began packing.[10] Mr. Parsons, the headmaster at Englewood, took the news of Bert's defection well, although in parting he pointed out that Professor Bell had picked him not on the basis of talent, but because Bert was Edwin A. Grosvenor's son.[11] Another person might have taken the comment as a rebuke; Bert Grosvenor embraced it. Family connections had brought him this far, and he would never hesitate to use them to advance his own interests or those of his kin.

Bert was proud of his family, which was far from wealthy or socially prominent. But in an era when class consciousness intruded into almost every aspect of daily life, the Grosvenors felt a sure sense of entitlement, based on their standing among New England's educational elite, not to mention the nobility of their background. Without a trace of self-consciousness, Bert Grosvenor could boast to an interviewer that "every drop of blood" in his veins came from English ancestors.[12] And, although they were too polite to do so publicly, the family kept track of where others stood in the social order, especially those who had risen from humble origins; none of the Grosvenors ever forgot, for instance, that the mother of the polar explorer Adolphus Greely had been their washerwoman in Newburyport, Massachusetts.[13]

Professor Grosvenor's children grew up confident of their place and standing, in part because they were reminded of it, but also because they received unqualified parental support for their initiatives, large or small. When, for instance, Lilian Waters Grosvenor learned that Bert was trying to grow his first mustache, she urged her son on. "Good luck and long life to it!" she wrote. "May it be as bushy and fine as Ed's last summer and of longer duration."[14] In countless ways, the Grosvenors displayed that particular closeness that sometimes accrues to families who live abroad for long periods, relying on one another for comfort in strange surroundings. When the family scattered, they remained close, producing a flood of correspondence that documents Bert's first days at *National Geographic*.

Reporting for work on April 1, 1899, he may have thought that Bell had sprung an elaborate April Fool's joke on him. Bert found the Society's one-room office in the Corcoran Building at 1330 F Street in Northwest Washington "a perfect pig sty," with the rough lumber walls of a boxcar, papers littering the floor and dirt on every surface. There was a spittoon but no desk. Hearing about the missing desk, Bell sent over his own big rolltop for his new protégé, who set about putting *National Geographic* into shape.[15]

The first order of business was for Bert to get the magazine out on time, then to have it reviewed to attract new members outside of Washington. The hard-pressed John Hyde was happy for the assistance, and Bert worked eagerly, toiling late at night, dogging contributors, and writing filler for the journal's back pages, which noted the progress of polar expeditions, archaeological surveys, scientific conferences, and census reports.[16] With deadlines threatening on one occasion, Bert addressed a month's worth of magazines by hand, bundled the issues into a bag, and lugged them to the post office on his back.

He and Bell began a public-relations offensive, nudging prominent members to write to newspapers and magazines for reviews of *National Geographic*. "Now as you know up to this time the magazine has hardly ever been noticed,"

Bert wrote to his father in April. "If I can get a dozen notices in such papers as *Outlook, Independent, Brooklyn Standard Union, Boston Transcript . . . Review of Reviews,* etc. I shall have made a good gain. Mr. Bell thinks that personal letters to diff editors will do a great deal. . . . Now I know you are very busy but if you could write a note to diff men you know . . . it would be of great value to me. . . . I would like these notes as soon as possible and shall be much obliged for them. Tell Ma I feel strong & well Mashallah. . . ."[17] Bert's wheedling letter found its mark. His father, a former contributor to several of the target papers, wrote to influential editors he knew, and the reviews began to appear. Bert also subscribed to a clipping service and built a file of favorable notices, which helped sell the magazine and documented his progress as editor.

With Bell's encouragement, Bert also began one of the first known direct-mail campaigns in magazine history, promoting likely members from the lists of the National Education Association, the American Association for the Advancement of Science (AAAS), Phi Beta Kappa, the National Academy of Sciences (NAS), the army and navy, and other organizations. Bell would sign letters to AAAS and NAS candidates; Adolphus Greely prepared notes for army officers; and Professor Grosvenor promoted to NEA prospects. Many years later, when the magazine began to operate in the black, a portion of each year's revenue was set aside to support expeditions and scientific research; by the year 2004, for instance, the National Geographic Society gave out $8.5 million in such grants. In Bert's day, though, the challenge was to stay afloat, and the conceit of membership—as opposed to subscription—was something of a marketing ploy, as Bert frankly explained to his father. "You see," he wrote, "we want to make the fact of an election to N.G.S. an honor so that people will think they are complimented by a nomination."[18]

Bell made other marketing innovations at *National Geographic* magazine, where he was one of the first to suggest inserting subscription blanks—nomination forms, in the polite language of the National Geographic Society—into the journal. "Every issue of the magazine should contain at least one printed blank for the nomination of members which would be torn out by anyone desiring to nominate a friend and sent to the secretary for confirmation by the board," he wrote Bert. "This I consider to be very important but I don't think it has been done in the magazine yet."[19] Bert got it done, and fast. He proved himself the ideal lieutenant for Bell, who appreciated the young man's industry. For his part, Bert practically worshiped Bell, who gave him all that he asked, provided a deluge of ideas and encouragement, and introduced Bert to "piles of big guns" in Washington society. "He is just fine," Bert said of his patron. "I wish I cd pay him back in some way."[20]

That repayment came soon enough, in the form of the membership gains that Bell had made the central aim of his administration. Within weeks of Bert's employment, the society had gained 250 nonresident members representing every state and parts of Canada.[21] A year later, Bell could boast that membership had almost doubled to 2,462. "It is obvious . . . that we are moving in the right direction," he wrote. "There is every prospect that attention to the Magazine and to the needs of our outside members will result in an increase in membership so great that we may hope in a few years to have thousands of members where we now have hundreds, and to establish on a lasting foundation a great national society of which we may all be proud."[22]

Bell impressed all who knew him with his big plans for the magazine. "I see no reason why our mag. should not be of great popular interest, and grow ultimately into a mag. as large as the *Century*," he wrote. "I may of course be sanguine as to the possibilities of the mag. But there is no harm in having a high ideal, even though it may be beyond the feasibility of actual realization. 'Hitch your wagon to a star' is a sentiment that just suits me."[23]

Bert Grosvenor had come under Bell's tutelage with big dreams of his own, and these were not only for the magazine. He had his sights set on Elsie May Bell, the inventor's eldest daughter, who had grown into a tall, handsome young woman of twenty by the time Bert arrived in Washington. She had her father's dark good looks and his childlike impetuosity, combined with her mother's frankness, flashing wit, and adventurous spirit; from both parents, Elsie acquired expensive tastes and a keen sense of independence.

"She is so thoroughly frank and unreserved that you see the best and worst of her at once," wrote Elsie's mother. "All her hardness, selfishness and want of sympathy, her absorption in herself and her pleasure in her own good looks, and at the same time her perfect sincerity and honesty, her great desire to do right, to be a good woman and conquer her faults."[24]

Bert Grosvenor displayed little of Elsie's fire or dash—in fact, one could search for a long while before finding two people less alike. Where she was bold and sure, he was understated, sensitive to criticism, and diffident. And with his bookish manner, thinning hair, soft chin, and glinting pince-nez, he could hardly be described as handsome. But evidence of another Bert flickered behind the lenses, something in his mild blue gaze suggesting a hint of hardness and a determination to prevail. Like many of the Grosvenor males, Bert was a driven man. He had to succeed, in love as in work, and he set about doing both within days of reporting to Washington. He was ruthless in handling social invitations, accepting offers from the Bells, Greelys, Hydes, and other important families, rejecting others. "I am going to refuse every bid except when I think there is good in going, I mean gain for this present position," he reported to his father.[25] Not long after this, he stepped up his campaign for Elsie's affections.

They went for picnics in the country and to dances in the city. They took carriage rides on warm spring nights serenaded by peepers. They chatted in the glow of Chinese lanterns at garden parties. They sat together at church. Bert took his first horseback ride in hopes of impressing Elsie, an accomplished equestrienne. His horse had different ideas. It ran away with him, threw him twice, and prompted him to bail out. He turned up his collar to hide the cuts and scratches and announced that he had had a beautiful time. Although his horsemanship needed work, both Mabel and Elsie gave Bert high marks for spirit. "I don't believe that Gilbert intends to be sat down on unnecessarily," said Mabel.[26]

Indeed, what Bert lacked in style, he made up in staying power, trailing Elsie like a spaniel. He became such a fixture around the Bell home that his presence was taken for granted. "Bert is here now," Mabel wrote Alec. "He is almost always here when he has nothing better to do."[27] This constant exposure gave Mabel, a sharp observer of the human species, ample opportunity to study the courtship of Elsie and Bert. "They are evidently good friends," she wrote, "very fond of each other, but it seems to me more the good camaraderie of congenial friends than the sentiment and shyness of lovers." Elsie was openly affectionate, grabbing for Bert's hand, resting a casual elbow on his knee, and cooing over him: "Isn't he a dear?" she asked Mabel in his presence. "Isn't he cunning? Isn't he a baby?" Bert squirmed, protested, and withdrew his hand from Elsie's. "He admires her and wants to be with her, but there is nothing loverlike in his attitude, he doesn't pay her compliments. . . . I don't know what to think," Mabel wrote, perhaps recalling the thermodynamic quality of her own courtship with Alec.[28]

Despite these misgivings, Mabel found other aspects of Bert's personality appealing. She liked the way he talked, and was impressed by his occasional boldness, as when he flourished a copy of *National Geographic* magazine, opened it to the masthead, and pointed out his name. "He said he thought it was the most aristocratic and handsomest name there," Mabel said. "He does not want for a certain kind of self confidence." On the other hand, she wondered if Elsie's growing attachment to Bert might be unwise, given her lack of experience with other men and Bert's sketchy financial prospects. "It will never do for her to marry a poor man and have to live in a small house, yet she seems to be drifting that way," Mabel wrote.[29]

To arrest that drifting, Mabel imposed a separation on Bert and Elsie, taking her daughter to Nova Scotia with the rest of the family for the summer of 1899. She forbade her to correspond with Bert—in hopes that she might lose interest in him. Lonely and disappointed, Bert nonetheless remained determined for Elsie, encouraged by his father. "Let me tell you, young man, that I never saw a finer girl than Elsie Bell and whoever wins her is a fortunate man," Professor Grosvenor wrote. "She is worth working for and waiting for as Jacob did for Rachel."[30] And so, like Jacob, Bert waited, hopeful that time and perseverance would carry the day. "I'm going to win that girl if it is within human power," he vowed.[31]

Then he buckled down to work at the magazine. He continued recruiting new members by the hundreds. He helped plant more reviews in leading papers. He corresponded with the prominent journalist Ida Tarbell of *McClure's,* with whom he worked out arrangements for trading free advertising space. He fielded letters from Bell, who still gushed with ideas, which afforded little time for Bert's languishing over absent love.

"Things are working up in the Transvaal," Bell warned.

"Bubonic plague is reported to have made its appearance on the American continent. . . .

"The lecture course should be taken in hand at once. . . .

"Some of the Aleutian Islands are reported to have settled from twenty to twenty-five feet. . . .

"I would have you specially bear in mind the advisability—NO THE NECESSITY—of building up our advertising pages. . . .

"Wherever there is anything of interest going on we should have geographic information to illustrate it. . . .

"Now would be the time to get an article from the Weather Bureau on the West India Hurricanes. . . .

"I hope you are looking out for a new location for headquarters for the Society. . . .

"The great danger we have to fear is the danger of preaching away above the heads of ordinary well-educated people. . . .

"I strongly recommend that you have a big map of the Transvaal made. . . .

"MAKE the magazine interesting to the general reader. . . .

"I wish we could get into a fire-proof building. . . .

"Dry and long-winded articles of technical character should be avoided. . . .

"Ask Mr. Hyde what he thinks of the possibility, feasibility, or advisability of the National Geographic Society publishing books which should appear in parts in the magazine. . . .

"Elsie says 'send my love', but I don't know that she means it."[32]

On occasion, Bell risked Mabel's wrath by sneaking such messages to Washington. Perhaps he remembered how painful his pursuit of Mabel Hubbard had been, which made him naturally sympathetic toward his stumbling young associate. Whatever his motive, Bell found a way around the romantic obstacles his wife had erected; he dictated letters to Elsie, who in turn wrote them out in her bold, slanting hand for Bert.[33] This way, at least the young lovers could maintain contact, however tenuous, through that long summer.[34]

While the Bells were gone, Bert cultivated other important allies in Washington, among them Adolphus Greely, with whom he dined on occasion, and Charles J. Bell, a prominent banker, a board member, and the treasurer of the National Geographic Society. Charles Bell and Bert played golf together in the rolling hills at Chevy Chase, Maryland, and Charles came to admire this purposeful young man for the same reasons his cousin Alec had done so.

Bert also made regular excursions to Twin Oaks, where he developed a cordial relationship with Gertrude Hubbard.[35] Although not active in the Society's day-to-day operations, she still wielded great influence with the board members who revered her husband's memory. Seeing that Bert was a regular worshiper at the Church of the Covenant, Mrs. Hubbard invited him to sit in the family pew for the Presbyterian services on Sunday mornings.[36] There, in the sanctuary where Gardiner Hubbard had received his final send-off, they sat under a stained glass window dedicated to his memory. The glass, glowing gold and green in the diffuse light, depicted a bearded man scattering seeds over the earth, just as Hubbard had done in life. He would have been proud of the way the National Geographic Society was growing into something bigger.

Others were not inclined to applaud Bell and Grosvenor's ambitious growth plan. It made for a shocking change from the intimate early days, when most of the Society's members had known one another and belonging had meant inclusion in Washington's cultural elite. Bell's push to make the National Geographic Society more populist—and its magazine more prominent—would, of necessity, lower the quality of the membership. In addition, Bell seemed to be perfectly happy to make the magazine more commercial—he had met with publisher S. S. McClure, who suggested that *National Geographic* magazine increase its advertising, boost newsstand sales, move to New York, and even remove the word "Geographic" from its name. Bell did not embrace all of these suggestions, but he helped create an atmosphere in which few of the old certainties remained. He made the old guard uneasy, and this was accentuated by his frequent absences from Washington. Where was Bell as *National Geographic* magazine groped its way from being a learned journal toward something new? In Canada flying kites—*literally* flying kites—which left the magazine in the hands of the overworked John Hyde and utterly green Bert Grosvenor.

Few of Bell's critics were eager to challenge him openly, but they were not averse to tweaking his newly hired associate. Hyde, the peppery editor in chief, had come to the conclusion early on that Bert was primarily interested in building up the membership—and his own reputation—and not in the tedium of magazine editing. The young man spent a lot of time writing letters to prospective members, but at first there was little proof that Bert's feverish mail campaign and grand ideas for the *National Geographic* magazine had produced results.

"By all means make the Magazine as attractive in the quality of its illustrations and the variety of its contents as it can be made," Hyde told Bert, "but do not let us embarrass ourselves and the Society, either for the personal gratification it would afford us to issue large and fine numbers, or in the mistaken belief that we shall get back in increased membership the additional money we may expend. As I recently said to you, these are the days of cheap Magazines. . . . As many publishers of large means have found out to their cost, to build up a circulation is either a tedious and expensive or very risky business. . . . ."[37] Bert swallowed Hyde's criticism, put more effort into helping him, and tried to maintain a respectful relationship. But he also continued the membership push, knowing that the board of managers would ultimately back Bell. "My job," he recalled years later, "was to get members so I proceeded to get about 800 from June to Oct., regardless of Hyde's objections."[38]

Without meaning to, Bell had planted the seeds of discord between the two editors. Hyde, who had been laboring without pay to get the magazine out, must have felt undervalued by the new arrangement, which initially seemed to provide relief from his grinding editorial duties, but in fact diluted his authority as editor in chief. As Bell's reliance on Bert grew, so did the young man's influence at the National Geographic Society. Even so, the magazine's cumbersome management structure made it difficult for the new man to operate very efficiently. At the top, the organization was supervised by a board of managers composed of twenty leading scientists and explorers, who were all loyal to Bell. An executive committee, headed by Henry Gannett, was detailed to run Society affairs from day to day, in consultation with a publications committee headed by John Hyde. Hyde, in turn, was expected to coordinate the magazine's operations with twelve associate editors, none of whom was paid, all of whom served as board members, and many of whom held decided opinions about how the magazine should be run.

Bert tried to steer among them, but a wreck was inevitable. For one thing, Hyde and Gannett felt that the practice of soliciting members was undignified; they preferred the more straightforward approach of subscriptions. Bert would be squeezed between the conflicting aims of the board of managers and the executive committee. In addition, the editor in chief seemed unimpressed with Bert's editorial performance. On several occasions Hyde had found it necessary to rework the young man's writing and editing. Instead of accepting this as part of his editorial education, Bert complained to Bell and to Elsie behind Hyde's back.[39] The executive committee, in turn, lodged what seemed like a good-natured complaint about Bert with Bell. According to Henry Gannett, Bert was so occupied with Elsie that he neglected his magazine work. Instead of alarming Bell, this news delighted him. He made a joke of it, summoning Bert for a discussion.

"What have you been up to?" Bell demanded, shedding his cheerful manner.

"You have a Harvard professor of geography on your board," a shaky Bert answered. "He doesn't like what we are trying to do. Perhaps he wants to change the editor."

Bell boomed with laughter.

"The executive committee thought the chairman should tell the president that Mr. Grosvenor was paying more attention to his daughter than to *National Geographic* magazine!" Bell sputtered, seemingly oblivious that Bert was headed for trouble.[40]

Bert's detractors had one thing right——he was devoted to Elsie. When she returned from Canada, they resumed seeing each other, floating through the winter and toward another spring, a pair of old friends adrift on an unruffled lake, but where were they going? The tepid nature of this courtship still baffled Mabel Bell, who predicted that Elsie would probably decide to marry Bert, "but she is so perfectly matter of fact about it that I am sure she is not in a particle of love with him. I wish we could put her in position to meet big young men. I am sure she ought to have some one stronger than Gilbert."[41] After months of badgering Alec on the subject, Mabel finally prevailed upon him to take the whole family to Europe for the summer of 1900——Elsie's last chance to meet the "big young men" her mother had in mind.[42]

Left to swelter in Washington, Bert resumed the role of Jacob, resigned to another summer of working and waiting. "As to Elsie and myself," he wrote his mother, "nothing is definite now. We are letting things rest, but I believe I have made progress these last few weeks and that she really cares for me now as she never has before. Meanwhile, I am going to have a good summer. . . ."[43]

As if to guarantee that prediction, before Alexander Graham Bell sailed for Europe he persuaded the board of managers to recognize Bert's outstanding service by ratifying the young man's position for another year and by suggesting a significant promotion. But when Bell pledged to subsidize Bert's salary again, John Hyde tried to dissuade him, citing the magazine's need to grow. The executive committee, he said, was "of the opinion that it would be unwise for the Board to take any action that might seem to commit it to Mr. Grosvenor, either when your generous contribution should cease to be available or when the increasing importance of the magazine should make larger demands upon its editorial management."[44] President Bell and the board of managers ignored Hyde's advice, unanimously approving Bert's promotion. Bell asked the executive committee to work out the details in his absence.

As soon as the Bells left Washington, Bert was plunged into the biggest crisis of his young career. Hyde and the executive committee sprang on Bert for a variety of lapses. He had allowed mistakes in the magazine. He had no business ability. He was lazy, incapacitated, and ignorant. He was ordered to stop soliciting members. Then Hyde presented Bert with the letter of agreement putting his recent promotion into effect. But instead of the promotion the board had approved, this document gave Bert a subservient role. His standing reduced to that of a clerk, he could be fired without cause on one month's notice. It looked like a prelude to Bert's dismissal.[45]

Isolated from his patron, Bert was devastated. He turned to his father, laying out his anxieties and documenting the first National Geographic power struggle in a series of letters over the course of the summer; the young editor turned to the right counselor in his hour of need.

Like his friend Alec Bell, Professor Edwin A. Grosvenor was a multifaceted character. Raised by a guardian after his parents died young, Grosvenor won honors as a student at Amherst, proceeded to seminary, became ordained as a minister, gravitated to Constantinople, mastered French and modern Greek, delved into archaeology, absorbed European and Asian history, contributed articles for leading newspapers from abroad, and eventually returned to New England to teach at Amherst. His reputation for political wisdom spread from that Massachusetts town, and he entertained a succession of politicians, reputable and otherwise, at an old maple table in the family kitchen.[46] There the professor puffed cigars and dispensed political insights to no less than seven presidents of the United States, which is why the table is still known in the family as the Seven Presidents' Table. Although a lifelong Democrat, he crossed party lines to vote for Calvin Coolidge, a former student. Despite his high connections and considerable learning, Grosvenor was an eminently approachable man, level-headed, short, and square. His lively blue eyes and permanently arched brows gave him an impish look, like someone about to tell a good story—something he often did. Having suffered through his own youth, he showed a special sympathy toward young men. This won him a popular following at Amherst, where students pronounced him "a peach" and affectionately called him Grosvie (pronounced "Grovie").[47]

Grosvie knew a thing or two about politics and he thought he understood the root of Bert's troubles at National Geographic. The boy was young. He was successful. The magazine showed promise, making its control suddenly attractive to John Hyde and the old bulls who had allowed it to drone along. The board of managers' recent recognition of Bert constituted the final insult for Hyde and Company, who now sought to drive Bert from the organization.

"It is in a certain sense a dethroning of Mr. Hyde," Grosvie wrote. "It is impossible that a young man, so much younger than the Associate Editors or the Editor in Chief, should . . . not arouse jealousies." Grosvie advised his son to hang on, to remain strong, to exercise self-control, and to wait patiently for Bell's return in the fall. Throughout that long summer, Grosvie wrote his son almost daily, counseling patience, steeling his resolve, soothing his frayed nerves. "You are all right, Bert, Mashallah!"[48]

Bert was not all right. Judging from the frequency and intensity of their correspondence, Bert was worried and frightened, and it took all of the professor's cajoling, advice, and flattery to keep his son from quitting. Grosvie was doing for Bert what he had done for countless youths in his care, building their confidence with kind words, teaching the sort of loyalty and quiet strength celebrated in the book of Job. "You are a newcomer and you are young," Grosvie told his son. "You are Elihu in the midst of Eliphaz, Bildad, and Zophar, men a good deal older. Elihu came out all right and they didn't and they had to make room for him."[49] Reinforced by this biblical wisdom, Bert held his temper, reported to work each day, and avoided blunders. But under the tension of that summer, he could not help complaining to his father—why had Mr. Bell not settled Bert's position before he left the country?

Grosvie shot back a warning. "Be sure in no way to doubt or distrust Mr. Bell. Count on him as on a rock. He had no idea of any such jealousy breaking out and supposed that every thing was practically settled. . . . Act and feel that you are fully backed by all the Bell influences. . . . Whenever you feel a doubt of Mr. Bell, Mrs. Hubbard or any of that blessed family say to yourself, 'What a fool I am! I ought to be ashamed of myself.'"[50]

As a student of history, Grosvie understood the power of dynasties, even one as young as the Hubbard-Bell reign at National Geographic. Having brought Bert under control, the professor began a long-distance campaign to rally the "blessed family" to his son's cause. First he corresponded with Gertrude Hubbard, briefing her on Bert's travails. To cement her alliance, he encouraged Bert to speak frankly with Mrs. Hubbard regarding Hyde,[51] who threatened the well-being of the National Geographic Society. She responded to Bert in belli-cose terms. "Load your guns and be ready to fire," Bert recalled her saying.[52] Charles J. Bell closed ranks with Mrs. Hubbard, followed by Elsie, to whom Bert wrote a detailed letter describing his difficulties. "Hyde knows that if he can turn me out, he is safe for years," Bert wrote. "He wants me out now so that when your father returns he can make it appear that I voluntarily withdrew. . . . Then a new man will have to be broken in & Hyde naturally is still editor." When he was done maligning Hyde, Bert shifted into romantic mode, which removed any doubt of his intentions toward Elsie. "My precious little sweet-heart, the prettiest, loveliest and noblest of all women," he wrote. "I am going to have you for my own, mine wholly, not just a slice of you, but your lips, your hair, your dear eyes, your whole self is going to be mine, do you know that girlie? You can't escape and you don't want to, do you, sweetheart?" The letter, written on National Geographic stationery, moved Elsie to sympathy.[53] With the rest of the Bell and Hubbard clan lined up, Grosvie then made his final ap-peal to the patriarch on Bert's behalf.

"Whether one be my son or not, I have an intense sympathy with a young man," Grosvie wrote to Bell. "It makes my blood boil when from some of those above him a young man receives the very opposite of appreciation after he has striven to do his work to the best of his ability. . . . I do not deem it worthwhile to make any reference to charges of ignorance, incapacity and laziness which Mr. Hyde brings against Bert. . . . The purpose seems to be to get Bert out of the Magazine before you return. Nor shall I regret greatly if he should prefer to resign rather than to stay on. But what makes me indignant is his being treated in this way. I think I shall advise to put up with the present disagreeables [sic], keep as cool as he can and wait for your return."[54]

Then both Grosvenors settled down to wait, one of them nervous for the future, one supremely confident. "When this state of affairs is all over and you are at rest and peace," Grosvie wrote Bert, "I can not help believing you will look back upon these days as the most valuable and fruitful portion of your life. . . . You are in fact master of the situation."[55] He was right. Within days, Elsie sent word from London that good news was coming. Then a telegram from Alexander Graham Bell arrived, saying that he would return to "see what I can do for my boy." Then, out of the blue, a surprise from Elsie: She wanted to marry Bert, whose troubles with Hyde and feverish love letter provided the final spark. Even Mabel Bell revised her assessment of Bert's strength. "We feel that Gilbert has proved his mettle in this summer's trials and deserves the reward Elsie wants to give him. . . . I doubt whether Elsie would have been as sure of her own mind if all her love and sympathy had not been aroused by her indignation at the attacks upon him."[56]

As promised, when Bell returned in September, he made a show of supporting Bert. Bell's loyalty to his friend Edwin A. Grosvenor was stronger than his loyalty to John Hyde, whom Bell had inherited when he reluctantly took over National Geographic. Since Bell had approached Professor Grosvenor with the intention of hiring one of his sons, he could hardly abandon Bert now. Bell had developed a sense of authorship regarding Bert, whom he had grown to like, especially when the young man stood up to those who had tried to thwart Bell's innovations at the magazine. Most important, Bert's fate had become a family concern—with the flourishing of his romance with Elsie, there was now no choice for Bell but to back his future son-in-law. Bell quickly persuaded the board of managers to affirm Bert's permanent position, to approve his promotion to the newly created title of managing editor, and to give him an $800 raise.[57] Never one for grudges or confrontation, Bell allowed the disgraced Hyde to remain as editor in chief, confident that he would cause no further trouble. Bell then turned his attention toward England, where his family added an autumn wedding to their European vacation. Bert suggested a date he considered auspicious—October 23, his parents' anniversary.[58] That was fine by Elsie, but she soon discovered a snag in the Church of England's ceremony, which required a bride to love, honor, and *obey* her husband. Love and honor she could promise, but Elsie drew the line at obedience to any man.[59] So the Bells shopped around London for a house of worship that would allow a nonconformist ceremony, and they found one at the King's Weigh House Church.

There the marriage of Elsie May Bell and Gilbert Hovey Grosvenor took place in a Congregational ceremony on October 23, 1900.[60]

But Bert's trials were not over. While the newlyweds began their honeymoon tour of Europe, the executive committee of the National Geographic Society remained busy at home. Without Bert's approval, the committee shifted the magazine's printing from Washington to New York, where S. S. McClure's company would also manage the journal's advertising. Bert returned home in November to find the new arrangements already in place. He was incredulous, certain that the deal would eventually lead to McClure's takeover of *National Geographic*. Armed with the first printing bills from New York, he marched in to see his new father-in-law, waved the bills around, and pointed out that the new arrangement had cost twice as much as the old one, with no increase in quality. How could Mr. Bell condone this?

"Well, Bert," said Bell, "the board made you the managing editor. You are responsible now."[61]

With that authority, Bert cancelled McClure's printing contract, returned control of the magazine to Washington, and methodically began to consolidate power. John Hyde quietly resigned in 1902,[62] removing any doubt about who controlled the National Geographic Society and its magazine. From Hubbard to Bell to Grosvenor, the line remained unbroken, defended by three families now acting as one.

CHAPTER FOUR
A NEW GENERATION

GILBERT H. GROSVENOR IN THE HEADQUARTERS OF
NATIONAL GEOGRAPHIC MAGAZINE.

When the anarchist Leon Czolgosz reached out of a crowd in Buffalo, New York, and fired two shots into President William McKinley, it announced the start of a violent new century. Yet the death of America's twenty-fifth president, in September of 1901, also cleared the way for a fresh start, symbolized by his exuberant replacement, Teddy Roosevelt.

The promise of the moment was not lost on Gilbert H. Grosvenor, who dashed off a note to his father as McKinley's funeral train lingered in the black-draped capital on its way home to Ohio. "It seems incredible to me that he can be dead," Bert wrote. "But in Roosevelt the country would seem to be very fortunate. The change will undoubtedly be very great in Wash., not at once but in a few months; it can't be otherwise for Roosevelt belongs to a different generation."[1]

Like the new president, the young editor had reason for optimism, even in the midst of national tragedy. Having just routed his chief detractor at *National Geographic* magazine and married into one of the country's most prominent families, Bert Grosvenor now relished the prospect of building up a popular new publication like no other. He had the full support of his new father-in-law, Alexander Graham Bell, a certified genius whose eagerness matched President Roosevelt's. Other family members, Gertrude Hubbard and Charles J. Bell, had just pledged more than $30,000 toward the construction of Hubbard Hall, which would soon take shape at Sixteenth and M Streets, a few blocks north of the White House. When finished, the new office building would assure the National Geographic Society a prominent, permanent presence in the capital. On top of these blessings, Bert and Elsie Grosvenor were expecting their first child in a matter of weeks. So even in the aftermath of McKinley's violent departure, life was about to reassert itself—this time in the bawling person of Melville Bell Grosvenor, born November 26, 1901.

Within hours, Melville Grosvenor was brought down the long curving stairs at 1331 Connecticut Avenue, where Bert and Elsie had moved in with her parents. Melville was tendered to a beaming Alexander Graham Bell, who displayed his first grandchild, squirming like a prized piglet, before admiring dinner guests. Then Grandfather Bell leaned close to the child's face and bellowed: "Baaa!" Melville answered with a satisfactory screech. "Aha! He has perfect hearing," Bell announced as the baby was bundled off to the safety of his third-floor nursery.[2]

Grandfather Bell's nocturnal demonstrations would be repeated at regular intervals, as Melville was plucked from his sleep and taken downstairs for another show, always producing waves of shouting and laughter in the parlor. "I have the greatest work to get anything like system, or regularity in the care of

Melville," wrote a dismayed Bert, who worried that the disorderly nights robbed his son of much-needed rest and overstimulated his delicate nervous system— just as they did Bert's.[3]

The young family soon moved out of the Bells' brick-and-stone mansion for the sanctuary of a rented house nearby. At the same time, they bought a lot on Eighteenth Street, backing onto their in-laws' property, and began to build their own place, which would be a close copy of the Bells'. Over the years, Bert and Elsie would fill their rambling house near Dupont Circle with plenty of company for Melville, as five sisters and a brother arrived.

As the family grew, so did the magazine, and thus the National Geographic Society. All three would remain intimately entwined for more than a century, identities and destinies interdependent. When Bell inherited responsibility for the National Geographic Society at the tag end of the nineteenth century, he gave the magazine new emphasis as a means of extending the organization's reach. In an effort to find new members, he had lowered dues from $5 to $2 a year and challenged his young editor to make the magazine irresistible for a broad national audience. Now it was up to Bert, the first paid employee of the National Geographic Society and the managing editor of its magazine, to put his father-in-law's ambitious vision into practice.

Bert had performed admirably since 1899, pushing the journal's subscribing membership up from the shaky one thousand it had been at his hiring to almost three thousand at the start of 1903. The National Geographic Society had become truly national in scope, with two out of every three members living outside of Washington and every state represented. When Hubbard Hall opened for business in the autumn of that year, Bert had been promoted to editor.[4]

He gathered the magazine's files and mismatched furniture from his shabby rented office in the Corcoran Building and moved them a few blocks north to Hubbard Hall, where the National Geographic Society finally had a home. The new office, a two-story structure built of buff-colored brick, had been designed in the Italian Renaissance style, with gracefully arched windows, detailed crowning cornices, a discrete balcony over the recessed entrance, and a heavy tile roof. And lest any visitor forget the founding father, the name GARDINER

GREENE HVBBARD marched across the building's stone entablature, rendered in bold neo-Roman letters. Intended to inspire, the stout and sturdy Hubbard Hall looked less like a temple of learning than it did a bank building—though it held very little money at the time.

"We did not even have the means to provide for its lighting or to take proper care of it," Alexander Graham Bell said of Hubbard Hall. "It threatened to be a white elephant."[5] But they opened the doors to their core constituency, those one thousand living in the Washington area for whom membership meant more than a magazine subscription. Local members went on seasonal outings touring Civil War battlefields, watching eclipses, enjoying picnics; they also flocked in their hundreds to the lectures Bert organized for them. Bell had added this assignment to Bert's other duties and he arranged talks on arctic exploration, the uses of radium, the Alaska boundary dispute, sand dune control, volcanic explosions, surveys of Siberia, and the basis of wealth and power in the United States (these being land, water, soil, industry, and mining). In his spare time, Bert played host to members who came to browse through the new library at Hubbard Hall, where books from Gardiner Hubbard's collection filled the shelves.

Beyond this flurry of local activity, the real focus of the National Geographic Society had turned outward, as Bell had planned, past the small circle of well-wishers in the capital to the nation at large. Now, instead of the National Geographic Society supporting its infant magazine, the adolescent journal was expected to keep its parent alive. Bert begged his father, Adolphus Greely, and other prominent friends of the National Geographic Society for more names and addresses. He sent out rafts of letters to prospective members and prayed for good returns. The new headquarters functioned primarily as a magazine office and a direct-mail center, its purpose being to find and keep as many readers as possible for *National Geographic* magazine.

In the meantime, Bert tried to make the journal less technical and more appealing for a popular audience. "If you can make the magazine interesting and entertaining to the general reader," Bell wrote to Bert Grosvenor, "there is no reason why the membership should not reach ten times ten thousand, in which case the magazine would become one of the most important and influential journals in existence. . . . Therefore I say form your ideal for the magazine and make it so high that you can only hope to approximate it."[6]

Bert took this advice, hired a small staff of clerical helpers, and set out to make the National Geographic Society bigger and better, gambling the organization's still scanty reserves for direct-mail campaigns to boost the magazine's circulation. By 1904, he would gain another few thousand members. It was not enough to put the organization in the black yet, but the numbers would keep climbing, reaching a circulation of more than ten thousand in 1905 and surpassing twenty thousand by 1907. Prospective members responded favorably to the polite solicitation letters Bert sent from the National Geographic Society, which gave the impression that one was being invited into a very special club. "I have the honor of advising you," read a typical letter, "that the Membership Committee extends you a cordial invitation to become a member of the National Geographic Society."[7] Such letters came on embossed letterhead and heavy stock, conveying the organization's importance and its dignity. Bert understood the importance of appearances. "People don't pay any attention to cheap letters, cheaply written letters," he said. "The same would be true . . . about the magazine. In the first place you couldn't get good pictures unless you used good white paper and good ink and good press work."[8]

Despite earlier pressures to transform *National Geographic* magazine into a strictly commercial enterprise, Bell and Bert insisted on maintaining it—and its parent Society—as a membership organization. There were no stockholders clamoring for dividends and all proceeds went back into operations. When the organization started to make money after 1905, it began to pay taxes and did so until the 1920s, when it was declared tax-exempt. All along, Bert Grosvenor was promoting the magazine—and membership in the National Geographic Society—as something extraordinary, beyond the reach of personal greed or daily commerce.

"As the magazine is not published to make money for any one," he wrote in an early solicitation letter, "but is the sole property of a great national organization which reinvests practically all receipts from the magazine in the publication itself, we can most strongly and disinterestedly ask you to recommend it to your friends."[9] Under National Geographic Society regulations, each new member had to be proposed by an existing one, who sent a nominating form to Washington. There, a membership committee (usually consisting of Bert, Bert, and Bert) reviewed applications and accepted almost all of them.

With a few notable exceptions, the standards for membership were rather loose; if one had a sponsor, a stated interest in geography broadly defined, and two dollars to spare, membership was automatic; if no sponsor was available, the National Geographic Society could usually find one. A Chicago resident named Al Capone was a member of the National Geographic Society until 1932. When he entered federal prison that year he was stricken from the membership rolls, but he continued getting the magazine as a subscriber, for which he was charged an extra fifty cents.[10] Capone's subscription arrangement was very unusual. Except for a few notorious cases, most individuals who wanted the magazine were automatically made members, while institutions such as libraries, prisons, and schools had subscriber status.

Letter by letter, issue by issue, Bert slowly created an identity for *National Geographic* magazine, which would come to be known as approachable, decent, authoritative, clear, optimistic, inquisitive, and all-embracing in its interests— much, in fact, like Alexander Graham Bell. Although Bell did not create a unified blueprint for the magazine, he sketched out the broad principles over several years in a series of letters to his son-in-law. Bert Grosvenor, in turn, converted Bell's multitudinous publishing ideas into print each month, enlarging upon them as he gained in confidence and experience.

He made the reporting accurate and reliable, tried to anticipate news developments, and kept a reservoir of articles on hand to publish subjects at the peak of public interest. He modernized the look of the magazine. He cleaned up the cover, reducing the crowded type and simplifying its appearance. He experimented with cover formats until, on the fifth try, he found one that stuck. Launched in 1910, this design featured a thin yellow outer border, an internal border of oak leaves representing strength and longevity, and a crowning garland of laurels symbolizing the arts of civilization. Cartouches depicting the hemispheres were inset at the cardinal points of the compass among the foliage—these maps reflected the magazine's intention to report stories from every quarter. The cover's sturdy frame was a window, inviting readers to look inside. There they found articles cleanly typeset, two columns per page, relieved by plenty of white space and the occasional halftone photograph. The new look improved the readability of stories that had formerly appeared in a single broad column, looking impenetrable and uninviting.[11]

Bert also began to weed out the long, technical articles that had characterized the magazine's early years. In their place he published shorter pieces on a dizzying array of subjects domestic and foreign, from U.S. coal-mining statistics to accounts of tribal customs in New Guinea. Some articles were excerpted from recently published books; others were lively originals from resilient contributors like Eliza Scidmore, who roamed Asia and reported on funeral rites in India, elephant hunting in Siam, art museums in Mukden, natural wonders in Ceylon, and a mountaintop necropolis in Koyasan, Japan, where "the humblest may freely go and cast a fragment of a cremated body into the well in the Hall of the Bones."[12] Other bylines appeared from foreign ministers serving in the United States; they wrote intelligent pieces designed to deepen cultural understanding of Mexico (Juan N. Navarro), France (Jean Adrien Antoine Jules Jusserand), Great Britain (James Bryce), Japan (Eki Hioki), and Peru (Alfredo Alvarez Calderon).

For the most part, the magazine approached its foreign subjects with a sensitivity that celebrated the dignity of humanity and the brotherhood of man. Emperor Haile Selassie of Ethiopia not only sat for an early photographic portrait in the magazine, he became an enthusiastic member who collected back issues, bound them in monkey hide, and complained when his magazine arrived late to Addis Ababa. The subjects of his empire, like the ordinary people of Chinese Turkistan, Russia, or India, generally received decent treatment in *National Geographic* magazine. Writing from Asia in 1910, for instance, contributor Melville E. Stone argued for an end to racial prejudice in that part of the world: "As a soldier . . . the man of color has shown himself a right good fighting man; in commerce he has, by his industry, perseverance, ingenuity, and frugality, given us pause; and before the eternal throne his temporal and his spiritual welfare are worth as much as yours or mine."[13]

Closer to home, however, the standard sometimes shifted. Blacks were most often ignored in the magazine's articles, even in southern states, where African Americans comprised a healthy share of the population. If they were mentioned at all, blacks were often portrayed as affable, shuffling stereotypes, like this man, delivering a load of Georgia melons to a busy New York market: "Lawdy, dere's a powerful lot of hauling goes on here. Ah'm plumb confused."[14] The author of this article, J. R. Hildebrand, was not some fly-by-night contributor but a valued senior editor who worked at the magazine for more than twenty years.

Overt racism surfaced occasionally. Writing for the magazine in 1912, Prof. Robert DeC. Ward of Harvard called for stricter limits on immigration to the United States. Reprising the antiforeign theme from the magazine's earliest issues, he worried that the flood of recent Europeans was diluting valuable "native" genetic stock. Some sort of eugenic screening, combined with physical and mental tests, was in order, he said. "Should we not exercise at least the same care in admitting human beings that we are now exercising in relation to animals?" he asked. "Yet it is actually true that we are today taking more pains to see that a Hereford bull or a Southdown ewe, imported for the improvement of our cattle, are sound and free from disease than we take in the admission of an alien man or woman who will be the father and mother of American children."[15]

Alexander Graham Bell would have disagreed with Ward's bigoted conclusion. But a lifetime's interest in genetic studies had led Bell—himself an idealistic democrat, an enthusiast for universal suffrage, and an outspoken foe of racial intolerance—into association with the eugenics movement, which held that the human race could be improved by selective breeding. Professor Ward obviously held that view, which also found support from the plant breeder Luther Burbank, from the biologist David Starr Jordan, chancellor of Stanford University, and from other prominent scientists of the early twentieth century. Without seeming aware of the racist implications, Bell loaned his prestige to the eugenicists by attending their conferences, by serving on committees, and by publishing papers on human genetics in journals embraced by the movement. In retrospect this pseudoscientific path seems absurd, but it was not out of character for an inventor whose mind was open to all manner of wacky ideas. After all, he thought humans could fly; he thought people could communicate across the barrier of death; he was temporarily captured by phrenology; and he spent more than twenty years breeding sheep with extra nipples in the mistaken belief that they might produce more milk. He pursued each of these—most of them lost causes—with his usual enthusiasm until they failed him, which the eugenics movement did soon enough. When he realized that anti-Semites and racists had seized control of the discussion, he went off to search for more creditable lines of inquiry.[16]

The same, alas, cannot be said of his son-in-law, who published bigoted views in the magazine from time to time and who pursued—with varying degrees of effort and success—a racist membership policy through the formative years of the National Geographic Society. Despite such exotic exceptions as Haile Selassie, the truth was that most black members were unwelcome in the National Geographic Society. This exclusionary policy was in place by at least 1921, when Bert Grosvenor noted the presence of members in "every community of 100 white people in the United States."[17] As that membership grew in subsequent years, Bert and other executives worried that without careful monitoring, the expansion of the National Geographic Society could bring unwanted new members into the family, particularly in the capital, where blacks might try to use the library, vote at annual membership meetings, or appear at lectures, alienating the loyal white initiates. African Americans were excluded from membership—at least in Washington—through the 1940s.

The magazine's public forays onto such disturbing terrain, while memorable, were infrequent. Most stories from the early Bell-Grosvenor collaboration appeared to be models of mainstream propriety—sober, respectable, and well scrubbed—and for good reason: They were often written by U.S. diplomats, consular officers, and government officials. Public servants provided their material for free, which made the articles attractive to a young magazine with tight budgets. This brand of contributor wrote clearly and, if not with style, at least with a worldly authority based on experience. Bert was particularly proud of pieces by then secretary of war William Howard Taft, a former governor of the Philippines, on U.S. policies in that island territory, which was often in the news as a result of the recent Spanish-American War.[18] In this same era, John W. Foster, a former secretary of state under Benjamin Harrison, prepared an informative background piece on the Boxer Rebellion in China, which had raised concerns about foreign commercial prospects in Asia; Foster provided an unsentimental assessment of China's future, which he believed to be promising, "if internal peace is preserved."[19] Lesser officials filled the magazine with articles on the weather bureau, aquaculture, forest preservation, the Persian Gulf, railroad building, the Panama Canal, pearl fisheries, Cuba, red ants, archaeology in Asia Minor, and irrigation schemes for western states. One such contributor, O. P. Austin, chief of the U.S. Bureau of Statistics and secretary of the National Geographic Society, produced a story entitled "Queer Methods of Travel in Curious Corners of the World," which charmingly described the decorated camel carts of Delhi, the dog-powered milk wagons of Belgium, the ox-drawn sleds of Madeira, and the inflated bullock skins of India, which were used for river crossings. Perhaps these glimpses of the world were tinted with a touch of cultural smugness, but they emphasized infinite variations on the theme of humanity.

So many bylines from so many bureaucrats and military officials appeared in the magazine that it sometimes seemed like an extension of the government. This coziness between the journal and official Washington, which might be unseemly by today's standards, bothered neither Grosvenor nor Bell in the least. They were not interested in uncovering scandal or criticizing policy but meant to celebrate "this great world of ours," as Bert once put it.[20] Bell encouraged his young associate in this direction and pressed for even closer cooperation with the government. "We have exceptional facilities for obtaining inside information concerning the work of the Departments," he wrote in 1901. "I would urge upon you the importance of frequently calling upon your associate editors and ascertaining what is going on in the Departments, so that the public press of the country may find out from you."[21]

Over time, the National Geographic Society's cultivation of Washington's power structure began to pay dividends. Presidents, senators, and other dignitaries were happy to attend the organization's annual white-tie dinners. In his dual role as editor and liaison to local members, Bert instituted the first such banquet in 1905, when William Howard Taft, Bert's cousin by marriage, was the central attraction. Aside from giving Washington members an added benefit, the dinners boosted the National Geographic's profile in the capital. Such events also gave Bert access to senior officials when he needed a favor. For instance, when Alexander Graham Bell was appointed by the Smithsonian Institution to escort the remains of its benefactor, James Smithson, from Genoa to Washington in 1904, Bert convinced President Theodore Roosevelt to provide a special escort. Roosevelt sent a naval warship, the USS *Dolphin,* to collect the funeral party upon its arrival in New York and conduct Smithson to the capital, where the entourage was met by a military honor guard.

Smithson's Washington finale went off smoothly—but just barely. Weary from his recent voyage, Alexander Graham Bell had tried to stay in bed that day when his valet, Charles F. Thompson, came to rouse the inventor, a reluctant morning person under the best of circumstances.

"Breakfast and newspapers here, Mr. Bell," said Thompson. "I can't give you another minute."

"Why am I to get up if I don't want to?" Bell asked.

"Because you are to be at the Navy Yard at 10 A.M. sharp to escort the remains of James Smithson to his last resting place."

"Nonsense," said Bell. "He's been dead 50 years."

"Can't help it, sir. He is in Washington and you brought him here."

Bell went silent for a moment. "What did he come back here to bother me for?"

With Mabel Bell's help, Thompson got the great man out of bed, dressed him in the appropriate formal clothes, and delivered him to the Navy Yard with minutes to spare.[22]

Within weeks of Smithson's resettlement, Bert Grosvenor was back at his desk worrying over the January 1905 issue of *National Geographic,* which was going to be late unless he found a way to fill a gaping hole in the contents. The printer was demanding eleven pages of copy—now—and Bert had nothing to offer. Feeling sorry for himself, as well as a bit panicky, he shuffled through the papers on his desk, found a bulky envelope, and opened it without much enthusiasm. When he saw the contents, his heart began to race. The packet held fifty crisply executed photographs from Lhasa, the mysterious Tibetan capital. Now strewn across the editor's desk were pictures of the Potala soaring white among gray storm clouds; an image of two dark-skinned women in windblown skirts trudging barefoot on their way to market; and other glimpses into a city that had been off-limits to visitors for centuries, taken by photographers from the Imperial Russian Geographical Society of St. Petersburg. The pictures, among the first ever from the forbidden capital, were being offered to *National Geographic* for free. Bert did not hesitate. He ordered plates, filled the gap in the January issue, and went home for the day. There he warned Elsie that he expected to be fired for publishing eleven solid pages of scenic pictures using the magazine's scarce resources to pay for the plate making.[23] But when the January issue arrived, his unprecedented use of photography proved to be a hit with readers, who stopped him on the street to express congratulations.

Bert had stumbled upon a popular way to tell stories with photographs, in addition to words. Although other prominent editors of the day—notably the respected Richard Watson Gilder of the *Century* magazine—shunned the use of such illustrations as vulgar, Bert instinctively understood the power and appeal of this relatively new medium, which placed previously homebound readers in Russia or Africa or any of the far-flung places *National Geographic* could now show them. From the time of his success with the Lhasa coverage, he explained, the word "photograph" became "as musical to my ear as the jingle of a cash register to a businessman."[24]

He heard that music again while meeting early in 1905 with William

Howard Taft, who mentioned that the government would soon publish its first census of the Philippines, along with an extensive collection of photographs that had been commissioned while Taft served as governor of the territory.

Could Bert borrow the pictures?

Of course, Taft said. This sent the editor scurrying to round up more than a hundred copperplate engravings from the U.S. Census Bureau, which had already paid for their preparation. That April, Bert filled most of *National Geographic* with photography from the Philippines. The coverage, which included 138 pictures, was well received, adding to the magazine's growing identification with photography. Although the quality of *National Geographic*'s pictures would remain uneven for some time, it hardly seemed to matter in those early years, when the immediacy and the novelty of the medium was an attraction in itself. With time, as Bert's sophistication grew, the magazine's pictures improved in quality as in quantity. By 1908, photographs occupied at least half the space in the eighty-page magazine, which would allocate an even greater share of its pages to illustrations in future years.

Alexander Graham Bell viewed his son-in-law's innovations with pride, but he never shied from suggesting improvements. Although Bell found the use of so many pictures impressive, he thought they looked like padding because they had not been thoughtfully integrated into the journal's editorial content. "The strange feature about the matter," Bell wrote, "is that in this case that which is intended as padding is the most interesting part of the Magazine. . . . I must confess to an occasional feeling of disappointment at the lack of connection, and it seems to me that one notable line of improvement would be, either to adapt the pictures to the text, or the text to the pictures. Why not the latter? Judging by myself, the pictures are the first things looked at."[25]

In these few words, Bell had distilled an editorial philosophy that would guide *National Geographic* magazine for most of its long life. The magazine would henceforth be organized along photographic lines, on the assumption that readers went first to the pictures, then to captions ("legends" in the language of *National Geographic*), and then finally to a complementary article, which might run to eight thousand or ten thousand words. All three parts of the story sup-

ported one another, with photographs providing emotional impact, text fur-
nishing the substance, and legends the connective tissue. Maps became an added
specialty.

The magazine's pictorial philosophy provides the most lasting evidence of
the fruitful collaboration between inventor and editor, who also discovered a
deep affection for one another. Bert often addressed Bell in letters as "My Loving Father" and signed himself "Your Loving Son." Bell matched the sentiment,
once scratching out the editor's title on a letter and replacing it with a scrawled
phrase of his own: "No, not the Editor but 'My boy.'" When Bell learned that
Bert and Elsie were suffering financially, he not only arranged for the family
trust to take over their mortgage, he also had the bank—his cousin Charlie's
bank—refund the principal the Grosvenors had invested and provided them
with $10,000 in stocks to begin their own portfolio.

"I realize the fact that the difficulty you have experienced in making both
ends meet is not due to extravagance, or anything unworthy of you," Bell wrote.
"Start the New Year without any debts, then my gift of $10,000 . . . will set
you on your feet fairly and squarely."[26]

In addition to these infusions of capital and emotional support, Bell continued showering Bert with ideas for the magazine—on the need for better photographs as "bait" for readers, on the unrest in Korea, on the Mad Mullah's threat
to Abyssinia, on erosion at Niagara Falls, on the health of Emperor Franz Josef
of Austria, on a new volcanic island in the Bay of Bengal, on polar exploration
by Amundsen, Peary, Cook—the world and all that was in it. Many years later
Bert suggested that the sheer volume of his father-in-law's ideas made it difficult to take them seriously. "Mr. Bell was so full of ideas that if I tried to answer
them all I'd just be standing on my head and getting nothing done," Bert said.[27]
In truth, he had relied upon Bell's creative powers and petitioned him to keep
the ideas flowing, especially in the magazine's critical early years.

As with most successful collaborations, the Bell-Grosvenor partnership
amounted to more than the sum of its parts. Bell provided confidence, enthusiasm, and ideas, which might have flown away on the wind if Bert had not been
there to catch them. With patience and doggedness, he brought Bell's scattered

notions into print each month. Neither man would have been nearly as effective without the other.

Nonetheless, for all the pleasure Bell took in charting a new course for the National Geographic Society, he was primarily an inventor, not an editor, and his greatest invention was long behind him. Now in his midfifties, he knew that time was slipping away, as Mabel Bell explained in a family letter. "What he is trying so hard to do now is creative work, which is essentially young people's work. He is now in the prime of life, but it is not more than this, and he has nothing more to look forward to."[28]

Bell felt the pull of Canada, where he had established a summer home and extensive laboratories and his flying experiments were languishing. With Bert in control of the magazine and membership climbing past three thousand, the organization was finally heading for solvency. Bell's fellow board members had been impressed by the young editor; although not yet a member of the board himself, Bert's influence was felt among the trustees on an almost daily basis as he solicited them for new members, helped them with articles, and arranged for them to lecture. Bell's duties as the National Geographic Society president had been largely ceremonial—presiding at meetings, giving speeches, dropping by Hubbard Hall from time to time to sit in his office—and he could see that Bert no longer required his inspiration or his protection. The boy would do fine on his own. So, having set things in motion for Bert, for the magazine, and for the young National Geographic Society, Bell announced in 1903 that he was stepping down as president. He left his son-in-law with a breezy valediction.

"Bert," he said, "you are competent to paddle your own canoe."[29]

CHAPTER FIVE
EDUCATING MELVILLE

MELVILLE BELL GROSVENOR AND
ALEXANDER GRAHAM BELL IN BADDECK, C. 1912.

At almost precisely the time that Alexander Graham Bell shed his title at National Geographic to resume flight experiments in Canada, a spindly biplane lurched over the sand dunes in North Carolina, wobbled aloft for twelve seconds, and clattered back to earth. On December 17, 1903, Wilbur and Orville Wright had beaten Bell to the sky.

The Wrights' success must have been a terrible disappointment for Bell. But he did not let it show. He praised the Ohioans for their accomplishment and expressed pride that Americans had led the world into the air. Then he set out to improve upon the Wrights' achievement. At Mabel's suggestion, and with her pledge of $20,000, he established the Aerial Experiment Association (AEA) in April of 1907, with the goal of producing a flying machine capable of sustained, stable flight.[1]

The association intended to develop a motorized kite that would carry a pilot. But the kites, which Bell gave fanciful names like *Cygnet* and *Frost King,* proved difficult to steer and therefore unfeasible. In the process of testing kites, however, Bell developed and patented the tetrahedron as a structural element, like the one employed years later in the design of towers, bridges, and buildings. And when the AEA gave up kites for biplanes, it was Bell who suggested that the tips of the upper wings be hinged to enhance lateral control. This aileron-like contrivance, which the group patented, became an essential element in modern airplanes and is still used today.[2]

Despite these advances, Bell never achieved the fame in aeronautics that he had in acoustics. Indeed, over a lifetime occupied with lofty goals, he had grown used to falling short of them, fighting through the ensuing (and usually brief) gloom, and sometimes comforting himself with the philosophy of Robert Burns: "The best laid plans of mice and men gang aft agley!" Bell would spout,[3] meanwhile erecting another formidable barrier to be surmounted. Even after decades of marriage, his resilience continued to impress Mabel Bell.

"I do so appreciate all the wonderful, unfailing, uncomplaining patience that you have shown in all your work and the quiet, persistent courage with which you have gone on after one failure after another," she wrote him. "How very, very few and far apart have been your successes. And yet nothing has been able to shake your faith, to stop you in your work."[4]

Even if Bell's aeronautical achievement could not compare to that of the Wrights, at least he had the satisfaction of seeing his own planes fly. With Douglas McCurdy, a local engineer, at the controls, the *Silver Dart* lifted over the frozen surface of Bras d'Or Lake at Baddeck, Nova Scotia, on February 23, 1909, and flew for half a mile at forty miles an hour. That flight, the first in the

British Empire, prompted a euphoric Bell to proclaim that Tuesday a "Red Letter Day!" in his diary.

More than that, the *Silver Dart* had flown where all of Baddeck could see it, which helped to vindicate Alexander Graham Bell in the insular Canadian community where he and Mabel had first settled in 1886. Seeking relief from Washington's torrid summers, the family found solace on this rugged fringe of the Cape Breton Peninsula, with its low blue mountains, tumbling salmon rivers, and pine-scented breezes. They bought fifty acres, gradually adding several hundred more from neighboring farms and woods until they owned most of the Red Head promontory across the water from town. Bell renamed the sea-wreathed headland Beinn Bhreagh (pronounced "Ben Vree-ah"), Gaelic for "beautiful mountain." At the foot of this mountain, within a scone's toss of the shipping lanes, the Bells built their Point House, a chateau-style mansion of cedar siding and stone, with round towers, massive chimneys, and verandas and balconies jutting out at odd angles. A grand structure, it looked like a cross between a citadel and a lunatic asylum.[5]

The hardworking farmers and fishermen of Baddeck might have regarded Point House with mild suspicion as they looked up from cutting hay or pulling lobster pots to see that unfamiliar hump of timber and shingles clinging to Red Head. The suspicion would have hardened when the Bells settled in and locals began to hear about queer doings at the Point House, where Mr. Bell worked by night and slept by day and came down the stairs hooting like an owl and drank his soup through a glass straw. He wandered the mountain in nocturnal storms, shielded against the cold by a black fox fur his wife had discarded.[6] Sometimes he lay nude on the roof of his houseboat sketching seagulls.[7] Sometimes he rode about in a wagon behind a great gray horse named Champ, which would stop and start as Mr. Bell clapped his hands. Once he tried to convince a local girl that he could mesmerize a chicken.[8] People got used to seeing him in town, dressed like an oversized boy in a Norfolk jacket and knickers and his favorite Pyrenean beret.[9] But they never grew accustomed to his talk of flying, which seemed quite mad to the no-nonsense locals with names like MacDermid and

MacDonald and MacFarlane who still observed the Sabbath with a stiff dose of Gaelic singing and Calvinist sermons and not much else.[10] They long remembered the day when Mr. Bell had floored the men's club by telling them, in that bold way of his, how a fellow would soon get on a flying machine in New York and step off in London a few hours later. The men had listened politely and adjourned their meeting and stood by the lake afterward shaking their heads as Angus Furguson spoke for all of them: "Did you ever hear such nonsense in your life?"[11]

It remained nonsense until that snapping cold February day in 1909 when the whole community turned out on ice skates and in horse sleighs to cheer one of their own, Douglas McCurdy, as he pulled the *Silver Dart* shuddering up into the air over the ice and Mr. Bell followed with his dark, searching eyes, which finally shone with boyish delight when McCurdy got airborne.[12] Mr. Bell was one of their own now too. He and Mrs. Bell had gradually won that honor by their dealings with the local people. They had invited villagers to parties and dinners at Point House, provided work for many in the community, and organized clubs to keep the men and women occupied through hard northern winters. One Christmas Bell had handed an envelope to Walter Pinaud, one of his workers and the father of eight. When Pinaud got home and opened it, he found the deed to his house inside, paid off in full. Once you got past the shock of Mr. Bell's eccentricities, he was a fine neighbor.[13]

As time went by, the Bells gradually increased their stays at Beinn Bhreagh, traveling north in March or April, returning to Washington after Thanksgiving.[14] Bell needed the solitude he found in Nova Scotia, where ordinary distractions did not intrude. He had been haunted by interruptions and social obligations in Washington. He had once locked himself in the bathroom there to avoid going to tea at Twin Oaks; Mabel armed herself with a Mississippi bullwhip, lashed at the window of Bell's sanctuary, drove him into the open, chased him through the house, and cornered him in the attic, where she found the great inventor hiding behind a stack of carpets. He went to tea.[15]

Other rules, looser rules, applied at Beinn Bhreagh, where it was understood that Bell would remain scarce. He could disappear with a satchel of notebooks and pencils for a solitary weekend of thinking and writing on the *Mabel of Beinn Bhreagh,* the old houseboat he had hauled up among the spruces in a secluded cove;[16] there, with a fire crackling in the woodstove and a pipe of tobacco for company, he worked in what he called the Roman style, lying prone on a worn horsehair sofa gradually molded to his form.[17] It was here that some of his best ideas took shape, including many of those he passed along to Bert at *National Geographic* magazine.

"You have a high standard of excellence to keep up and it is difficult to suggest improvements," Bell wrote to Bert in 1909. "It is a great pity, I think, that you don't continue writing articles on your own account. You were making a name for yourself, why not continue in that line of work? There is little in the magazine to bring you personally before the public. . . . In a word my recommendation to you is WRITE."[18]

For once, Bert did not take his father-in-law's advice. "I can't give up my editorial work to write," Bert told Mabel Bell, who often became the conduit for disagreements between Bert and her husband. "The success of our Magazine depends on the personal attention I give it and on the business direction by me of the Society. . . . This sitting down and dashing off articles in an hour or so exists only in imagination. No successful writer ever made a reputation by scribbling. Writing requires thought and time, neither of which to my great regret, can I now afford. Besides I have nothing to write about."[19] For the rest of his career, Bert would stick to editing.

As the magazine's circulation grew—from a meager one thousand in 1899 to more than three thousand in 1904 and past ten thousand in 1905, Bert's confidence and independence grew accordingly. Impressed by his performance, the board of managers boosted his annual salary from $2,000 to $3,500 in 1905. At the same time, the board brought Bert onto their six-person panel, and gave the twenty-nine-year-old editor a new title. Now he was director of the National Geographic Society—the parent organization that, through its board of managers, set policy, collected dues, and arranged annual meetings—in addition to his position as editor of its official journal. Bert's influence was spreading. "Consider yourself the National Geographic Society," Bell told his son-in-law.[20]

Buoyed by his promotion and recent success, Bert began advising Alexander Graham Bell on how to deal with the Society's board and launched his own initiatives, proposing awards for polar explorers like Robert E. Peary and Roald Amundsen. But he went too far in advocating National Geographic's highest honor, the Hubbard Medal, for the Wright brothers in November of 1908. Without clearing the idea with Bell, Bert took the proposal to the board of managers, which voted to recognize the Wrights for "discoveries which will greatly promote exploration and geographic research."[21] When Bell heard of this action, he objected, as did Adolphus W. Greely, still a board member. Greely felt that the Wrights' achievement lay "outside the domain of geography."[22] Given that Greely had previously served as the National Geographic Society's vice president for the geography of the air and had written articles on aviation for the magazine, his objection seemed curious; most likely he was acting out of loyalty to his old friend Bell, who had been a rival of sorts to the Wrights. For his part, Bell may have objected because of his deep friendship with the late Samuel P. Langley, the scientist who had tried—and twice failed— to launch manned airplanes in 1903. Bell's note to Bert offered no explanation: "Don't think Hubbard Medal would be appropriate for Wright Brothers."[23] That was enough. The award was withdrawn by the Society, which never recognized the Wrights' success but subsequently awarded the Hubbard Medal or its equivalent to Richard E. Byrd, Charles A. Lindbergh, Anne Morrow Lindbergh, Amelia Earhart, John Glenn, and several other astronauts and aviators. The slighting of the Wrights represented a rare instance when Bell's generosity failed him.

For the rest of his life, Bell remained influential in National Geographic Society affairs, even from his faraway Canadian retreat. There his days were crowded with so many experiments and family interests that he was constantly threatened with derailment. He avoided it by carefully regimenting his time at Beinn Bhreagh, where his schedule read as if it might have been written by the author of Ecclesiastes: There was a time to rise (10 A.M.), a time to dictate "Home Notes" (11 A.M.), a time to sing with the grandchildren (12 noon), a time to take the carriage down to the Kite House (12:15 P.M.), a time to dictate "Laboratory Notes" (4 P.M.), a time to hold hands with Mabel and walk home for tea (5:00 P.M.), a time to read to the grandchildren (5:30 P.M.), a time for dinner (7 P.M.), and a time for singing and playing the piano before the family went to bed (9 P.M.).[24] Then came a time for silence, the blessed silence settling like a benediction on Point House, where Bell worked alone until the sky began to lighten. That signaled a time for bed, for which Bell prepared by wrapping a towel around his eyes to shield them against the coming of the day. Then he folded his hands on his chest and descended into profound sleep.[25]

When the Bell grandchildren began to swell the family roster—there would eventually be ten of them—summer visits to Beinn Bhreagh became an essential facet of their education and a focal point for the clan's identity. "It's what held the family together," said Mabel Grosvenor, the second daughter of Bert and Elsie. She spelled her first words in Grampy Bell's study ("c-a-n-d-y p-l-e-a-s-e" and "t-h-a-n-k y-o-u"). She learned to look directly at people when she spoke to them. "We took it for granted that we had to look at Grandmother when we spoke," she said. "If we didn't she would take our faces in her hands and point them toward her. We thought she did that because we had been impolite. I didn't learn until I was ten that it was because she was deaf." Mabel Grosvenor learned to sail with her cousins and siblings on Bras d'Or Lake—the same place they learned to swim by Grampy Bell's unorthodox method, which involved taking a child to the end of the dock, tying a rope around him, and tossing him into the water. "My sister never quite got over it," Mabel Grosvenor recalled more than ninety years later. "But he didn't frighten me. I thought it was fun."[26]

Equipped with a childlike view of the world, Bell exerted a pied piper's influence on children, who usually found the inventor irresistible. "He was very good with children," Mabel Grosvenor said. "He could almost think like a child himself."[27] He developed a particular affinity for his first grandson, Melville Bell Grosvenor, a lively boy whose first memories were of sitting on Bell's lap and looking up at a jolly man with a Santa Claus beard.

"Pull my nose, Melville," said Bell. Melville obliged, provoking a rousing "Bow-wow-wow" from his grandfather. "Pull my hair, Melville." Melville pulled. "Baaaa!" said Bell. "Pull my beard, Melville." Melville yanked it, prompting a leonine growl that rattled the windows. Although Melville saw this routine repeated when each of his siblings arrived, he never grew tired of it.[28]

Melville begins to romp barefoot through family photographs about 1908, when he can be seen hanging on to his grandfather's hand, dancing along a line of men flying an experimental kite, shadowing Bell across a narrow bridge, flopping on the floor between Bell's protective legs. Even on the flat paper of an old picture, one senses a special tenderness between them, a bond explained by Mabel Bell: "I am almost jealous of the attention Mr. Bell is giving Melville," she said in 1914. "He did not devote himself so regularly to his own daughters. But he says there is a difference—Melville is a boy.* I wonder what would have happened if his own boys had lived."[29]

The boy and his grandfather became faithful companions. They shared a sweet disposition and a forgiving nature, which allowed both to overlook the failings of others. And they had in common the gift of enthusiasm, which occasionally led them astray. Unable to pass a certain bakeshop near Bell's office in Georgetown, the two often went inside, ploughed through generous helpings of apple pie, and tried to keep it secret from Mabel Bell, who forbade her diabetic husband such food at home. "Don't say a word to your grandmother," Bell warned.[30]

Melville stayed with his grandparents for extended periods in Canada when the rest of the family had decamped for autumn. As long as the boy kept quiet, he was always welcome in his grandfather's study, where he tried to help the older man, fetching books, looking up words in the encyclopedia, helping to organize reports. They had discussions from time to time—about how to measure the height of a mountain with trigonometry, about the Latin origin of English words such as "magnitude" and "agriculture,"[31] about geometry, using the twin dials of a barometer to illustrate various angles. They talked about news

*Two of Alec and Mabel's children, both boys, died in infancy.

items that had caught Bell's eye and found their way into his vest-pocket note-book. They read newspapers or books aloud to each other every day.[32]

The story of Robinson Crusoe became a favorite one summer. It remained so through the following winter, as man and boy laid plans to replicate the Daniel Defoe adventure when they returned to Baddeck. When summer finally arrived, they went into the woods like shipwreck survivors with no food or weapons, not even a knife, wearing their bathing suits. "We were castaways," Melville recalled. "Grampy insisted that I call him 'Rob' and he called me 'My Man Friday.'" It began to rain. "A fine Scotch mist!" Bell explained. The air grew cool. After a few hours in the woods, hunger drove the pair to Bell's houseboat, where they got the stove going, ate a can of beans, and considered the effect of latitude on climate. "Robinson Crusoe was lucky," Bell told his grandson that day. "His island was tropical."[33]

Without really knowing it, Melville was being educated by a master teacher who had never lost his gift for pain-free instruction. Having been an indifferent student in his own youth, Bell was more than critical of regimented educational methods, which took little notice of students' individual needs. He employed what he called the "indirect method" in teaching Melville. "He is under the im-pression that my eyesight is a great deal worse than it really is and that he is therefore saving my eyes, by reading to me things that are of no earthly interest to him. I have therefore . . . developed a great interest in fairy tales and stories that interest young people. . . . There has been a marked improvement in his ability to read naturally in short phrases, and with frequent pauses between."[34]

Bell sent Melville to look through the shelves of his extensive library on the pretext of needing a particular book. "I asked him to read the titles on that shelf and I would tell him when he came to the right book. In the course of his search he read the title of Hutchinson's 'Extinct Monsters.' He was interested in that title, glanced inside and gave a whistle of astonishment when he came across a *megatherium*. I said nothing but asked him to pass me the book I wanted. I im-mediately dipped into it leaving Melville to himself. As I had anticipated, he went right back to the bookshelves, took down Hutchinson's work and spent a happy half-hour in looking at the pictures and reading about them."[35]

Bell took these extraordinary pains with Melville not only because he loved his grandson but also because the family knew that the boy was in trouble and needed special attention. Like his grandfather, Melville was described as a "late bloomer" who had difficulty concentrating in school, perhaps because of an undiagnosed learning disability. Teachers at the exclusive St. Alban's School in Washington informed Bert that Melville would never make it to college, which must have been a keen disappointment to a stellar student raised among university professors.[36] "The education of Melville is the hardest problem I ever tackled," wrote Bert, who had taken his son out of class at age ten for instruction at home. There Melville got very little schooling in grammar or arithmetic, subjects his father felt were better suited to older boys.[37] But when Bert noticed that Melville had an interest in writing, he bought the youth a typewriter. Pounding away on the keys seemed to dissipate some of Melville's excess energy, and it helped him to organize his thoughts. He soon became so attached to the machine that he could not write without it. "I can rattle the keys off like the wind," a proud Melville wrote his father in 1914. "I am beginning to typewrite without looking at the keys."[38]

While Bert was at work, he had Melville compose stories of up to four hundred words four or five times each week. They reviewed the results together. "I rarely point out a misspelled word or a grammatical fault, as he usually discovers them himself," Bert wrote. "I also never laugh or poke fun at any of his expressions, *as I want him to be perfectly natural. . . .*"[39]

Melville also received special instruction in photography from his father, an accomplished camera-man who gave his son the same advice he did the professionals at *National Geographic*—shoot plenty of film. "Of course, the films are expensive," Bert wrote, "but one photographs to get a good picture, and it is more satisfactory and cheaper in the end to sacrifice three or four if thereby one good photograph is secured. So I have been encouraging Melville to take a lot of photographs, as experience is the best teacher."[40]

Although such training would boost Melville's later editorial career, Bert still worried about the gaps in his son's education, which was compromised by the boy's learning difficulties and by his emotional fragility. By age eight or nine, he had begun to demonstrate what his grandmother called "trouble of the mind" and Grandfather Bell termed "a disordered nervous system." Like others in the family, they worried over the youth's extreme timidity and hypersensitivity.[41]

The slightest criticism seemed to flatten Melville. He developed an unreasonable aversion to the dangers inherent in swimming, sailing, or riding a pony for fear of an accident. And he acquired a particular compulsion—that of twisting an unruly lock of hair on the back of his head—which his Grandfather Bell took to be the outward manifestation of deeper trouble. "I tell him there is no harm in fiddling with his hair," Bell wrote Bert, "but when he does it hundreds and hundreds of times a day he wants to check this habit. . . . He went into town yesterday to have his hair cut and I found out that one of the reasons the poor fellow had in his mind was the desire to get rid of that obnoxious lock of hair with which he was constantly fiddling. He told me he thought he could not fiddle with it any more *if it was not there.*"

Melville's grandparents kept him busy at Beinn Bhreagh, where the salt air and sympathetic treatment gradually helped him to conquer his constitutional fearfulness and, to some extent, his hair-pulling compulsion. And much to his Grandfather Bell's delight, Melville developed into an intrepid rider and a bold sailor in all weathers. "There is no question about it, he is getting braver," Bell reported to Bert. "He has been a great comfort to me here and quite a companion. . . . I can only say that I love the boy and feel proud of him."[42]

Melville and his grandfather remained close long after their time together at Beinn Bhreagh, an experience that served to underscore their kindred traits, genetic as well as learned. Melville became fastidious in dating his neatly typed letters, a practice his grandfather had stressed as a result of his patent cases. And, like his grandfather, Melville began to sign his letters with a flourish, forming a little switchback pattern under his initials. "I think Melville was more like my Grandfather Bell than he was my father in many ways," recalled Mabel Grosvenor. "He was terribly enthusiastic and impulsive. Father had to think things over and look at every angle. Sometimes Melville just got carried away—and that was *very* like my Grandfather Bell."[43]

After graduation from Bell's study, Melville resumed his unconventional education—at Dr. Henderson's progressive school in Samarcand, North Carolina, where students rode horses between Latin classes and ran naked among the pines; at an army-navy "cram" school in Washington, D.C., where Melville earned top marks in arithmetic* and almost flunked geography;[44] and finally at the U.S. Naval Academy in Annapolis, Maryland, where the young sailor became an expert one, trained in leadership. But even the combined attentions of Alexander Graham Bell, Dr. Henderson of Samarcand, and the U.S. Naval Academy could not iron out the essential Melville Grosvenor, a man who remained sweet for life, eager to please, and easily wounded. And in times of stress Melville the boy returned again and again, reaching back to fiddle with that lock of hair.[45]

---

*At this "cram" school, which prepared candidates for Naval Academy entrance exams, Melville scored 100 percent in arithmetic, 90 in history, 65 in geometry, and 65 in geography. The passing grade was 62.5:

CHAPTER SIX
THE FIRST HERO

EXPLORER ROBERT E. PEARY BOUND FOR
NORTH POLE ON *ROOSEVELT*, 1908–1909.

Back when a gentleman's word counted for something, Gilbert H. Grosvenor arranged for Comdr. Robert E. Peary and Dr. Frederick A. Cook to share the stage at the National Geographic Society's annual dinner in Washington. The date of the white-tie event—December 15, 1906—is memorable because it was the last time the two explorers, among the best-known travelers of their day, would appear together in public. After this, the two friends would become rivals and they would come to hate each other.

At the time of the 1906 dinner, more than twenty years had passed since Peary had stood with Gardiner Greene Hubbard and received from him an American flag, along with the charge to plant it as far north as Peary could reach. In all the years since, the naval explorer had been doing just that, edging a bit closer to the North Pole with each season, first from Greenland, then from the Canadian Arctic, but never quite attaining his objective. In the process, he had sacrificed seven toes, most of his meager income, and his youth to the crusade. Now at age fifty, he was by his own admission an old man—old, at any rate, for the rigors of arctic work—but he was still trying.[1] During the season that had just passed he had pressed to a claimed latitude of 87°6', farther north than any other explorer had managed, which put him not only in the history books but also within striking distance of the Pole.[2]

Cook, meanwhile, had been striving to make his own mark in the uncharted regions, swarming back and forth between the extremes of north and south like an Arctic tern.* He had traveled to Greenland with Peary in 1891 and had become enchanted with the region. He popped up on later expeditions to the Arctic before joining a foray to Antarctica that never reached that continent. Then he shot north again, shifting his sights to Alaska's Mount McKinley—at 20,310 feet the highest peak in North America.[3] In his last try at McKinley, the Brooklyn physician announced that he had reached the summit on September 16, 1906. He claimed to be the first person to do so.[4]

If their assertions proved true, both explorers had earned success worthy of recognition. By planting the Stars and Stripes in the most remote of places, they exemplified the grit and fortitude of the United States, which was demonstrating its presence in the world as the new century began. Under the stimulus of President Teddy Roosevelt, the United States had routed the Spanish from Cuba and rebuffed Colombia's objections to the Panama Canal, which would soon link the Atlantic and Pacific. Roosevelt had also built up the navy, about to

*The Arctic tern migrates some twenty thousand miles a year, from the Arctic to the Antarctic.

launch its Great White Fleet on a muscle-flexing global cruise. And for broker-
ing an end to the war between Japan and Russia, the president had just been
awarded the Nobel Peace Prize. "The equilibrium of the world is moving west-
wards," a French observer lamented in 1903.[5] That shifting balance caused the
public mood to swell with national pride on this side of the Atlantic, where
John Philip Sousa's concert tours were playing to packed houses. His music
made the perfect score for Manifest Destiny, which held that the United States
was ordained to dominate the Western Hemisphere, if not the world. Such a
world needed heroes in the mold of Roosevelt, Peary, and Cook—and so did
the National Geographic Society.

The organization had grown steadily under Alexander Graham Bell and
Gilbert H. Grosvenor. By 1905 the National Geographic Society crossed an important threshold, enrolling more than ten thousand members, thanks almost
entirely to a solicitation campaign by Bert Grosvenor, who had boosted the staff
of the organization to nine souls and put them to work at Hubbard Hall, where
they cranked out more than fifty thousand invitational letters to prospective
members.[6] The resulting flood of membership dues—at $2 a head—put the
National Geographic Society into the black for the first time in many years.
With a lordly surplus of $3,500, the organization could begin to sponsor scien-
tific research and expeditions, such as Robert E. Peary's ongoing quest for the
North Pole. He received the first grant—in the amount of $1,000—that the
National Geographic Society had given any explorer. The organization's largesse
would distinguish it from other organizations in future years, when its scattered
membership of armchair travelers would send adventurers across the globe to
gather stories for the magazine.[7]

By such gestures, Bert Grosvenor, Alexander Graham Bell, and their allies
on the board sought to transform the National Geographic Society into something more than a successful publishing house. "What we want," Bert explained
to Ida Tarbell, an editor and writer who had become a friend, are "not subscribers to a magazine but members of a society. . . . They will become members because they get two things, the distinction of membership in a well-known
society and also a good monthly journal. . . . The *Magazine* exists not for it-

self, like your *Popular Science Monthly*, but is the means, the tool by which we plan to build a society having thousands and thousands of members."[8] By linking its fortunes with Peary, Cook, and other adventurers, the National Geographic Society would become *the* authority on matters of science and exploration. In turn, these well-publicized explorers would lift Bert's organization from relative obscurity into national prominence. The explorers appreciated National Geographic's stamp of approval, which helped them to raise money—always a prime concern for an expedition. Both sides benefited from the arrangement.

With the first chill of October, when word reached Washington that Cook had scaled Mount McKinley, Bert invited him to be the guest of honor at National Geographic's annual banquet. Cook's triumph on McKinley also boosted him into the presidency of the Explorers Club of New York, where he replaced the departing president, Adolphus W. Greely.[9] Cook's moment of glory faded quickly, however, when November brought the news of Peary's Arctic achievement. At Bert's urging, National Geographic double-booked its guests of honor for December. It speaks to the innocence of those times that nobody at National Geographic made an effort to verify either explorer's claim—one's word of honor was proof enough.

Peary's recognition at the Washington dinner would be the greater, for he would receive the first Hubbard Medal, the National Geographic Society's highest honor, named for and endowed by the organization's founding family. Bert, chairman of the dinner committee, somehow snagged President Roosevelt to make the presentation. Roosevelt already admired Peary's swashbuckling style and he was predisposed to like an explorer who had named his expedition steamer the *Roosevelt*. With weeks remaining before the December dinner, Bert hastily had the solid-gold medal designed and struck by Tiffany & Company. On one side, the four-inch disk showed the Society's seal, a map of the Western Hemisphere with latitude and longitude lines converging on the poles; on the obverse, an engraved citation to Peary "For Arctic Explorations, Farthest North," crowned by a blue sapphire star. The medal arrived just in time for the banquet, which helped to inaugurate the capital's social season.[10]

More than four hundred guests—including the city's leading diplomats, scientists, politicians, and reporters—converged on the ornate gray castle known as the New Willard Hotel, where lights from the lobby resisted the winter darkness along Pennsylvania Avenue. Diners shuffled through a receiving line headed by Willis L. Moore, then National Geographic Society president and head of the U.S. Weather Bureau, and proceeded to the tenth-floor banquet room, where ministers from Norway and Bolivia mingled with those from Colombia and Switzerland. The ranking diplomat, Ambassador Baron Mayor des Planches of Italy, settled in with the Japanese ambassador at the head table, framed by palms, roses, and hollies.[11] The affable Dr. Cook, blunt-featured, smooth and dark, took his place beside Commander Peary, whose granitic face had been cracked and creased by decades of Arctic punishment; he radiated unblinking confidence and natural authority.[12]

"I was tremendously impressed by the man," Bert recalled of Peary. "I never met a man who so captured me by his evident kindliness—he was the ideal all young Americans had of a naval officer. . . . I was so impressed with Peary that I tried in a humble way to help him all I could."[13] Bringing Roosevelt and Peary together in Washington was the first step in that campaign.

From his inconspicuous perch at Table D, Bert swept the room with an apprising glance, making last-minute checks as the time moved toward 7:30 P.M. The Hubbard Medal was nestled in its padded box, ready for President Roosevelt's after-dinner arrival. The U.S. Marine Corps Band formed ranks, brandished their instruments, and made the chandeliers tremble. John Philip Sousa settled in to listen to his own music at Table G. Alexander Graham Bell, whose phosphorescent white hair and beard made him visible clear across the room, chatted happily with a congressman's wife at Table A. Finally the Senate chaplain droned to the end of his invocation, and waiters glided out through the banquet hall with bottles of sauterne and platters of oysters. The evening was up and running.[14]

It would be Bert's evening, from Potomac bass to roast quail. A born organizer, he loved planning such events, with their seating charts and diplomatic subtleties and thousands of shifting details, but they taxed his nervous system and his increasingly delicate stomach. Before occasions like this one, he usually required a predinner meal of bacon and eggs at home, followed by infusions of strong coffee, which ironically seemed to settle him.[15] Once suited-up and plunged into action, he appeared the picture of calm, but those who knew Bert best could tell when he felt nervous—he would reach for the back of his head and begin fiddling with a lock of hair.[16]

There was ample cause for fidgeting at the dinner that night. Peary, whose ego was prodigious, did not like sharing the stage with anyone; he tolerated it this evening because he did not consider Cook a serious threat to his North Pole ambitions and, besides, Peary had displaced Cook as the main attraction. For consolation, Cook would be introduced by Alexander Graham Bell, whose star quality might assuage any lingering disappointment the doctor felt.[17]

Peary not only suffered through such dinners, he seemed to relish them, perhaps in the knowledge that heroes were made as much by these tribal rituals as by their deeds in the field. The long evening wore on through chicken gumbo and sweetbreads, through bottles of St.-Estèphe and pudding, through fancy cakes and cigars. Perhaps to Peary's discomfort, his nominal boss, the secretary of the navy, rose to offer a toast. Like other old navy hands, Charles J. Bonaparte saw Peary as something of a showboat who never served at sea and took long leaves for his arctic excursions, which were private affairs, not official navy expeditions. Bonaparte spoke at great length about the importance of a strong navy and respect for men in uniform—never mentioning Robert E. Peary at all. The omission could not have been an oversight. Ten o'clock came and went with no sign of President Roosevelt. Bert kept an eye on the door. Alexander Graham Bell heaved himself up, crossed the room, and mounted the stage to speak for Frederick Cook. Never one for half measures, Bell was effusive in his praise for Cook, whose achievement seemed to rise higher than Peary's for a few tense moments. "We have had with us, and are glad to welcome, Commander Peary of the Arctic regions, but in Dr. Cook we have one of the few Americans, if not the only American, who has explored both extremes of the world, the Arctic and the Antarctic regions," Bell intoned. "And now he has been to the top of the American continent, and therefore to the top of the world. . . ."[18]

Top of the World? Peary must have felt a twinge of jealousy at that. Cook took the spotlight and began to spin a spellbinding tale of his McKinley ascent, which set off steady applause. Cook climbed higher and higher through snow and ice and biting wind until he was interrupted by a rustling outside the banquet room. Alexander Graham Bell announced the president's arrival and Roosevelt bowled in at high speed, with Bell puffing behind him. The president squirmed through a standing ovation and motioned for Cook to finish his talk. Cook hurried along, broke off, and yielded to the president, who launched into a sermon on hardiness and persistence in the American character.[19]

"Civilized people usually live under conditions of life so easy that there is a certain tendency to atrophy of the hardier virtues," Roosevelt said in his high, squeaking voice. "And it is a relief to pay signal honor to a man who by his achievement makes it evident that in some of the race, at least, there has been no loss of the hardier virtues. . . . I feel that in an age which naturally and properly excels, as it should excel, in the milder and softer qualities, there is need that we should not forget that in the last analysis the safe basis of a successful national character must rest upon the great fighting virtues, and those great fighting virtues can be shown quite as well in peace as in war. . . ."

Roosevelt turned to Peary. "You did a great deed that counted for all mankind, a deed which reacted credit upon you and upon your country, and on behalf of those present, and speaking also for the millions of your countrymen, I take pleasure in handing you this Hubbard Medal, and in welcoming you home from the great feat which you have performed, Commander Peary."[20]

To sustained applause Peary stood on his ruined feet, crossed the stage to meet Roosevelt, and took the Hubbard Medal in his strongly freckled hands. The president's eyes glistened. Peary's muscular chest strained at the studs of his boiled white shirt. His pale eyes took in the admiring dinner guests as he stood above them with the president and nodded his thanks, looking as sound as ship's oak, every bit the man of virtue Roosevelt had just described.[21]

Peary began to speak, thanking Roosevelt, thanking the National Geographic Society, and thanking the millionaires from the Peary Arctic Club for supporting his many expeditions. He announced that he would try for the North Pole one last time. "To me the final and complete solution of the Polar mystery which has engaged the best thought and interest of the best men of the most vigorous and enlightened nations of the world for more than three centuries, and today quickens the pulse of every man or woman whose veins hold red blood, is the thing which should be done for the honor and credit of this country, the thing which it is intended that I should do, and the thing that I must do," Peary said. The commander's words rolled over his audience in a soothing wave, a slow, round baritone that sounded more presidential than Roosevelt's urgent barking.[22] From the moment Peary pledged to go north again, there was no turning back—not for the commander himself; not for President Roosevelt, who had endorsed him; and certainly not for Gilbert H. Grosvenor, who had brought the two men together and arranged this solemn honor bearing the family name.[23]

By the next morning, Peary's polar interests had become inalterably entangled with those of President Roosevelt and the National Geographic Society. "President Praises Peary's Hardy Virtue—Fighting Qualities Needed in Peace as Well as War," the *New York Times* reported. "Must Find the Pole—Peary Declares He Will Solve Arctic Mystery," proclaimed the *Washington Post*. "From the Hand of the President," the *Washington Star* reported, "Medal of Geographic Society Awarded to Peary." Even the London *Times* weighed in: "In the presence of a distinguished audience of diplomatists, statesmen, naval and military officers, and men of science, the President last night presented the first gold medal offered by the National Geographic Society for extraordinary achievements to Commander Peary."

From such headlines were national heroes made. The *New York Times* arranged for exclusive rights to Peary's North Pole story. Roosevelt ordered the navy to extend Peary's leave for three years.[24] National Geographic stepped forward with its modest but symbolic $1,000 toward Peary's fourth—and final—try for the Pole. The organization had given advice, moral support, and publicity to many explorers before—but never money. The now invisible Frederick Cook, barely mentioned in the accounts of the Hubbard Medal ceremony, quietly departed Washington to resume his medical practice in Brooklyn. There the doctor spent a fitful winter, soon finding himself bored with the office routine, which "seemed like the confinement of prison."[25] Perhaps impressed by the popular attention Peary had received,[26] Cook found his thoughts turning northward again. Without fanfare, he closed his Brooklyn practice and made his way to Etah, Greenland, on that island's northwestern coast, in the summer of 1907. There he began to lay in supplies for a long journey. He engaged the services of Inuit guides and hunters and struck out for the North Pole in February of 1908, just as the long arctic night began to fade.*

Unaware of Cook's intentions, Peary rode the wave of his own luminous publicity into 1907. He began a hectic lecture tour to raise money for his next expedition, which he hoped to field by the summer of that year. But this was not to be—the Roosevelt, damaged by its most recent battle with the ice, needed extensive refitting that would delay Peary's departure for another year. As the summer of 1908 approached, he gathered supplies, including 25,000 tons of pemmican, supervised the building of new dog sledges and special stoves, and ultimately learned of Frederick Cook's plans.[27] The news seemed to infuriate Peary. He complained that Cook had appropriated *his* Inuit and *his* dogs. Cook was using *his* staging area in Etah, where the local people had learned *his* methods of polar travel.[28] Of course the Inuit belonged to no one, their dogs were their own, and, while it was true that Peary had adapted Inuit methods for his expeditions, it was he who had learned from them, not the other way around. For someone who had dismissed Cook's chances for success, Peary seemed unduly rattled by the prospect that the doctor might beat him to the Pole.[29] Having obsessed over that prize for two decades, having sacrificed his health and fortunes for it, the commander was not inclined to yield the territory to any rival. Besides, the certainty of Peary's success had been sanctioned by President Roosevelt, by the National Geographic Society, and perhaps by an even higher

*Inuit, meaning "the people," is the preferred term for the Arctic people known in Peary's day as Eskimo, a usage now considered offensive among the Canadian Inuit.

authority. "I believe it is written in the eternal book that I shall win next time," he wrote to Bert in the weeks just before sailing.[30]

By July 6, 1908, Peary was finally on his way, steaming out of New York Harbor on the *Roosevelt,* cheered by crowds, hailed by sirens, and saluted by the rhythmic boom of guns. The city receded astern and the *Roosevelt* turned northeast for the president's summer estate at Oyster Bay, where Roosevelt had arranged a farewell luncheon for Peary. Despite a punishing heat wave, the president bounded aboard his namesake vessel for an inspection. Racing around in his white linen suit and Panama hat, he probed the lower hold, scrutinized the sailors' berths, examined the engine room, and demanded to know the name of each person on board. When he came to Peary's cabin, he found an etching of himself on the wall, overlooking a pianola outfitted with rolls of music for marches and ragtime. Captain Bob Bartlett, the bluff Newfoundlander and ship's master, who had not foreseen such a thorough assessment of his vessel, heaved a sigh of relief when Roosevelt pronounced the craft "Bully!" Neither the president nor the commander mentioned Cook when they shook hands and parted company that day.[31]

"I believe in you, Peary," said the president, "and I believe in your success— if it is within the possibility of man."[32]

That endorsement, like a crisp following breeze, hurried the *Roosevelt* north, where she shipped whaleboats in New Bedford and a spare rudder in Maine. Then it was on to Sydney, Nova Scotia, for the bittersweet moment when Peary's wife, Josephine, and their children left the ship. "Brave, noble little woman!" Peary wrote. "You have borne with me the brunt of all my Arctic work. But somehow, this parting was less sad than any which had gone before. I think that we both felt it was the last."[33]

Among the crowd cheering from the Sydney docks, Peary made out the familiar thin form of Gilbert H. Grosvenor, who had met the *Roosevelt* coming the other way three years before with the same words of encouragement. Then as now, Bert had driven over from Baddeck to show his support. His commitment to Peary was no less—even with Frederick Cook still abroad. The doctor had not been heard from since the early winter of 1908, when he departed Etah with a full complement of Inuit and sled dogs; he was a tough, resourceful traveler with ample polar experience, and he was not to be easily dismissed. Cook had a full year's start on Peary, and if he came back with a competing claim to the North Pole, Bert's attachment to the commander might put National Geographic's credibility at risk. But Bert was nothing if not loyal: He had cast his lot with Peary and he remained confident that the Hubbard laureate would finish first. "He seemed to have everything necessary," Bert reported from Sydney. "A fine crew of Newfoundlanders took the place of the crew who had brought the ship up from N.Y. Peary was very appreciative of what the Nat. Geog. Soc. had done for him. . . . I also obtained some good photos of the ship etc. which may be of service to the Magazine when he returns."[34]

As Peary sailed off for the Arctic, Bert returned to the familiar routine at Baddeck, with its blend of family imperatives, long-distance editorial duties, and soul-calming beauty. For two or three months each summer, Bert set up office here in a military-style tent outfitted with a heavy desk, filing cabinets, and a gnarled bentwood chair. Between occasional sailing forays and visits to town, here among the pines he pored over magazine articles, worried about budgets, and corresponded with readers and colleagues. Bert claimed that this Spartan summer office freed him from the distraction of visitors and family, which now included four little Grosvenors and a fifth on the way.[35] A photograph from this period tells another story: Bert sits deep in the tent on the edge of his chair in dim light, straining to focus on his work. Alexander Graham Bell, poised to speak, looks over his right shoulder; a barefoot Melville Grosvenor leans against his father's chair with an arm draped over the back; and Bert's eldest daughter Gertrude squats comfortably on the ground, blocking the tent's entrance. The editor looks boxed in, beset by official duty in one direction and family obligations in the other, and indeed these years proved worrisome for Bert. He began to show the strain of overwork, money troubles, and the problems of a growing family. He developed ulcers and had periodic bouts of fatigue. He became testy.

As the Grosvenor family grew, Bert discovered that his wife had inherited her father's independence and nocturnal habits. Thus it was Bert who rose early

and made sure that his children had breakfast and were dressed for school—
while Elsie slept late. Although she was by all accounts a devoted mother and
wife, she showed very little interest in the tedium of homemaking, preferring
to spend her spare time riding horses, playing bridge, visiting friends, organizing clubs, and campaigning for women's suffrage. Despite his own upbringing,
which emphasized a traditional role for women, Bert supported Elsie's outside interests. But he also yearned for the regular home life the perfectionist always wants but seldom gets, which became a source of tension between Bert
and Elsie.

Had Bert been less obsessive about work and family life, he might have
avoided the ill health that began to plague him in his thirties. With his burgeon-
ing responsibilities, the vigorous young man grew frail, which prompted alarm
from loved ones and colleagues. "I am anxious that you should stay away until
you get well rested, for I could see before you left that your nervous energy had
been a little overtaxed," wrote Willis L. Moore, National Geographic's president, addressing Bert just after he had seen Peary off.[36]

Bert had been ordered by doctors to rest that summer. But by now he
seemed powerless to break the pattern; for much of his career, he would work
himself to exhaustion, rest when forced to do so, and return to the office, there
to begin the cycle anew. He did this in the belief that the success of *National Geographic* relied upon him alone. "I realized no one else could develop the vision
I had," Bert said, recalling those formative years. "I realized even then the possibilities ahead for the NGS. . . . Mr. Bell thought that anyone could fill the
pages of a magazine. Anyone could fill pages. But only a very astute man could
choose things that would sell a Geographic Magazine."[37] This assessment was
largely true in the magazine's early years, when Bert functioned as a one-man
band, arranging the lecture program, selecting photographs, buying articles,
commissioning maps, rewriting stories and laying them out.

Bert's hard work paid off for the National Geographic Society, which blossomed under his care. From 1906 through 1908, memberships jumped from
10,000 to 20,000. By 1911, circulation had reached 84,000, representing, in
Bert's phrase, "the best culture and the biggest brains in the United States."[38]

To service those brains, Bert's staff had grown with the times: By 1911, the National Geographic Society employed about fifty people. Almost all of them worked in Hubbard Hall, which would soon be expanded to accommodate still more employees. Even when it was clear that the National Geographic Society and its journal had become a success, staffed with sufficient talent and flush with money, Bert remained active in every aspect of the operation and the last authority on all matters pertaining to the magazine—in short, Bert was a micromanager long before that unlovely term entered the language. So thoroughly did he impose his own identity on National Geographic that it was soon impossible to think of the organization without envisioning Gilbert H. Grosvenor. "He is the Society," said Comdr. Robert E. Peary.[39]

Peary had emerged from the Arctic in the summer of 1909 and steamed south toward Labrador with exciting news. On September 5, he reached the first port equipped with telegraphic facilities, left the *Roosevelt,* and rowed to the village of Indian Harbor. From that rocky outpost, he began sending cables.[40]

"Have won out at last," Peary wired Gilbert H. Grosvenor on September 6. "Pole is ours. Regards to yourself and Mrs. Grosvenor. Peary."[41] On the same day, he sent dramatic lines to the Associated Press and the *New York Times:* "Stars and Stripes Nailed to the North Pole," he reported.[42] He wired his backers in the Peary Arctic Club with a request: "Kindly rush following: wire all principal home and foreign Geographic Societies of all nations including Japan, Brazil, etc., that the North Pole was reached April sixth by Peary Arctic Club's expedition under Commander Peary."[43] He wrote to his wife, Jo, in Maine: "Have made good at last. I have D.O.P." When a reporter asked what "D.O.P." meant, Mrs. Peary replied: "Damned Old Pole."[44]

Almost as an afterthought, Peary remembered to cable the new president, William Howard Taft. "Have honor to place North Pole at your disposal," he wrote on September 8, two days after the rest of the world had read about his triumph. "Thanks for your generous offer," President Taft replied. "I do not know exactly what I could do with it."[45]

Taft's jocular response may have been an attempt to defuse a developing controversy, for less than a week before, the president had received a competing North Pole claim from Frederick A. Cook, who declared that *he* had reached 90° north almost a year before Peary. Their dispute over the discovery, which would come to be known as the Polar Controversy, would produce the biggest news story of the year, with the *New York Times* and the *New York Herald* backing rival camps and the public hungry for each new morsel served up to them. The Cook-Peary dispute would inspire countless cartoons, several books, and hours of acrimonious congressional hearings. And it would draw the National Geographic Society into the argument, creating rifts among its leaders and taxing Gilbert H. Grosvenor's character and diplomatic skill to the utmost.

The first news of Cook's claim hit the papers on September 1, 1909, just days before Peary's announcement hummed down the wires from Labrador. In brief, Cook declared that he had wintered over 1907 at the limit of navigation in the Smith Sound region of Greenland, where he prepared food and equipment for a seven-hundred-mile march to the Pole. When the sun reappeared that winter, his party of eleven men and 103 dogs struck north with heavily loaded sledges. He claimed to have reached the Pole two months later, on April 21, 1908, accompanied by two Inuit companions, the others having turned back along the way. In the September 1 dispatch, Cook described his moment at the North Pole for his sponsoring newspaper, the *New York Herald:* "An endless field of purple snows," he wrote. "No land. No life. No spot to relieve the monotony of frost. We were the only pulsating creatures in a dead world of ice." On their harrowing return journey, Cook described how his party had been thwarted by drifting ice and open water and forced to pass the winter of 1908–09 in a cave on Cape Sparbo north of Baffin Island, Canada. With the spring sunrise, Cook made his return to Greenland, arriving near Etah on April 15, 1909.[46]

Peary's first report, published in the *New York Times* on September 12, 1909, described how he had concentrated his personnel and supplies at Cape Columbia on Ellesmere Island, and how he had sent relays of men, dogs, and sledges out in late February to establish a series of camps. This line of camps stretched across almost five hundred miles of ice between Cape Columbia and the Pole. On April 2, from the northernmost of these support camps, at 87°47', Peary made his final dash to the Pole, 136 nautical miles over the frozen ocean (156 statute miles) away, with teams of fresh sled dogs, four of the strongest Inuit guides, and his longtime associate, Matthew Henson. Henson, a black Washingtonian who had accompanied Peary for almost two decades, had begun as the commander's valet, but he became an indispensable companion whose survival skills made it possible for the crippled Peary to continue exploring after he had lost most of his toes to frostbite. Henson, an expert sledge driver, skier, and hunter, also became fluent in the language of the Inuit, whose cooperation was crucial to success; Peary never learned to speak their language. Despite his long friendship with Henson, Peary did not wake his friend for the last march to the Pole on April 6. As Peary himself later testified, he did not want to share the glory. So as Henson slept, Peary made the final approach accompanied by two Inuit. By abandoning Henson, Peary not only wounded a loyal colleague of eighteen years' standing, he also denied himself a witness to the greatest—and the most hotly disputed—claim of his career.

"The Pole at last!!!" Peary wrote on a loose page from his bound diary for that day. "My dream and goal for twenty years. *Mine* at last. I cannot bring myself to realize it."[47]

Nor could Peary bring himself to believe the news of Cook's success. Had it not been written in "the eternal book" that Peary should be first to the Pole? And who would remember the *second* man to claim that prize? The doctor had to be discredited. In very short order, witnesses stepped forward to begin that process. On September 6—the same day Peary emerged in Labrador—the *New York Sun* reported that Cook had faked his Mount McKinley climb three years before. According to two guides who had accompanied Cook on the mountain, the expedition went no higher than five thousand feet and produced photographs of smaller peaks to deceive the public. All subsequent evidence supported this allegation—Cook's McKinley story was a hoax from the start. If Cook had lied about Mount McKinley, said Fred Printz, one of the guides interviewed by the *Sun,* how could anyone believe his North Pole story?[48]

Like a raptor sensing weakness in his prey, Peary moved in for the kill. From Indian Harbor, he tried to scuttle Cook's claim to the North Pole. "Cook's story should not be taken too seriously," Peary wired the Associated Press on September 7. "The two Eskimos who accompanied him say he went no distance north and not out of sight of land." Like the charge that Cook had faked his McKinley climb, this one also had a basis in fact: The Inuit on Cook's North Pole journey had given affidavits to Peary's men in Etah, where the commander had prudently collected this evidence on his way home. Peary also launched an attack in the *New York Times,* which had sponsored his expedition. "Do not trouble about Cook's story or attempt to explain any discrepancies in his statements," Peary wrote in a letter published on September 11. "This affair will settle itself. He has not been at the pole on April 21st, 1908, or at any other time. He has simply handed the public a gold brick."[49] Then, in a move that would return to haunt him, Peary demanded that Cook submit proof of his claim to "some geographical society or other reputable body." At this point, Peary had been so busy criticizing Cook that he had yet to provide the details of his own story, much less any proof to support it.

To those accustomed to a genteel tradition among explorers, for whom good faith and truthfulness had been the standard, Peary's aggressive treatment of Cook violated a high-minded code. Outraged by Peary's churlish behavior, Arctic veterans like Adolphus W. Greely rose up in indignation. "Peary has given a black eye to exploration, very much to my regret," wrote Greely, who had introduced the commander to the National Geographic Society in 1889 and still served on its board. The general warned Bert to proceed with caution in the Cook-Peary controversy "if a one-sided row can be called a controversy."[50] Sharing Greely's misgivings, Alexander Graham Bell seemed more upset by Peary's outburst than by the possible evidence of Cook's fraud. "There was glory enough for both," Bell wrote, expressing regret at the cynical atmosphere surrounding both claims. "Peary has spoilt his own reputation as a broad and liberal-minded man, by showing himself to be narrow and ill-liberal. The whole controversy, therefore, is most unfortunate to the characters of both men. . . . I cannot believe that there is any man living, who could dare to impose upon the world to the extent Cook is claimed to have done. . . . Don't let us be too cocksure of the results until all the facts are in."[51]

But Bert was sure. He knew that Peary had reached the Pole. Like former president Roosevelt, he believed in Peary, and that belief did not allow for the possibility that Cook might surpass him. Bert never wavered throughout the controversy, even as he sought to maintain the public appearance of objectivity. The October 1909 issue of *National Geographic* magazine carried North Pole accounts by both Peary and Cook with no commentary, except for a note from Bert encouraging both explorers to submit proof of their claims to suitable authorities. To those National Geographic board members who worried that the organization's association with Peary might damage the society's image, Bert offered reassurance. "We are trying to take an absolutely impartial position in this Polar matter," he wrote Greely on October 6.[52] "The Society is ready to honor Cook the moment he submits proof," he told Bell.[53]

After waiting a few weeks for Cook's proofs, the National Geographic Society's board of managers assembled a panel of experts to examine Peary's claim. The panel was composed of Henry Gannett, chief geographer of the U.S. Geological Survey and vice president of the National Geographic Society; O. H. Tittman, superintendent of the U.S. Coast and Geodetic Survey; and Rear Adm. Colby M. Chester, former superintendent of the U.S. Naval Academy and an expert on navigation. Gannett was appointed as chairman. All three men were highly capable and respected—but all three operated under conflicts of interest. Tittman and Gannett had both been founders of National Geographic and active on its board for most of the organization's history; Chester was a board member at the time of his appointment; since the National Geographic Society had sponsored the Peary expedition, how could a committee appointed by its managers and comprised entirely of its board members render an honest assessment of Peary's claim?

With these questions lingering, Peary arrived in Washington on November 1, 1909. He met with members of the tribunal at Admiral Chester's house that afternoon. The group went through the commander's books and papers. With a map at hand, they discussed his navigational readings. After dinner, by which time it was dark, the subcommittee retired with Peary to the train depot at Union Station, where his equipment trunk had been stored for the day. Crowding into the baggage room, Peary's inquisitors peered into the trunk in the dim light. They did not remove or examine any equipment. They said their good-byes to Peary, who left Washington on the Midnight Express for New York, taking all evidence with him.[54]

For a scientific panel, the Gannett committee behaved in an exceedingly casual manner. They did not interview Matthew Henson to corroborate Peary's navigational readings, nor did they check Peary's story with Captain Bartlett and other members of the expedition. Peary was their sole witness. In examining his diaries, they should have been curious about the blank pages appearing about the time he is supposed to have reached the Pole. They made no comment on this. Nor did they express concern about the loose page inserted into his diary on April 6, 1909, the day of his great discovery: "The Pole at last!!! The prize of 3 centuries . . ." This anomaly was never explained—not by Peary, not by the Gannett panel. The experts seemed satisfied by Peary's navigational readings, which appeared on loose sheets inserted with other records. Why were these essential observations not part of his bound diary? No explanation.

The subcommittee accepted Peary's unprecedented travel times, which had him sledging from the camp farthest north, at 87°47', to the Pole and back in eight days. This round-trip of 296 nautical miles (340 statute miles) was possible, even over pressure ridges and gaps in the frozen surface, but hardly credible without further confirmation.* Peary never supplied it, nor was he asked for it. Then there was the problem of drifting ice: The North Pole is not a physical feature anchored on land like a mountain or a lake, but an invisible point sitting on plates of shifting ice. To fix its location requires precision with a sextant and a chronometer that Peary might have possessed but did not demonstrate. He made no longitude calculations during his march to the Pole—this despite violent easterly winds that would have been pushing the ice (and therefore Peary) westward as he raced north. Peary's ignorance of his longitude makes it

---

*An international team of five explorers, using huskies and wooden sledges like those Perry employed, reached the North Pole in record time on April 26, 2005. Their achievement, which shaved five hours from Peary's reported traveling time, does not vindicate his much-disputed claim, but it mitigates some of the doubt.

most likely that he arrived some 30 to 60 nautical miles (35 to 69 statute miles) *west* of his objective, according to Sir Wally Herbert, a modern explorer who has made four journeys to the Pole and has examined the commander's original records.[55]

There on the frozen ocean, with the April wind screeching around him, Peary may have experienced a rare moment of doubt. When Matthew Henson stepped forward to celebrate their triumph, Peary seemed to crack, according to Henson: "I ungloved my right hand . . . to congratulate him on the success of our eighteen years of effort," Henson wrote, "but a gust of wind blew something into his eye . . . and with both hands covering his eyes, he gave us orders to not let him sleep for more than four hours. . . . From the time that we knew we were at the Pole Commander Peary scarcely spoke to me. . . . It nearly broke my heart," Henson said.[56] The commander remained curiously subdued when he finally rejoined the *Roosevelt* with Henson at Cape Columbia. Peary never mentioned the Pole until a crew member asked about it, and the commander responded with marked understatement: "I have not been altogether unsuccessful."[57] This was a man who had spent hours in camp sketching plans for a mausoleum to himself, dreaming about his admiral's pennant "with a diagonal white bar," and planning new publicity photographs—"Have Borup take a 5"x7" to 4ft. focus portrait of me in deer or sheep coat (face unshaven) with bear roll, & keep on till satisfactory one obtained."[58] For one of Peary's soaring ambition, his behavior during this period seems out of character.

How to explain it? Perhaps he was simply exhausted, worn down by two decades of nonstop travel. Perhaps he was depressed by the death of a young crewman who had fallen through the ice. Perhaps, having finally gained the prize he had sought for so long, he was deflated. Or perhaps he knew that he had come tantalizingly close to the Pole—but that he could never prove it. Whatever the cause of Peary's behavior, whatever the reason for the troubling gaps in his records, his backers at National Geographic remained steadfast, as if their role, like his, had been "written in the eternal book."

"He came, he saw, he conquered," the *Washington Post* declared the morning after Peary had presented his proofs to the National Geographic Society.[59] "The committee had everything it wanted," one of its members told the *New York Times* within hours of Peary's visit.[60] The Gannett panel unanimously confirmed the commander's claim in a report to National Geographic's board of managers on November 3.[61] The board ratified the report that day, declaring that Peary had reached the North Pole on April 6, 1909. For good measure, they voted to award him another gold medal. No mention was made of Frederick Cook, who would plunge into disgrace that year.[62]

Early in December, just as Peary was returning to Washington for another victory banquet at the National Geographic Society, Frederick Cook turned over the long-awaited documentation to officials at the University of Copenhagen, a premier center of polar studies, where experts had agreed to review the doctor's claims. Cook's seventy-seven-page report was another disappointment. Instead of containing authentic proofs, it consisted of a long narrative that repeated information from newspaper accounts of his journey, and purported copies of his notebooks, which contained no original astronomical observations.[63] Tired of waiting for the evidence that never arrived, the Copenhagen commission closed the door on Cook on December 21, 1909. His proofs, said the commission report, were "of such an unsatisfactory character that it is not possible to declare with certainty that the astronomical observations referred to were actually made; there is [sic] likewise lacking details in practical matters—such as sledge journeys—which could furnish some control. The Commission is therefore of the opinion that the material transmitted for examination contains no proof whatsoever that Dr. Cook reached the North Pole."[64] He was allowed to keep his medal (awarded previously by the Royal Danish Geographical Society) and his honorary doctorate (conferred by the university), but he never proved his claim, which has since disintegrated under the scrutiny of scholars. He was convicted of fraud in the 1920s for having misled investors about oil reserves in Texas. He was still serving part of a fourteen-year sentence in federal prison when, in a final twist to an already strange tale, oil was discovered on that land. Pardoned by President Franklin Roosevelt in 1940, Cook died bitter and broken that year, still proclaiming his honesty in all things.[65]

Peary was promoted to the rank of rear admiral in March of 1911 and retired from the navy at $6,000 a year. But this came after contentious hearings in the House of Representatives, where a sharply divided Committee on Naval Affairs raised serious doubts about his North Pole claim.[66] Ironically, the most damaging of these questions might have been resolved by Matthew Henson, had he been allowed on Peary's final dash to the Pole; Henson was never called to testify. The official congressional resolution, which had originally recognized Peary as *discoverer* of the North Pole, was changed to say that he merely *reached* the North Pole. It was not a perfect victory, but it was official recognition and that was enough for Peary, whose name would forever be associated with the achievement, not the scandal. Soon after he retired, Peary joined the board of managers at the National Geographic Society, where he remained a firm friend and ally of Gilbert H. Grosvenor.

The same could not be said of Peary's strained relations with Adolphus Greely, who continued to serve on National Geographic's board. The explorers remained at odds, in part because Greely never believed that Peary had reached the Pole. The old man was obsessed by this, laboriously clipping every newspaper article, every report, every memo on the Cook-Peary controversy and pasting these hundreds of pages into bound journals. In these books, he underscored each criticism of Peary in ink.[67] "I have always praised the ability, courage and tenacity which brought him to the head of American explorers," Greely wrote of Peary, "a rank he will undoubtedly retain. Though he reached an unprecedented high latitude, his attainment of the Pole must be doubted by nearly every one who studies the documents instead of listening to others. Official records show that Peary lacked the power of accurate observation."[68]

For his part, Bert Grosvenor remained unfazed by the bruising character of the Cook-Peary argument. Having declared Peary a hero before any proof was in, Bert stubbornly ignored the gaps in the commander's records, his incredible traveling times, his shoddy navigation, his lack of witnesses, and his exceedingly odd behavior on his return from the ice cap. Against all of this, Bert took Peary at his word, along with the glory he had gained for the National Geographic Society. When Peary died in 1920,[69] Bert arranged a monument for him at Arlington National Cemetery, just as he did within the National Geographic Society where, in the museum known as Explorers Hall, the tattered American flag Peary had carried north was proudly displayed. There, Peary was elevated as an iconic figure. Above reproach or criticism, if not above doubt, he was the first hero.[70]

CHAPTER SEVEN
OFF TO WAR

ARCHAEOLOGIST HIRAM BINGHAM
AT MACHU PICCHU, 1912.

Among the unforeseen casualties of the Polar Controversy was the National Geographic career of Willis L. Moore, a meteorologist who failed to predict the consequences of the Cook-Peary storm. When it abated, Gilbert H. Grosvenor emerged in undisputed control of the magazine, while Moore was compelled into an abrupt resignation as National Geographic Society president.

The trouble began when president and editor lined up on opposite sides of the Cook-Peary argument, with Moore backing Cook and Grosvenor backing Peary. By the time National Geographic's board officially cast its lot with Peary in November of 1909, Moore had shifted sides—but by then it was too late. He had annoyed Bert Grosvenor and fumbled several high-profile incidents arising from the polar dispute. The most embarrassing of these came when Moore led a delegation of executives from National Geographic Society to the State Department in a clumsy effort to influence deliberations at the University of Copenhagen, where officials were reviewing Frederick Cook's arctic records. Unfortunately for Moore, this diplomatic sortie not only became public, but so did the Danes' wilting rejection of National Geographic's offer to assist them.

Moore revealed himself to be painfully out of step with the Bell-Grosvenor family, which held National Geographic's membership concept as an article of faith. Seemingly ignorant of this, Moore suggested that National Geographic abandon the membership notion in favor of subscriptions. This brought an amiable but firm rebuff from Bell, who warned that such a change would "strike a blow at the very cause of our financial success, upon which all things else depend. People prefer to be members rather than subscribers," Bell said.[1]

Finally, Moore made the mistake of trying to exert more control over the magazine, which Bert Grosvenor interpreted as an intrusion. "He wanted to assume, as president, authority over me, which had not been given him," Bert said. "He assumed the president was boss. . . ."[2] Yet Moore kept testing Bert—by attempting to place new controls on the editor's spending; by scheduling meetings of the executive committee when Bert was absent; by trying to fire Bert's handpicked lieutenant, John Oliver LaGorce. When the editor resisted each of these moves, Moore complained that he was not showing proper deference. Relations between the two men, already frayed by months of disagreement, reached the breaking point under the strain of the Polar dispute.

Mabel Bell calmly observed the power struggle from the sidelines, gathering evidence from Bert and from Charles J. Bell, Alec's cousin and a board member, before finally sounding the alarm in November of 1909. By then she believed that Moore was threatening her family's stewardship at National Geographic. "I feel as if you ought to be here rather than Baddeck, now," Mabel wrote to Alec from Washington. "Bert is on Mr. Moore's black books and he is trying to eject Bert, while a great many people feel that Mr. Moore has made himself almost impossible. He is responsible for the Danish muddle and his speech at the Peary lecture was of such character that it was hard to listen to it with any sort of patience. It was entirely unworthy the President of a scientific society. Charlie [Bell's cousin Charles] is about aroused to action, but of course it is difficult to act when it is one's relation that is in question."[3]

The dispute boiled over the next day, when the board met to discuss changes in the bylaws of the National Geographic Society, which would have given Moore greater leverage over the magazine. "Mr. Moore got mad when he found the whole Board against him," Mabel reported to Alec. "He arose and said that as they seemed able to conduct the discussion without him he would go. Then as he put on his coat he returned to say that he had not meant to tie things up so that Mr. Grosvenor could not buy a photograph without his authority, but he did wish to have some authority to act while Mr. Grosvenor was away on his three months vacation, or was it four months this summer? He said as President he would like to be able sometimes to come into the building and have a letter written, as things now were he couldn't do it without Mr. Grosvenor's approval!"[4]

Within weeks, the board had quietly replaced Moore as president of the National Geographic Society, just as Editor John Hyde had been eased aside to make room for Bert years before. By now the lesson was clear: Any assault on Bert constituted an attack on the founding family and would not be tolerated, least of all by Bert himself, whose courtly style masked a tough-minded autocrat who disliked sharing power and who proved to be a dangerous opponent when threatened. Moore's replacement as president, Henry Gannett, grasped the realities of his position and demonstrated proper esteem for the editor and the dynasty he represented. The pair got along famously, and from the time of Moore's leaving, Bert never again wrestled with the board. He ran the magazine with a free hand, and it prospered under his innovations.

The most notable of these, which separated *National Geographic* magazine from other periodicals for many years, came in photography, where Bert's willingness to take risks and spend money soon began paying dividends. The magazine's first extensive color story, "Glimpses of Korea and China," walked in the door under the arm of a world traveler named William W. Chapin of Rochester, New York. Chapin, a tall, florid man with money to burn, presented Bert with a stack of black-and-white photographs, many of which had been painstakingly colored by a Japanese artist according to Chapin's instruction. Bert published them in November of 1910, long before color film came into widespread use. The article, also written by Chapin, contained twenty-four painted photographs in thirty-nine pages. By today's standards the tinted photographs look posed and unnaturally pale, like faded Polaroids, but Chapin's story created something of a sensation, providing visions of workaday life in Asia, where men carried live swine to market on their backs, bent almost double by the weight; where a Korean peasant, looking like a haystack with legs, shielded himself from the rain in a green thatch cape; where two Chinese prisoners, bound together with heavy neck chains, awaited release from jail; where four elegant Manchu women in sculptured headdresses and turquoise silk drifted down a dappled Beijing lane, transforming it into an opera set.[5]

"No magazine had ever made such a display of color," Bert recalled. "It made a tremendous hit with the members."[6] Although the color printing cost almost four times the normal amount, the positive response, along with a surge in new memberships, made Bert's gamble pay off. Until he could afford to use color in every issue, Bert saved his best color stories for each November, a key month for renewals. Chapin, a close friend of Kodak pioneer George Eastman, continued traveling and feeding Bert new ethnographic material, as did a flock of contributors from all quarters. Having witnessed Chapin's success, the resourceful Eliza Scidmore soon brought her camera to bear on Japan, where she prepared an exquisite series of hand-tinted photographs. "Herewith 31 pictures of 'Women and Children,' mostly children, as you see," she wrote Bert from Kyoto in 1912. "I have had them made uniform in size and strongly colored, so that you can cover yourself all over with glory with another number in color and thereby catch a few thousand more subscribers."[7]

Such coverage helped establish ethnography as a staple for *National Geographic* magazine. At the same time, Bert's photographic advances accomplished exactly the effect Miss Scidmore had predicted, the number of readers growing with each passing year. New members brought new money, which Bert reinvested in the magazine. He established the organization's first photographic laboratory in 1910, appointing a seasoned photographer named Charles Martin as its director in 1915. Martin's charge was to keep abreast of the latest color technology and to apply each improvement at *National Geographic*. Bert also splurged on groundbreaking work by George Shiras 3d, a Pennsylvania congressman who produced a remarkable series of wildlife photographs, including nocturnal studies made possible by trip wires and illuminating magnesium powder. Bert demonstrated an almost boyish gusto for technically difficult displays, an enthusiasm that also led him to fold an eight-foot-long panorama of the Canadian Rockies into the magazine in 1911. These innovations caused critics to wonder if Bert's photographic fervor occasionally went too far, giving his magazine the haphazard look of a scrapbook crammed to the bursting point with curiosities. Reading the magazine in those days was seldom boring, but it could induce a kind of vertigo as it lurched from subject to subject with hardly a pause for breath—from grasshoppers to the Panama Canal, from the Los Angeles aqueduct to acorns, Persian women, oysters, the temples of Teotihuacán, cocoa plants, the lost wealth of Midas, and a Liberian board game called *Kboo*.

All subjects, it seemed, were fair game. As long as an article promised some visual interest, Bert's definition of geography could be wonderfully elastic, as the case of Hiram Bingham proved. Bingham, a young assistant professor of Latin American history at Yale, approached National Geographic in 1908 about sponsoring his archaeological projects. Bert demurred, citing an oversupply of South American stories in inventory. But when the editor saw Bingham's photographs from Venezuela and Colombia a few years later, Bert suddenly changed his mind. Not only did he feel that there was room for Bingham's work in *National Geographic* magazine, Bert also invited the professor to give lectures at the Society, where his talk on pre-Columbian civilization won an enthusiastic response. When Bingham's excavations at the Peruvian site of Machu Picchu produced promising results in 1911, Bert encouraged him to submit a proposal for the next season—this despite strong feeling among the board of managers that archaeology had no place in *National Geographic*. "I believe that this work pertains to the Archaeological Societies," said one of the doubters, Adolphus W. Greely. "If such allotments of our funds are legal, they are not sound and defensible policy. They open wide the door for exploitation of our funds for other purposes . . . for which there seems to be no warrant in our charter as a Society."[8] Warranted or not, Bert overcame Greely's objections, and in the spring of 1912, the research committee voted a $10,000 grant for Bingham's continuing fieldwork at Machu Picchu, the ancient Inca citadel that proved to be one of the most spectacular archaeological finds of the twentieth century.

The National Geographic grant, combined with a similar contribution from Yale, allowed Bingham to mount a well-equipped study at Machu Picchu that year. Battling snakes and hacking through thickets, members of the survey team—which included local guides, mapmakers, an osteologist, a surgeon, and a small army of assistants and mules—edged its way to Peru's mist-shrouded heights to probe the mysteries of the lost city. Most of Bingham's theories about Machu Picchu did not withstand scrutiny—that it was the last redoubt for Incas resisting the Spanish Conquest; that it was a sanctuary for sacrificial Virgins of the Sun; that it was the birthplace of Inca civilization. Later scholarship suggests that the Andean site was most likely a retreat for Incan royalty, a sprawling South American version of Camp David.[9] But these particulars hardly mattered when news of Bingham's find crashed into the headlines. The sight of those high, lonely ruins, deftly rendered in Bingham's black-and-white photographs, still evokes an involuntary shudder. Who were these people? What led them to this forlorn mountain? Who was the Ozymandian ego behind those massive works of stone?

Such questions touched a responsive chord in Bert Grosvenor, who devoted the entire April 1913 issue to the Yale–National Geographic Peruvian Survey.[10] Bert packed the magazine with maps, drawings, and no less than 250 photographs. As if that were not enough, the editor folded a breathtaking panoramic picture of the entire site into the issue. For an organization that had so recently debated the suitability of archaeology, the National Geographic Society had traveled a long distance indeed. Even Adolphus Greely grudgingly came to admire the Andean project: "*Caviare* to our readers," he wrote Bert, "but they can stand it occasionally."[11]

The popular response led Bert to support Bingham's work over several seasons, which contributed to the professor's renown and gradually added to an understanding of Inca civilization. After the 1915 season, for instance, the team botanist Orator F. Cook could report that the Incas had built their elaborate mountainside terraces not for military defense, as originally supposed, but for agriculture, at which they excelled, producing almost as many species of plants for cultivation as all of Asia.[12] This long-term commitment to winning subjects and proven talent became typical for Bert, who made extended investments in the Binghams and Pearys of the world; and, like many of the practices Bert began, this one became the magazine's approach with *National Geographic* contributors who would become famous in later years, among them Jacques Cousteau, Louis and Mary Leakey, Jane Goodall, and Bob Ballard.

Bingham's exposure in the magazine, along with his charming speaking style and aquiline good looks, helped launch his political career in Connecticut, where he served as lieutenant governor, governor, and United States senator through the 1920s and 1930s. Like Peary, Bingham maintained his links with Bert Grosvenor, who occasionally called on him for political favors. Bert's experience with Bingham also convinced him to embrace archaeology as a regular subject for his magazine. Throughout the twentieth century, the National Geographic Society would contribute thousands of dollars toward archaeological projects, most notably in the pueblos of the Southwestern U.S. and in the Maya regions of Mesoamerica. Today, archaeology remains an important area of research and magazine coverage for the organization.

While Bert lost himself in the mysteries of vanished empires, his family embraced the era's progressive movement, with its call for government regulation of trusts, reform of political machines, conservation of natural resources, and regulation of the food industry. After some children became ill from drinking tainted milk in Washington, Elsie Grosvenor joined other women campaigning for new dairy standards. She also gathered her own children and, encouraged by her parents and by Bert, joined thousands of other marchers in a March 1913 parade for women's suffrage, which developed into one of the biggest political demonstrations Washington had ever seen.[13] It made Alexander Graham Bell very proud: "We are a united family on the subject," said Bell, who endorsed suffrage long before most of the Bell and Grosvenor women did.[14]

The women of the National Geographic Society generally fared well during Bert's tenure. His magazine was among the first businesses in Washington to hire women in large numbers; albeit as clerks and secretaries, they took positions traditionally held by men. "I found them much more efficient in handling machines and typewriters,"[15] said Bert, expressing a sentiment that would land him in hot water today, but reflected progressive thinking when he and the twentieth century were young. Less progressive were the dining arrangements he approved for National Geographic's Sixteenth Street headquarters, where men and women ate in separate rooms, perhaps in the belief that this would inhibit lust in the workplace (which it did not). Dining by gender, like other antiquated practices at National Geographic, did not end until the 1960s.

Bert encouraged and published many articles by women, whose sensitivity, he believed, often made them better reporters and writers than men. But he never published as many feminine bylines as he would have liked. One wonders if more women might have found their way onto the editorial staff or the pages of National Geographic magazine if Bert had pressed harder for them. Most came to work over the objections of his key lieutenants, John Oliver LaGorce and Franklin L. Fisher, who believed that women belonged at home, not in the office. Indeed, Fisher was so inhospitable to women that he resisted having photographs of them in the magazine, a quirk that Bert tried to correct. "I had to insist that they put some women in," Bert said. "Mr. Fisher didn't have a very high opinion of women and Mr. LaGorce never did either."[16] Although Bert sometimes felt it necessary to bring his lieutenants up short, the truth was that he deferred to them on most matters, gave them wide discretion, and trusted them to keep the office running when he and Elsie traveled abroad each year.

These journeys gave Bert a chance to gauge the pulse of the outside world, provided ideas for future articles, and offered opportunities to test the latest photographic equipment and film under field conditions. By keeping pace with the latest technology, *National Geographic* magazine kept its edge over other periodicals just awakening to the promise of photography. For himself and for those who shot pictures for *National Geographic* magazine, Bert spent happily for the latest cameras and equipment; he also encouraged contributors to travel widely, to shoot plenty of film, and to anticipate the events that would make future headlines.

Such conditions were brewing in 1913, when Bert and Elsie headed for Russia, then on the eve of the convulsions that would change the old empire forever. Using a new folding Kodak camera and roll film—much easier for a traveling photographer than the bulky format cameras and glass plates then generally in use—Bert recorded the twilight of Czarist Russia, characterized by horse-drawn droshkies, devout peasants, street corner shrines, and rough-riding Cossacks—all immortalized in the November 1914 *National Geographic*. Among other subjects, Bert photographed one of the empire's recently shuttered vodka shops, or saloons, in Nizhniy Novgorod. The prohibition order signed by Czar Nicholas II "was one of the surprises of Russian history," reported Bert, who had given up smoking and drinking in his twenties. "Everywhere it was received with acclaim, and there were such widespread and universal evidence of the approval of the government's stand . . . that it soon became evident to the Tsar that what he had intended as a temporary measure could be made permanent."[17]

Bert's prescience proved more reliable on the harbingers of World War I, which smoldered in the Balkans that summer, threatening the alliances that had so far held off a European conflagration. In June of 1913, Bert and Elsie got a foretaste of the turmoil to come when the Second Balkan War broke out, prompting banks to close in Britain and stranding the Grosvenors in Edinburgh without funds for three days. Based on that experience, along with the war rumors that dogged him all summer, Bert made plans for the great conflict. As soon as he returned to Washington that fall, he arranged to print three hundred thousand maps entitled "New Balkan States and Central Europe" so that *National Geographic* readers could follow the conflict when it came. By the start of 1914, the freshly printed maps arrived. Instead of folding them into the next issue, Bert stored them in the basement at Hubbard Hall. There the maps stayed through June 1914, when the heir to the throne of Austria-Hungary, Francis Ferdinand, was assassinated in Sarajevo, touching off war between Serbia and Austria-Hungary and gradually drawing in Germany, France, Russia, and Britain later that summer. From his office in London, Sir Edward Grey, Britain's foreign minister, saw the dusk settling in St. James Park and uttered the most famous line of his career: "The lamps are going out all over Europe; we shall not see them lit again in our lifetime."[18]

When war broke out in August 1914, the new maps appeared with every issue of *National Geographic,* and such previously unfamiliar places as Liège, Mons, the Marne, Ypres, and Lemberg were made real and fixed on paper, satisfying a portion of the sudden demand for cartographic information in the United States. "I tried always to publish what the members would find helpful and what they would like, knowing that . . . was the only way to get increased membership,"[19] said Bert, who brought another one hundred thousand readers to *National Geographic* on the strength of his 1914 war map. At such times of change and conflict, Bert discovered, the public hungered for authentic geographical material, and now he knew how to provide it. By July 1915, he would publish a new European map, followed by a detailed "Theatre of War" map in May 1918. Based on these successes, Bert created the National Geographic Society's first cartographic division, which would continue printing detailed map supplements as world events demanded.

As a monthly, *National Geographic* was less successful at covering the rapidly breaking war news, which was thoroughly reported by newspapers. Alexander Graham Bell perceived the problem as the first weeks of fighting began. "It is obvious that the results of the war should form a principal topic," he wrote Bert, "but by the time you come out much of the war news that would be interesting now, will be stale by then."[20] Bert responded by commissioning articles on the historical and cultural antecedents of the nations in conflict. With few exceptions, these pieces read as if they might have been lifted from an encyclopedia or prepared by press agents for allied governments. The magazine did little to improve its war coverage when the United States entered the conflict in 1917. *National Geographic*'s war stories were upbeat and bloodless— hardly surprising since many were written by government officials, army officers, and others associated with the U.S. war effort. Herbert Hoover, then U.S. food administrator, contributed "The Weapon of Food"; Maj. Granville Fortescue, "Training the Armies of Liberty"; Gen. George P. Scriven, "Recent Observations in Albania"; and Charles M. Schwab, director general of the U.S. Shipping Board Emergency Fleet Corporation, "Our Industrial Victory." Bert also pulled together articles on the American flag, on armed forces insignia, and other patriotic subjects. Although the magazine's war coverage would win few journalism awards, it proved to be highly popular, perhaps because it provided a sense of hope, as well as relief from the bleak daily fare readers found elsewhere.

Even as thousands lay dying on the battlefields of Europe, Bert codified his own upbeat editorial philosophy that would guide his magazine through most of the century:

1. The first principle is absolute accuracy. Nothing must be printed which is not strictly according to fact. The Magazine can point to many years in which not a single article has appeared which was not absolutely accurate.

2. Abundance of beautiful, instructive, and artistic illustrations.

3. Everything printed in the Magazine must have permanent value, and be so planned that each magazine will be as valuable and pertinent one year or five years after publication as it is on the day of publication. The result of this principle is that tens of thousands of back numbers of the Magazine are continually used in school-rooms.

4. All personalities and notes of a trivial character are avoided.

5. Nothing of a partisan or controversial character is printed.

6. Only what is of a kindly nature is printed about any country or people, everything unpleasant or unduly critical being avoided.

7. The contents of each number is [*sic*] planned with a view of being timely. Whenever any part of the world becomes prominent in public interest, by reason of war, earthquake, volcanic eruption, etc., the members of the National Geographic Society have come to know that they will obtain the latest geographic, historical, and economic information about that region, presented in an interesting and absolutely non-partisan manner, and accompanied by photographs which in number and excellence can be equaled by no other publication.[21]

Like other principles Bert brought to the magazine, guidelines 1, 2, 4, 7, and possibly 6 show the influence of Alexander Graham Bell, a voracious reader often disappointed by the partisan bickering and frequent errors he encountered in his newspapers each morning. When he and Bert set out to reanimate *National Geographic* magazine, Bell saw an opportunity to improve upon the truthfulness of reporting, as well as the tone. His native optimism, generosity, and high-mindedness found their way into the soul of the magazine, which also came to reflect Bert's better traits—among them his fairness, his appreciation for learning, and his polite style.[22] The blending of these attributes resulted in a unique species of journalism at *National Geographic* magazine, which excelled at celebrating and explaining the world, especially in the areas of natural history, archaeology, ethnography, exploration, and science, which it knew best. But the magazine seemed to stumble when addressing the grim realities of the twentieth century, in part because of rules 5 and 6 on Bert's list; these kept vast regions of the globe off-limits through war, revolution, and other controversy. The Soviet Union, for instance, vanished from *National Geographic* for almost forty years because Bert could think of nothing positive to say about that country.

When the war finally ended in November of 1918, emissaries began converging on Paris for the peace conference that would recast the balance of power in Europe, carve new states out of the old Ottoman Empire, and advance

President Woodrow Wilson's proposed League of Nations. Because what happened in Paris would transform the maps of the world, Gilbert H. Grosvenor took a professional interest in the proceedings.

Initially supportive of Wilson's proposal for the League of Nations, Bert promoted the concept in the January 1919 *National Geographic*. As delegates gathered in Paris, he published "The League of Nations, What It Means And Why It Must Be," based on a speech by former president William Howard Taft. In the piece, Taft made strong arguments on behalf of the League, which he believed would promote peace and stability in the aftermath of the recent conflict. (Perhaps feeling that Taft's lengthy text needed spicing up, Bert managed to introduce into the middle of it a photograph of bare-breasted African villagers—that signature bit of prurient ethnography, by now a staple eagerly anticipated by many young *National Geographic* readers; the picture showed a group of colonists, in this case Cameroonians formerly under German rule, who might benefit from the new world order.) "Germany," Taft thundered, "has forfeited her right to the colonies by her mistreatment of them in the past."[23]

Almost as soon as Bert published Taft's article, he regretted it, having become skeptical as the details for the prospective League emerged. Like others confronted with President Wilson's grand-sounding but somewhat vague notions about self-determination, Bert grew concerned about their implications.[24] For one thing, he was afraid that giving small countries—such as New Zealand and Yugoslavia—an equal say in world affairs would diminish the influence of the United States.[25] For another, he fretted over the growing danger of nationalist and regional urges threatening the old stability. "If this doctrine was followed to its logical conclusion," he wrote, "it would mean the break-up of the British Empire. It would justify the establishment of a black republic in the United States, and it seems to me it would also justify the secession of our Southern States. The doctrine has been used to incite rebellions in Ireland, Egypt, and India, and I have felt is one of the principal reasons of the present turmoil throughout the world."[26]

Bert, who would worry over any problem until it was reduced to fine powder, grew increasingly fidgety about the fate of Russia, now in the grip of Bolshevism. This strange new force in world affairs prompted a rare show of emotion and a bout of political scapegoating in Bert. Vacationing in Atlantic City, New Jersey, where he had joined the family for Christmas, he picked up the *Newark Ledger* for December 24, 1918, and seemed to snap. What set him off was a front page photograph of Woodrow Wilson arriving in Paris with George E. Creel, the president's controversial chief of wartime information.

"Wilson was strangely influenced by this man," Bert wrote to his father that day, providing a clipping from the *Ledger*. "Creel is a violent hater of all things British and his absolute control of all the news from Paris explains the constant emphasis in the press dispatches of every condition that will arouse anti-British feeling. Creel is a Jew. How I hate that race; they are responsible for all the Bolshevik's propaganda everywhere and Bolshevik horrors."[27]

Bert's distaste for the Bolshevik horrors was understandable, given that his magazine had just documented, in the November 1918 issue, their desecration of Russia's cathedrals and cultural treasures, many of which he had photographed and admired before the war.[28] More surprising—even alarming—was his violent expression of anti-Semitism, all the more striking from a man who carefully considered his words, who forbade his children to criticize other people, and who usually promoted cross-cultural sensitivity in his magazine. For one who also insisted on accuracy, he had chosen a poor target—Creel's mother was Episcopalian, his father was Catholic, and Creel himself was a deist. But what was the source of Bert's anti-Semitism? And how did it affect *National Geographic*?

Bert's father, Edwin A. Grosvenor, whom he credited as the greatest influence in his life, publicly expressed admiration for the learning and adaptability of Jews and sympathy for the persecution they had suffered for centuries. But in private correspondence with Bert, it becomes clear that the elder Grosvenor regarded Jews as another tribe, strange and apart, not equal to his own social order. The polar explorer Frederick A. Cook would not be taken seriously because his father was a Jew, Professor Grosvenor wrote. A diploma from Bryn Mawr or Vassar was best for a young lady, he advised, but he warned that both schools "are crowded with Jewish Young ladies."[29] Hardly as venomous as the remarks his son had uttered, the professor's views suggest a dismissive and suspicious attitude toward Jews, which may have provided the foundation for Bert's own latent but occasionally noxious anti-Semitism.

Returning from Russia in 1913, Bert tried to explain why it was necessary for that country to enforce an anti-Semitic code. "The Jews in Russia have had a very hard time of it for generations," he wrote in *National Geographic* magazine. "An alien race prospering where the native race goes hungry naturally arouses bitterness, and that is what has caused the Russian government to adopt such strenuous restrictive measures against the Jews. Instances of this repression is [*sic*] written into every chapter of Russian law. There is a double tax on Kosher meat itself; there is a tax on religious candles used by the Jews; the head of the family must pay a tax for the privilege of wearing a skull-cap during prayers; not more than 10 percent of the students of a university may be Jews. . . . Russia feels that domestic policy requires these restrictions of the Jews. Without them, and unfettered, the wide-awake Jew would be too much for the lethargic Slav, say such authorities as Samuel Wilkinson."[30]

Bert's opinions about Jewish influence may also have been shaped by what he read—and edited—in *National Geographic*. In 1907 he published a story by William Eleroy Curtis, a popular travel lecturer and journalist who addressed himself to the stirrings of revolution in Russia. Curtis reported, with some justification, that Russia's leading revolutionaries had Jewish roots—some, including Leon Trotsky and Rosa Luxembourg did—then Curtis took a great leap: "Whenever a desperate deed is committed it is always done by a Jew, and there is scarcely one loyal member of that race in the entire Empire. The great strike which paralyzed the Empire and compelled the Czar to grant a constitution and a parliament was ordered and managed by a Jew named Krustaleff. . . . Every deed of that kind is done by Jews, and the massacres that have shocked the universe, and occurred so frequently that the name 'pogrom' was invented to describe them, were organized and managed by the exasperated police authorities in retaliation for crimes committed by the Jewish revolutionaries."[31]

This article and others like it violated Bert's own editorial rules 5 and 6—to avoid controversy or unkind comment. The editor's application of these standards, like his interpretation of geography, was flexible, providing a hospitable atmosphere for anti-Semitism in *National Geographic*. Bert would soon be wounded by it.

CHAPTER EIGHT
THE CHANGING ORDER

GILBERT H. GROSVENOR ON A MISSION TO SAVE THE SEQUOIAS, 1915.

Throughout the war years, Bert depended on John Oliver LaGorce, a barrel-chested, moon-faced man who had strutted into Hubbard Hall in the autumn of 1905 and asked for a job. Bert hired the former telegraph operator on the spot, paying LaGorce $60 a month to address magazines, field letters from members, and sell a few ads. Soon he was also handling administrative chores. "We made no contract," Bert recalled. "I asked for no reference. Here, I knew instinctively, was a man who shared my deep unquenchable faith in The Society's destiny."[1]

The pair had been inseparable in all the years since, with LaGorce's duties growing with the National Geographic Society. He eventually had responsibility for all advertising, and for as many executive and editorial duties as he could claim. He established himself as Bert's alter ego, the man to see when the boss was away. Over a half century of collaboration the two established a comfortable relationship, each contributing something to the partnership that the other lacked.

While Bert was understated, deliberative, and classically educated, LaGorce was a proud, blunt, decisive extrovert unencumbered by a college degree, the first in a long line of self-made people who would thrive under the meritocracy Bert established beneath himself. LaGorce's modest background, the blue-collar world of Scranton, Pennsylvania, provided a bridge from the magazine to the thousands of middle-class Americans Bert sought for his membership rolls. LaGorce also had more entrepreneurial flair, and perhaps better business sense. Although cursed with a sometimes debilitating stutter, Jack LaGorce brimmed with confidence, displayed shrewd political instincts, and made friends easily—among them the artist N. C. Wyeth, the heavyweight champ Gene Tunney, and the tycoon Carl Fisher. LaGorce plunged himself into Washington's social life, establishing high-level contacts in the business and political worlds, which would prove useful for the National Geographic Society. The loyal lieutenant never overplayed his hand or threatened Bert's supremacy in any way. For some at the magazine, LaGorce came to be seen as a bully, the enforcer unafraid to step on toes or to wield the hatchet for his patron. Bert, meanwhile, remained gratefully above the fray. "Certainly no two men ever worked together more closely or more harmoniously," he said of his colleague.[2]

With LaGorce on duty in Washington, Bert could spend months in Nova Scotia each year, keeping in almost daily touch with the magazine by telegram; the blizzard of correspondence was hardly necessary, as each man understood the other one perfectly well. LaGorce's presence freed Bert for regular travel abroad, and the lieutenant was always there to support the family in times of crisis. When Alexander Graham Bell Grosvenor, Bert and Elsie's fifth child, died at age six in 1915, LaGorce comforted his colleague and took over at the office for him.[3]

Overwhelmed by young Alec's unexpected death following what should have been a routine operation, Bert, overworked and worried about wartime pressures, broke down that spring. Leaving Jack LaGorce to manage the magazine, he retreated to the rolling Maryland countryside, where the family had recently acquired a farm they called Wild Acres. With Elsie and the surviving children, Bert immersed himself in the age-old rhythms of planting the garden, collecting eggs, and watching the cows grow fat on new spring grass. He took comfort at the sight of warblers and swallows returning on their flights north, leading the way for phoebes, which would soon begin their nests on the porch. Nothing could replace the loss of their son, to be sure, but the Grosvenors found healing in the countryside that spring, a solace as old as John Muir or Henry David Thoreau. The family slowly regained its poise.

Bert's devotion to conservation increased as he grew older. He used the magazine to celebrate natural wonders and to stress their preservation, themes that would be repeated down through the generations of *National Geographic* magazine. He promoted a vital piece of legislation known as the Federal Migratory Bird Act of 1913, which restricted commercial hunting and gave federal officials greater authority for conservation enforcement.[4] That same year, Bert represented the National Geographic Society at the Fifth National Conservation Congress in Washington.[5] Soon he became a logical recruit for the national-parks movement, which in 1915 struggled to bring some semblance of order to the country's scattered collection of underfinanced, unappreciated, and poorly protected federal lands. A wealthy Chicago businessman named Stephen T. Mather, recently appointed assistant secretary of the Interior for National Parks, stepped forward to address the problem.

He began by organizing a junket to the most spectacular lands of the High Sierra in California, where he hoped to build political support for parks. Mather's deluxe pack trip, set for July of 1915, included an assortment of opinion leaders and political figures, among them Henry Fairfield Osborn, president of the American Museum of Natural History; Burton Holmes, a well-known travel lecturer; Ernest O. McCormick, vice-president of the Southern Pacific Railroad; F. B. Johnstone, a member of the Sierra Club; Rep. Frederick H. Gillett of Massachusetts, ranking Republican on the House Appropriations Committee; and Gilbert H. Grosvenor. Bert leapt at the chance to join the Mather Mountain Party, as it came to be known.

Meeting in the town of Visalia, California, the nineteen-man group took horses and mules up through Sequoia National Park, down Kern River Canyon, and around Mount Whitney, all the while washed by bracing air and uplifted by rugged scenery, which spoke for parks as not even Stephen Mather could do. Mather subsidized the trip with $4,000 from his own pocket to ensure his guests' comfort—they slept under the stars on air mattresses, which sturdy guides inflated to preserve the lungs of gentlemen campers.[6] The party dined in comfort, sitting at folding tables spread with feasts of trout, porterhouse steaks, fresh cantaloupe, fried chicken, venison, and fresh-baked sourdough bread—all prepared by Ty Sing, a legendary camp cook of the region.[7]

For Bert, who had traveled abroad but had never ventured west of Ohio, the new country provided a welcome dose of vitality. Initially dubbed the Tenderfoot by fellow campers, Bert soon put to shame those who had given him that nickname by rushing past them on the march to Mount Whitney's 14,494-foot summit where thin air made such antics at once admirable and annoying.[8] When the party camped among the ancient trees of the Giant Forest, in the moss-muffled heart of Sequoia National Park, Bert was deeply moved by the dreamlike atmosphere. There he left the main party and insisted on sleeping among the ferns at the base of a mammoth tree, despite Mather's warning that he might be battered in the night by falling limbs.[9] The editor came away from the Western experience with a lifelong warmth not only for Stephen Mather but also for national parks, which would get frequent coverage in *National Geographic* magazine from that time forward. Indeed, Bert's enthusiasm for parks went beyond the stories he planted in his magazine; over a number of weeks, he worked behind the scenes with Mather, Horace Albright, and Frederick Law Olmsted, Jr., to draft legislation in 1916 establishing the National Park Service. The law, signed on August 25 of that year, consolidated the administration of all parks under one federal department, giving Mather the political clout and financial standing to make the U.S. system a model for the world.[10]

Bert also helped to protect a critical tract of Sequoia National Park that was vulnerable to logging. "Whoever has stood beneath these towering giants of the forest feels a reverent love for these grizzled patriarchs!" Bert wrote in *National Geographic,* with more enthusiasm than usual. "The oldest living thing! A thousand years may not bring them to their full stature, but a few days may wipe them out forever."[11] Unless money could be found to buy a 167-acre inholding (known as the Giant Forest, where Bert had slept beneath the trees), those private lands might be exploited. Bert and his allies of the Mather Mountain Party lobbied Congress for the appropriation, which came through in 1916. But the $50,000 sum, while gratifying, fell $20,000 short of what was needed. Bert immediately pledged to make up the difference with research funds from the National Geographic Society. The trees were saved. In gratitude, the park service designated a section of the newly acquired land as the National Geographic Society Grove.[12]

What makes Bert's largesse all the more remarkable is that it happened in a year when paper prices had risen by 20 percent because of the war, adding a $40,000 expense to the organization's budget. And Bert produced the sequoia funds over the objections of Alexander Graham Bell, who applauded his son-in-law's tree-hugging instincts but felt that Bert was misusing the money. "This would be a magnificent thing to do, but I must confess that I find a little difficulty in justifying the expenditure from a fund to promote Geographic research," Bell wrote from Baddeck. "I have no special objections to the appropriation if the research committee approves, but wish that it was more consistent with the objects of the Society."[13] Despite Bell's misgivings, the committee followed Bert's lead. The old order was changing.

This was fine by Bell. Now in his seventies, he had little time for diversions in Washington, and so he happily resumed scientific experiments in Canada, where he was building and testing hydrofoils, high-velocity boats which lifted above the water and skimmed along at speeds of up to seventy miles an hour.[14] Like those boats, Bell soared above the rough waters of life on an indomitable spirit, which he often kept aloft by willpower alone. "Smile and look cheerful, and pretty soon you'll feel so," he was fond of saying.[15]

For someone of Bell's luminous disposition, Gilbert H. Grosvenor's Coolidgesque reticence posed an occasional trial. Bell seldom criticized anyone, but he was known to feel distress at Bert's heartless streak and he once chided his son-in-law for his constitutional pessimism.[16] Certainly the older man came to know how Bert resented the unbreakable bond between the Bell daughters, Elsie and Daisy, who always lived close to each other, visited often, and occasionally aroused Bert's jealousy.[17] The sniping must have pained Bell, who retreated from controversy, especially when it involved loved ones. For his part, Bert was the most conventional of men, and he sometimes lost patience with his genius father-in-law's odd ways. And after *National Geographic* began to thrive, Bert grew snappish when Bell got credit for it. "I was the creator of the NGM," an elderly Bert scribbled in a note to himself when preparing for an interview in 1963. "Your adoration for Bell must not confuse you about his part of the NGM," he wrote to an imaginary interrogator. "I was the creator of it."[18]

Like all living organisms, families evolve to accommodate the demands imposed upon them by changing circumstances. With the Bells' growing attachment to Beinn Bhreagh came long absences from Washington, and although Bell remained on the board at National Geographic, he gradually withdrew from daily involvement there. Over time the psychic distance between Bell and Bert increased in parallel with the physical distance separating them. A coolness seeped into a relationship once defined by its open affection and warmth. Letters became more formal, less exuberant, and occasionally wary. The old spark was gone.

This became clear when the National Geographic president at the time, Rear Adm. John E. Pillsbury, died suddenly in 1919. Seeing opportunity in the vacancy, Bert began a backstage campaign for the job. Yet when he approached Bell to enlist him in lobbying, the older man resisted—this from the patron who had plucked Bert from obscurity, planted him in the editor's job, encouraged his romance with Elsie, deluged him with story ideas, and defended him when his position at National Geographic was threatened. As often happened when unpleasant news was brewing, Bell delivered his message to Bert indirectly, through Mabel Bell.

She tried to ease the impact. "Of course Father and I are very proud of the work you have done in building up that wonderful and successful organization known as the National Geographic Society," she wrote Bert from Canada.

Of course you are the natural, logical, and only possible successor to Admiral Pillsbury. . . . Everyone knows that you are the Society—the only thing to do is to admit it frankly honorably and publicly. That is our view—the other of course is that it is an honor which you have won by all those years of faithful service. The Society honors itself by conferring it on you—I read your letter to Father. Of course he agrees that you should have the position but he feels, as of course he would, that he could not vote on the proposition. I don't see why the fact that a man is your relation—your very own son or husband even—could prevent a person from voting for awarding merited honor on another. . . . You won't need Father's vote anyway. It's much better that the nomination & election should be the work of persons outside the family. The honor is greater.[19]

Three days later, Bert telegraphed Canada that he had won the election without Bell's help, and by a unanimous vote of those in Washington. What was more, he had decided to continue as editor. For the first time in its history, National Geographic's executive and editorial power now resided in one person, in whom it would remain for more than three decades. Bell, ever the democrat, disapproved of this concentrated power,[20] but he did not make an issue of it, for which Bert was grateful. "I can see clearly the difficulties of being President," he told Bell. "However, the complications of continuing any longer as the Morganatic President seem even greater. . . . Many thanks for your message."[21]

Bell went back to inventing boats, while Bert hustled to shore up his new position. Within weeks of his election as president, he was consulting a well-regarded Wall Street lawyer—his twin brother, Edwin Prescott Grosvenor—about changing the bylaws.[22] Bert and other board members had been alarmed by language in the National Geographic Society's original charter, which could allow a handful of the 750,000 members to take control of the organization, turn out its board and officers, and make off with its estimated assets of $4 million.[23] To avoid this possibility, the Grosvenor twins drafted new bylaws that abolished National Geographic's annual meetings and empowered a twenty-four-member board of trustees to elect officers, change bylaws, and fill vacancies on the board; the new arrangements also gave board members tenure for life and extended the National Geographic's charter in perpetuity, far beyond the one-hundred-year term envisioned by Gardiner Greene Hubbard and other founders. Finally, under Bert's revisions, the president-and-editor was given day-to-day responsibility for running the National Geographic Society. The board unanimously adopted the amendments on March 15, 1920, at the same time filling the trustee's slot left vacant by the late Admiral Pillsbury; John Oliver LaGorce was his replacement. Members of the new Society would still be required to contribute dues, but they would no longer have power—a change that troubled Alexander Graham Bell, who had been absent when the bylaws were approved and now refused to sign the new documents of incorporation.

"I am only sorry that the conversion of the Society into a Trust Corporation should have resulted in a form of organization in which the members have no voice in the selection of new Trustees," he wrote Bert.[24]

Bert tried to soothe his father-in-law. "By the old condition," he explained, "the changes in the by-laws could only be made by a meeting in Washington, at which less than .00005 per cent of the membership were wont to come.

To give a right of proxy to the 750,000 members outside of Washington would have been cumbersome and unsatisfactory, besides being exceedingly expensive to operate. . . . A person who thinks only casually might say that the National Geographic Society is no longer a Society, inasmuch as its members do not participate in the administration of its affairs. Herein, however, I differ. The members do not participate in the financial or business administration, but they do participate most actively and enthusiastically in the promulgation of the Society's activities for the increase and diffusion of knowledge, and prove their participation by the tens of thousands of new members which they are constantly recommending. The meaning of the word 'Society' is not limited to a smoking club or a 'talking fest'. . . . Unlike other educational and scientific institutions, such as the Carnegie and Smithsonian, the National Geographic Society is not founded and supported by one wise and financially fortunate individual. Instead, there are 750,000 wise and financially fortunate founders of the institution, all of whom are equal participants and promoters of the wonderful idea for which the Society stands."[25]

Whether Bert's argument helped to convince Bell is unknown, but the inventor eventually agreed to endorse the new charter after Mabel Bell interceded, reminding her husband of their son-in-law's new standing: "That's the president of the Society," she reportedly said, "and he recommends it, and you ought to agree." So threatened, Bell signed the papers of reincorporation, probably hissed to himself as he did when upset, and returned the documents to Washington. He was the last trustee to do so.[26]

The National Geographic Society's new leader proved to be both conscientious and effective. He and other officers worked very hard, lived comfortably but without extravagance, poured surplus funds into the magazine and important research projects, and saw the organization grow far beyond even the predictions of Alexander Graham Bell, who had so recently foreseen a membership of *ten times ten thousand*; after twenty years on the job, Bert had pushed that membership very near *one hundred times ten thousand*. His new position of influence agreed with Bert, whose past troubles at National Geographic had always mounted when his authority was divided or questioned; now with all the levers of control at hand, a loyal staff surrounding him, ample resources to draw upon, and a board of trustees packed with friends, relatives, and admirers, he could afford to be magnanimous. Among his subordinates Bert came to be known as the Chief, a distant but gentlemanly figure who often remembered anniversaries and birthdays at the office, never raised his voice in anger, praised notable performance, and established a collegial atmosphere that made

National Geographic seem less like a monthly magazine than a cloistered faculty club. He was particularly admired for his consideration of others. Once when a colleague from the north appeared for a summer dinner in a white linen suit and black shoes, Bert, similarly attired except for the shoes, excused himself without comment, disappeared for a few minutes, and returned in black footwear.

"My family gets the questionable benefit of all the temper which must be suppressed at my office," he confided to Mabel Bell about this time. "Here I must always be calm, cool, unprejudiced, affable, and temperate in language. I regard my job as editor very much as a judge regards his, except that I can exercise initiative, which a judge can't."[27]

Bert's initiatives in these years helped raise the organization's profile. National Geographic continued to improve its expertise at mapmaking, which won good reviews from members as well as discerning consumers at the State Department, which displayed National Geographic's "New Europe" map in its press room. Bert also sent regular National Geographic bulletins containing maps and abstracts of articles to more than two thousand U.S. dailies and weeklies eager for up-to-date geographical information. During this period the organization's research committee financed important long-term scientific surveys in the rugged Mount Katmai region of Alaska, mounted studies of northern glaciers, and contributed the major financing for archaeological research at Chaco Canyon in northwestern New Mexico; there Neil M. Judd, curator of American archaeology at the U.S. National Museum, directed five seasons of work documenting the spectacular communal ruins, among the most significant pueblo settlements of North America.[28] While pressing these initiatives, Bert raised $55,000 from members to add another 640 acres to Sequoia National Park.

The magazine provided readers with groundbreaking stories from contributors like George Shiras 3d, who created extensive natural-history coverage of the Lake Superior region, and from Ross Smith, who produced a firsthand account, running to more than one hundred published pages, of his 1920 flight from London to Australia in an open-cockpit airplane.[29] Bert would continue to acquire such articles as they became available, and he would stockpile manuscripts and photographs to fill the reservoir for future issues. But he also realized that *National Geographic*'s staff needed to grow if he was to provide readers with the distinctive, timely coverage they now expected.

By 1920, Bert began to fill the headquarters building with the artists, writers, editors, and photographers who would shape the future personality of his magazine. Most of the organization's four-hundred-plus employees were split between offices in central Washington and a converted brewery in Eckington, a Washington suburb, where clerical personnel kept track of correspondence and subscriptions. The editorial staff was housed at headquarters on Sixteenth Street, where the original Hubbard Hall had been expanded in 1913. The new wing, prosaically named the First Administration Building, accommodated a growing staff. Among them was a talented young artist named Hashime Murayama, who not only hand-tinted photographs for *National Geographic*'s early color issues but also produced exquisite watercolors for stories on frogs, birds, tropical fish, and other natural-history subjects. College men joined the magazine's ranks alongside high-school dropouts, a former diplomat, an ex-army sergeant, several well-traveled newspapermen, and a former Baptist missionary from Kalamazoo with a Columbia journalism degree and a triple-barreled byline conveying instant gravitas: Maynard Owen Williams.

Williams, a shambling six-footer with steel-rimmed glasses, an eternal smile, and a knack for languages, became the magazine's first foreign correspondent. A writer-photographer who dressed like a scoutmaster in knee-length boots and khaki shirts, always with a tie, Williams spent months and years abroad, during which he recorded the opening of King Tutankhamen's tomb, roamed the deserts of Tunisia and Arabia, lived among the Inuit of northern Greenland, sailed around the Celebes, drove across Hindustan, and retraced Marco Polo's overland journey from the Mediterranean to China. Williams was happiest on the road, living out of dusty steamer trunks, wrestling with camera equipment, and losing himself in foreign cultures. He pretended to know less than he did. "I get around the world by being a nice little lady," he said. "I never carry a gun. I tell everyone I am helpless, and that I cannot speak their language."[30] Thus he made friends from Prague to Singapore. Many American readers, who at this time had never ventured abroad, got their first glimpse of the outside world through this exuberant and often innocent man, who neither drank nor smoked nor ever uttered a curse stronger than "Oh, Thunder!"[31]

Like those who had founded the National Geographic Society, Williams looked upon the world as a benign place, full of hope—this, despite his previous experience covering the Russian Revolution, his duty as an army intelligence officer in the lawless provinces of China, and his painful months of organizing relief for the Armenians of Lake Van, where he saw hundreds of children starve and die as a result of war. He somehow emerged from these events, and from his arduous travels, without a trace of cynicism. If only people could meet and talk, Williams believed, the barriers of culture and politics would melt away, peace and understanding would follow, and the world would be better. During three decades of writing and photographing for the magazine, Williams never preached this sermon directly to readers, but it was always there, straining beneath the surface of his stories.

"This was a family, a small family," said a staff member who recalled Williams as one might a favorite uncle. In the same vein, Gilbert H. Grosvenor was regarded as the undisputed head of the National Geographic clan, and he watched over his growing staff with the same attention he devoted to his real family.[32] "This was typical of him," said Luis Marden, who had begun his National Geographic career in a basement laboratory. "He noted the sparrow's fall."[33] Marden still had the note Bert had sent him almost fifty years before, upon the occasion of his marriage. "I've been intending for several weeks to extend to you my congratulations on your marriage but have omitted to do so," Bert wrote. "However, here they are and as you have enjoyed the married state for some time now, you will now appreciate even more deeply than then how unfortunate bachelors are, and how much you ought to be congratulated."[34]

Such gestures earned Bert a special loyalty among those who worked for him, just as he demanded it in his own family. "My chief personal principle," Bert said, "is to be loyal to my family as I regard disloyalty, especially to one's family, as the basest of faults."[35] Mabel Grosvenor vividly remembered an object lesson from her youth. "It was on a Saturday morning and we were at Wild Acres picking up branches from a recent storm. Father made up a bundle of sticks and asked me to break them. I couldn't. Then the other children tried and they couldn't. Father pulled out individual sticks and we took them one at a time and they were easy enough to break. The bundle, Father said, was like the family. If we were loyal to each other and stayed together, it made us stronger. . Father was very big on loyalty."[36]

Among Bert's most faithful supporters was a young botanist named David Fairchild, whom the new National Geographic president invited to join the board of trustees. Fairchild had married Marian "Daisy" Bell, Alec and Mabel's younger daughter. But quite apart from that, he had established a distinguished career as a scientist, combing the globe in search of plants for the Department of Agriculture. Before he was forty years old, Fairchild came to roost as chief of the department's Bureau of Plant Introduction, from which he directed a worldwide network of botanists collecting plants to improve agriculture, industry, medicine, and the beauty of lawns and gardens in the United States. From this band of roving plant collectors, Fairchild spotted promising contributors for National Geographic magazine, among them Orator F. Cook, whose research at Machu Picchu advanced understanding of ancient Incan culture, and Joseph F. Rock, a self-taught linguist and pistol-packing botanist who lived for thirty years in China's remote southwest, where he traveled with mounted guards, a large format camera, a folding bathtub, and an ample supply of botanical gear. Rock collected thousands of plants, discovered new species, and documented the lives of border tribes retreating before that country's dominant Han culture. Rock was working there in 1923 when his funds from the Agriculture Department dried up. Fairchild approached Bert, who wangled a $17,000 National Geographic grant to keep Rock in place for two more seasons, during which he performed double duty for the government and the magazine.

Closer to home, Fairchild provided his own stories for the journal, most notably his May 1913 article on backyard monsters. That piece brought National Geographic readers face to face with a photographic gallery of yellow jackets, grasshoppers, beetles, flies, wasps, and other insects magnified up to twenty times their normal size. To accomplish this, Fairchild improvised his own macro-camera with a homemade lens twelve feet long. This brought his six-legged subjects into focus, their eyes popping, armor gleaming, hairs bristling; his work revealed an aspect of the natural world previously unknown to thousands of readers who benefited from yet another of the magazine's photographic innovations.

Friends and family were never far away in the National Geographic board-room, where Bert was surrounded by trustees who were, for the most part, ea-ger to prove their loyalty. Among them were Charles J. Bell, a steady ally dating to Bert's earliest days as an editor; David Fairchild, Bert's brother-in-law; William Howard Taft, cousin and ex-president; and such steadfast friends as Gen. Adolphus W. Greely, George Shiras 3d, Rear Adm. Colby M. Chester, Rear Adm. Robert E. Peary, and Frederick Coville, whose son would eventu-ally marry one of Bert's daughters. Jack LaGorce, who owed his job and his po-sition on the board to Bert, honored his patron not only in unflagging allegiance, but by naming his first-born son Gilbert Grosvenor LaGorce. When Gil LaGorce came of age, he went to work at National Geographic. The most prominent board member, Alexander Graham Bell, though erratic, could be counted upon to back Bert in a real crisis. In addition, Bert listed his brother Edwin on the masthead "of counsel" to National Geographic in 1919, and brought him on as a trustee as soon as a suitable vacancy appeared. Stephen Mather, Bert's camping partner from Sequoia days, soon joined them. Almost all of these board mem-bers contributed articles, for which they were paid handsomely. Fairchild re-ceived $1,000 for his article "The Monsters of Our Backyards." Bert's father got $1,200 for his 1919 piece "Races of Europe."[37] Bell happily accepted com-pensation for several articles, including those appearing under the byline of H. A. Largelamb, an anagram of his name.[38] Before he joined the board of trustees, Bert's brother Ed took a $2,338.88 fee in 1921 for his legal work at National Geographic.[39]

Such arrangements met the classic definition of nepotism, a tradition at Na-tional Geographic since Alexander Graham Bell inherited the organization's presidency from his father-in-law. The practice of preferential treatment for relatives and chosen friends became easier—and more frequent—after Bert's 1920 revision of the bylaws, which diluted the influence of members and increased that of the board. Bert essentially decided on the replacements for board vacancies, which were routinely approved by self-selecting trustees. He paid contributors and spent money as he saw fit. He ran the magazine with no outside check on his power. By today's standards, these practices would be questionable—but in Bert's day nepotism and other conflicts of interest were widespread, especially in family publishing companies such as Scribner's, the *New York Times,* the *Washington Star,* and various Hearst papers. Did nepotism harm National Geographic? Probably not—in fact, the opposite seems to be the case. Family influence helped to transfer important institutional values from one generation to the next, and it generally led to an orderly transition of executive power. It is undeniable that the National Geographic Society thrived under the Grosvenor family, which enlarged the organization's value and reach, set aside surpluses for research grants, and greatly improved the magazine. During his first two years as president, for instance, Bert increased the number of pages in the magazine by 10 percent; the paper weight by five pounds; the number of map supplements by 10 percent; the number of pages devoted to four-color illustrations by 85 percent; and the text pages by 25 percent.[40] With all proceeds going back into operations, the organization could afford such spending for its journal, which stood as the prime symbol of National Geographic's quality.

Had Bert been less scrupulous as president or his board less principled, the arrangements might have been a recipe for disaster. In practice, however, the organization came to reflect the solid, decent values passed down the line from Hubbard to Bell to Grosvenor. Almost without exception, the trustees around Bert were outstanding scientists or public figures who believed deeply in the National Geographic Society's altruistic mission. Bert's twin brother, Edwin, for instance, was a leading lawyer of his day, a former assistant attorney general under both President Theodore Roosevelt and, later, President Taft. In that role, Ed Grosvenor successfully prosecuted night riders in a Kentucky hate crime and broke the "bathtub trust" in a prosecution of plumbing manufacturers, which reached the Supreme Court and established important new principles of antitrust law. As an advisor to National Geographic, Ed appears to have been thorough and conscientious in his handling of the organization's legal concerns.

Even as family clustered around Bert at National Geographic, he felt the stirrings of disquiet at home, where the nest was beginning to empty. The first of the Grosvenor children was going away. "Melville is to be home only one week more," Bert wrote to his father in June 1919, as Melville packed for his first year at the U.S. Naval Academy. "I shall miss him very much, needless to say."[41] Bert felt the loss keenly, worried about the harsh treatment Melville received in his plebe year at Annapolis, and brooded when a persistent infection landed his son in the infirmary, where he was hospitalized for several weeks. To make matters worse, the Grosvenors' eldest daughter, Gertrude, was seriously injured in an automobile accident in 1922 and placed in a body cast. She became obsessed about being left alone and developed an anxiety disorder requiring medical treatment. All of this, on top of the usual pressures at work, had the whole Grosvenor clan on edge by midsummer.

Knowing how much Bert and Elsie needed a vacation, Mabel Bell encouraged them to go. All was well in Baddeck,[42] where she and Alec were enjoying an unusually quiet summer. So Bert and Elsie sailed for Brazil on July 15, accompanied by Melville and his sister Lilian. Back at Beinn Bhreagh, Bell, seventy-five, had several projects going as usual—experimenting with hydrofoils, tabulating studies on his twin-bearing sheep, and searching for ways to make seawater potable. On occasion, when the rest of the house slept, he would steal into the kitchen for a feast, piling his plate high with Smithfield ham and other forbidden treasures from the refrigerator. Then he would carefully wash the plates and clear away the evidence before creeping off to bed—only to be betrayed when indigestion swooped down a few hours later. The doctor, summoned to Point House after one such nocturne, extracted a confession from his patient, who had been diagnosed as diabetic since 1915. "That meal might have put an end to you, sir," the doctor scolded. "Well, as it is," Bell replied, unrepentant, "the game was worth the candle. It was the best meal I've enjoyed in an age."[43]

Despite such feedings, Bell began to lose weight. He turned pale and lost energy. Doctors diagnosed anemia. They prescribed iron injections, which did little good, in part because Bell was such an indifferent patient, and in part because years of diabetes had probably done irreparable harm to his pancreas and liver. When Bell fainted in the laboratory one morning, he revived in time to catch one of his workers, Walter Pinaud, reaching for the telephone to call for help. Bell grabbed the phone away, hugging it to his body so that Pinaud could not wrest it from him. "Never phone a doctor when this occurs again," Bell ordered, "because no doctor can save me now. It's too late."[44]

Before anyone else, Bell sensed that it was time to leave and he was determined to do so with the dignity that had characterized his life. As July faded toward August, he found that he could not get up from his sofa; he was moved to a sleeping porch, with its splendid view of the mountain he loved. He read late into the night and slept more than usual by day. But he remained alert. He awoke on July 30 to find Daisy and David Fairchild newly arrived from Washington. They came and went from Bell's room, taking turns with Casey Baldwin, Bell's closest colleague in Baddeck, and Mabel to keep vigil. Once, when the others had left the room, Bell put aside the novel he had been reading and fell into scientific conversation with David, who wondered about Bell's shortness of breath. "I asked him if the weakness might not be electrical in character," David recalled, "and he shrugged his shoulders and said as he always did when you asked him a question about which he had thought deeply but regarding which he had formed no conclusion, *'Je ne sais pas, Monsieur. Je ne sais pas.'*"[45]

Bell slept again, murmuring to himself and waking to ask David if Bert and Elsie knew of his illness. David told him that they were traveling; he sent word to Brazil, but knew it would be too late. Bell worried whether Charlie Bell, his beloved cousin, would arrive: "I am afraid he won't get here in time," Bell said. He called for his secretary, Catherine MacKenzie, to dictate some final notes. For once he fumbled with words. Someone told him not to hurry. "I have to," he said, trying to name all of those he loved.[46] "Mrs. Bell and I have both had a very happy life together, and we couldn't have had better daughters than Elsie and Daisy or better sons-in-law than Bert and David, and we couldn't have had finer grandchildren." He asked his family to help Casey Baldwin. "We have to stand by Casey as he has stood by us," he said, ending the session. Mrs. MacKenzie went away.

Bell tried without success to eat a little something for Mabel's sake. "His last hard service to me," she said. He squeezed her hand, dozed off, and woke on the afternoon of August 1 to find Daisy sitting beside him on the sleeping porch. "How beautiful it is here—the air is so fresh," he said, breathing in the cool fragrance of balsam and salt air. Night brought the stars wheeling over Bras d'Or Lake, where moonlight streaked the waves. David lit the bedside lamp. Bell slept again, calling for his cousin from the depths—"Charlie! Charlie!"—and when he awoke late in the night, Mabel was kneeling beside him. "Don't leave me," she pleaded. He smiled and squeezed her hand, transmitting their old signal for no. Still gripping her hand he closed his eyes for the last time, sleeping until his breathing stopped about 2 A.M. on August 2.

Two days later, John MacDermid brought a buckboard to a rolling stop just beneath the portico at Point House as he had done so many times before. But on this Friday afternoon, instead of blowing his police whistle to call Mr. Bell downstairs, MacDermid simply waited for his neighbors to load a rough pine box into the wagon. The casket, made just the day before by the laboratory men, was built tight as a dory and fixed with iron handles forged in the Beinn Bhreagh workshop. As a final tribute to their old boss, the workmen made his casket plenty wide and lined it with airplane linen so that he could rest easy inside, dressed in his favorite gray corduroy jacket and knickerbockers for this last trip up the mountain.

Casey Baldwin and one of Bell's grandsons, Graham Fairchild, led the way on foot, followed by twenty or thirty mourners who trudged up the long gray road bareheaded despite the cool mist wreathing the promontory that day. More than one in the funeral party remembered that it was just the sort of weather Mr. Bell loved best. John MacDermid swung his wagon in behind the walkers. He drove slow, followed by four cars. When the gravel road climbed out of the woods and onto the meadow the procession stopped, and Mabel Bell emerged from the first car. She wore a white dress and a soft white scarf around her neck, having decided that mourning clothes would have been a disservice to her husband's memory; she also knew that she would never be able to change from black if once she put it on. With her son-in-law David Fairchild supporting her, she walked the last few yards beside the wagon, crossing the meadow with her hand resting on her husband's casket. She kept her hand there through the brief service on the mountain, through the singing of verses from Tennyson, through Stevenson's *Requiem,* through "Bringing in the Sheaves" and the Lord's Prayer, maintaining contact as long as she could. Then she stepped back and watched them lower Alexander Graham Bell into the grave. David Fairchild stepped forward: "Mrs. Bell wants me to thank you all for coming," he said. "She knows you loved him." With the rest of the crowd, she who loved him best turned away and went down from the mountain to face life without him.[47]

CHAPTER NINE
THE CRASH

MIDSHIPMAN MELVILLE GROSVENOR, EDWIN A. "GROSVIE" GROSVENOR, AND
GILBERT H. GROSVENOR IN NATIONAL GEOGRAPHIC OFFICE, C. 1920.

I did not know your father was going until he was gone." Thus wrote Mabel Bell to Elsie Grosvenor on the day Alexander Graham Bell died. Even before her husband's funeral arrangements had been completed, Mabel, though stunned with grief, reached out to her absent daughter to ease any regrets Elsie might have felt for being so far away. "I was worried about him when you left but not much more so than I had often been before & anyway I couldn't have you all stand around waiting for something that I was bound wouldn't happen," Mabel wrote. "I am not reconciled. I never will be."[1]

Returning from Brazil, Bert and Elsie stopped in Washington to collect Gertrude, just emerging from almost eight months of convalescence, and the reconstituted Grosvenor clan converged on Baddeck that September. A family contingent trekked up the mountain to pay their respects at Bell's grave. The widow described the day as both lovely and horrid, with lashings of rain and wind straight off the sea, as if the weather had been ordered up by Alec himself. Back inside Point House with her grandchildren racing up and down the stairs, Mabel perceived that the shock of Bell's death had worked some change in Gertrude. "I believe the emotions consequent to her Grandfather's going have had just the effect I believed they would, and she is stirred to prove herself worthy of him," Mabel wrote.[2] Perhaps this was true; for whatever reason, after months of struggling with physical and emotional trauma, Gertrude edged toward recovery that autumn. The family scattered again. Gertrude moved to New York and enrolled in the Art Students' League, where she would study painting; Mabel Grosvenor returned to boarding school; Melville shipped south for his last year at Annapolis. Bert, having settled Elsie with her mother in Baddeck, returned to Washington, happy to be free of family troubles for a spell. He relished the routine at National Geographic, played golf two or three times a week, and took pleasure in the unaccustomed calm at home. "To be frank," Bert wrote Elsie in early October, "I am getting some rest & quiet which I need. . . .

I want to be exempt from family responsibilities until Nov. 1 at least. I am better off for the present alone. I love you, my dear wife, but it is good for married people to be separated occasionally and I feel that separation is very desirable just now for you & me. Since Jan. 1 I have been practically a head nurse to the Grosvenor family, to Gert, to Mel and to you and I want to be alone and pull myself together and get back into my office interests and plans. I am a young man and should have the best years of work before me but another year like this will wreck me. Give me until Nov. 1 by myself & I shall be rejuvenated. . . . Under no condition must you come down now.[3]

Bert's admonition might not have been as selfish as it appears; he outlined other reasons for Elsie's staying in Baddeck. There was the question of her own health, which had eroded during the family's recent crisis, and which Bert thought would benefit from Baddeck's soothing atmosphere. And since Mabel Bell had declared her intention to stay put in Nova Scotia, she needed Elsie close by. "She needs your bigness, your calm strength and unselfish views; you can supplement what Daisy lacks," he wrote.[4]

Elsie stayed through the fall, helping her mother to keep Bell's work alive. Mabel pored through the records of his sheep-breeding experiments before culling the flock; she encouraged Casey Baldwin to continue Bell's laboratory work; she gave some thought to a biography of Bell. "She must be doing *Papa's* work to be happy just now," Elsie wrote Bert. "It is the way she feels him nearer."[5] From the distance of Washington, Bert got the impression that Casey Baldwin was steering his mother-in-law toward laboratory work at the expense of the prospective Bell biography. The latter, Bert felt, would be more likely to perpetuate Bell's name. "I hate to see all this emphasis placed on the laboratory as if that was first & the most important to continue," Bert wrote Elsie. "The laboratory is of course the easiest to continue but it is not so far reaching or so permanent in its influence as will be the written story of your father's life which if properly handled will affect millions & millions for centuries *provided* it can be started right away."[6]

As she had often done in their marriage, Elsie calmed her husband, assuring him that Casey Baldwin was to be trusted and that her mother would focus on the biography soon enough.[7] When that happened a few weeks later, it was clear that Mabel had definite ideas about the project. She wanted a book that would present Bell with his faults, not a hagiography. "You must see that the biography does not picture Father as a perfect man," she wrote Bert, who was gathering material and suggesting writers for the prospective work.[8] While Mabel was content to have her son-in-law supervise the project, she made it clear that she did not want him to write the book. That, she said, would be like a doctor treating his own family—too close for a critical perspective of the subject. "All this does not imply any depreciation of your ability as an Editor, or indeed as a man generally," she wrote Bert. "I am a horrid old thing, but my husband is so much to me that I know the very best account of him that could be written will seem to me wrong in some way. . . .

> I would hate to have things attributed to him that were not so. He is a big man, very imperfect, lacking in things that are lovely in other men, but a big good Man all the same, broad-minded and generous and tolerant in some things beyond the comprehension of most, and then curiously the opposite in others. I do not know a person he has not fought some time or other, and some times I, his wife, have thought him badly in the wrong, and at others entirely in the right. But I could never say this publicly, it would seem disloyalty, and none of us would either I think, which would mean that the book would be inadequate. . . . Please forgive me and remember that the whole thing touches me in my tenderest part, and I can't be as nice as I would if I didn't care so much.[9]

Bert shook off the disappointment, found a spare office at the National Geographic Society, and installed an editor to catalog Bell's papers. He also put out word that the family was arranging an authorized biography so that publishers would not "rush immature and feeble ones into circulation."[10] He urged Mabel Bell and other family members to write their reminiscences of the inventor, and Mabel threw herself into the task, writing a few fragmentary accounts of her life with Bell. But she was forced to put it aside in November 1922, when she fell ill, developed jaundice, and lost all taste for food. At the family's insistence she left Baddeck in December for a medical examination at Johns Hopkins Hospital in Baltimore, which revealed cancer of the pancreas.

She faced this last crisis with the same blend of strength and levelheadedness she had brought to bear on previous ones. "Wasn't I smart to stay well until Alec no longer needed me?" she asked her granddaughter and namesake that winter. Mabel Grosvenor smiled at the memory. "All of us thought that it was a blessing that he went first," she recalled. "Grandmother was better able to cope alone."[11] She did not have long to do so. Brought to Washington from Johns Hopkins, Mabel Bell died at David and Daisy Fairchild's home on January 3,

1923, less than six months after Alec. The family carried her ashes up the mountain at Beinn Bhreagh that August, and laid her down with Alec to the familiar strains of "Bringing in the Sheaves."[12] At Mabel Bell's request, the service took place at 5 P.M., the hour she and Alec used to meet at the laboratory for their walk home together.[13]

Among her papers, Bert and Elsie found a recently written account of her first meeting with Bell, when she was a girl of almost sixteen striding up Beacon Hill. Bert gathered up the loose sheets of this memoir, which ended in midsentence, and added them to Bell's papers in Washington. That collection, to which Bert controlled access, would continue to grow over the years, eventually filling 183.2 linear feet of library space. Yet the urgency for a biography of Alexander Graham Bell seemed to die with Mabel. Not until the 1940s did a prospective writer step forward to take on the book—one who violated Mabel Bell's warning about family involvement, for the author was Lilian Grosvenor Jones, Bert's third daughter. When Lilian, a gifted writer who had profiled Bell for *The New Yorker,* asked her father for access to the Bell papers, he agreed to cooperate, but on the condition that the family review her manuscript before publication. Lilian, who had inherited her mother's independent streak, refused that stipulation, and the project collapsed.[14] Bert and Elsie would later cooperate in the production of *The Story of Alexander Graham Bell,* a movie starring Don Ameche and Loretta Young, in 1939, but no definitive life of Bell appeared until 1973, long after Bert died.

In some part, the delay in coming to terms with Bell reflected Bert's complicated feelings about his father-in-law. He loved Bell, who had given him the job that came to define him and furnished critical financial support. But Bert, a proud overachiever who wanted to be known for his own accomplishments, also resented living in the great man's shadow, which continued to dominate family affairs long after Bell's death. "We were always aware that Gramp chafed a bit at being Alexander Graham Bell's son-in-law," recalled Grosvenor Blair, one of Bert's grandsons. "One had to be very careful not to be too Bell-minded," Blair said, noting that this wariness had infected several generations of Grosvenors. "We're a little like Gramp in some ways. I hate being introduced as

Bell's great-grandson. I really hate that. When you grow up surrounded by Bell and his achievements, you still want to be known for your own."[15]

Bert remained touchy about the Bell legacy into old age, long after it seemed reasonable, given the strength of his own accomplishments. Under his guidance, National Geographic had become one of the country's biggest magazines by 1926, with more than one million subscribers and property valued in excess of $5 million; faced with this embarrassment of riches, Bert moved to strengthen National Geographic's position as a tax-exempt nonprofit institution under federal law.[16] In correspondence with the Internal Revenue Service commissioner, he crisply laid out his argument, noting how the organization functioned as an educational institution, served the public interest, and put excess revenues into scientific research and other programs—all requirements for the tax-exempt status he sought. By March of 1927, the IRS decreed that National Geographic qualified for nonprofit treatment, which the organization has retained ever since. "You would have made a fine lawyer," wrote Bert's brother Ed when informed of the IRS decision.[17]

With its surplus funds, the National Geographic Society had ample means to send explorers, writers, and photographers across the globe in search of adventure. In the process they discovered new mountains, animals, glaciers, plants, and lakes, often naming them after their patrons in Washington. Thus did the Grosvenor name begin to appear in obscure scientific papers of the twentieth century describing a seashell from western Greenland (*Mararites grosvenori*); an aphrodisiac plant from China (*Momordica grosvenori*); a blind fish from Peru (*Bryconamericus grosvenori*); a black-masked thicket warbler* from the Melanesian island of New Britain (*Cichlornis grosvenori*); and a rufous-chinned laughing thrush** from Nepal (*Garrulax refogularis grosvenori*). New maps and atlases sprouted the family name as well, from the Grosvenor Glacier of Peru, to Gilbert Grosvenor Island in the Canadian Arctic, to the Gilbert Grosvenor Mountain Range of Antarctica, to the Grosvenor Arch of Utah, to Mount

*Now known as the Whiteman mountain warbler.
**Now known as the variegated thrush.

Grosvenor and Grosvenor Lake in Alaska. An object named Grosvenorfjellet, or Grosvenor Mountain, emerged east of Norway's forbidding Spitsbergen Island. (By comparison, Alexander Graham Bell was immortalized in just one piece of geography, Graham Bell Island in the Canadian Arctic, although his name is still attached to millions of telephones.)[18]

For Bert, seeing his own name identified with far-flung places and newly described species must have been bracing. He was a relentless promoter who nourished a lifelong reverence for scientific discovery. But he was also a modest man, and proved this by telling an interviewer from *The New Yorker* that he had not bothered to keep track of the many plants, animals, and places named in his honor.[19] This might have been true at the time of his 1943 interview, but it was not so in 1925, when he got word that the U.S. Board on Geographic Names was going to wipe Lake Grosvenor, Mount LaGorce, and Lake Coville off the map of Alaska. Those place-names, from the remote Mount Katmai region, had been proposed by Robert Griggs, a scientist with the U.S. Geological Survey who had been sponsored by the National Geographic Society Committee for Research and Exploration (Frederick V. Coville, chairman). Over several seasons Griggs had explored the Alaskan backcountry, filling blank spots as he went. This cut no mustard with the Board on Geographic Names, a committee of fifteen at the U.S. Department of the Interior, which was responsible for standardizing place-names and following certain policies. Although explorers like Griggs were free to propose names for places they discovered, the board made the final call, giving preference to existing native names for physical features, which the Alaskan candidates apparently had. In addition, the board discouraged the naming of places after living people—unless they happened to be particularly distinguished. The committee, concluding that Grosvenor, LaGorce, and Coville did not carry sufficient weight for the Alaska map, rejected Griggs's proposal. Then, grossly misjudging their opponent's tenacity, they turned down Bert's appeal.[20]

Bert struck back, taking the matter straight to the president of the United States. The indignant editor marched into the Oval Office with Jack LaGorce on March 27, 1925, and asked an undoubtedly bewildered Calvin Coolidge to drop everything and intervene in the naming of one Alaskan lake and two mountains. Nothing less than the dignity of National Geographic was at stake. "When I remarked to President Coolidge that for the United States Government to eliminate all the names given by the National Geographic Society to the area it had explored, would reflect on the reputation of the Society, President Coolidge said, 'I think such action by the Board on Geographic Names would discredit the Government rather than the National Geographic Society,'"[21] Bert recalled. He came away with a favorable decision. Coolidge countermanded the Interior Department panel, which was lucky because, by the time of Bert's White House visit, National Geographic had already put into circulation more than one million maps with the new nomenclature.[22] In gratitude for Coolidge's action, Bert invited the president to join National Geographic's board as soon as he left the White House in 1929. Coolidge did.[23]

In this incident of the place-names Bert had used a sledgehammer to dispatch a mosquito, but it was because he sensed that the dignity of National Geographic had been threatened, a menace he never failed to meet with maximum force. Suggest that the organization's membership concept was a cover for subscription pitches, and you might get threatened for libel.[24] Make insulting comments about an editor, and you might be recalled from China to apologize.[25] Enemies to the dignity of the National Geographic Society even appeared from within on occasion: A young file clerk named Carolyn Bennett Patterson was reprimanded "for walking too fast down the hall."[26] Jack LaGorce scolded two employees who appeared in front of Hubbard Hall at 12:45 P.M. on Sept. 12, 1924, with a young man who "chased them up the steps to the entrance." An irate LaGorce asked the women's supervisor to "notify them that if anything of this kind occurs again in front of the offices of The Society they will be discharged."[27]

If scampering on the steps amounted to assault upon the decorum of National Geographic, then questioning the achievements of the organization's number one icon, Rear Adm. Robert E. Peary, constituted a felony. The distinguished explorer Roald Amundsen—discoverer of the South Pole and recipient of the Hubbard Medal—learned this to his dismay when he unwittingly re-opened the Polar Controversy in January 1926. On a lecture tour of the United States to raise money for a planned flight to the North Pole, the Norwegian stopped to visit his old friend Frederick Cook at Leavenworth Penitentiary, where Cook was serving his sentence for land fraud.[28] Following the jail visit, perhaps touched by the sentiment of the moment, Amundsen politely answered a reporter's question about the North Pole. "I have read Dr. Cook's story," he said,

> . . . and I have read Peary's. In Peary's story I have not found anything of consequence that Dr. Cook had not already covered. I am not only unconvinced that Dr. Cook was a faker, but, on the contrary, I am of the opinion that his story of the discovery is just as plausible as was Peary's. It is possible that neither of them actually reached the Pole, but, regardless, it seems to me that Dr. Cook's claims were just as sound as Peary's.[29]

These rather mild if ill-considered comments set off an explosion on the Potomac, where Peary's death six years before had done nothing to diminish his standing with Bert and other National Geographic officers. An attack on Peary was an attack on the organization that had made him famous, and the perpetrator would be punished.[30]

Jack LaGorce led the way, steeling his boss to cancel a lecture Amundsen had planned to give at National Geographic that spring. "Amundsen is a rat hearted, jealous, disappointed snake," wrote LaGorce. "He hates Americans & America yet takes their money & laughs. He laughs at us & uses us as best he may. . . . What do we care for the Danes," LaGorce thundered, in his anger misplacing the Norwegian's nationality. "Amundsen should be kicked out of the country. . . . The more I see of Arctic explorers the more respect I have for prostitutes!" Although Amundsen protested that he had been misquoted, it was too late. Bert cancelled the lecture, which had been an important stop on Amundsen's tour.[31] Constantly strapped for cash because his government provided little money for expeditions, Amundsen needed the income as well as the publicity a National Geographic appearance would have provided. In the end, he barely covered expenses; moreover, his dispute with National Geographic, which attracted the attention of the press, reduced demand for his South Pole movies, which he had been renting out to raise money. He sailed for home in financial trouble, leaving Peary loyalists to gloat over his misfortune. "You handled the Amundsen matter just right," Ed Grosvenor wrote to his brother in Washington. "He backed down all right."[32]

Despite this setback, Amundsen scraped together an expedition that spring. On May 13, 1926, he drifted over the North Pole in the dirigible *Norge I,* marveling at the benign white landscape that had broken so many explorers before him.[33] Yet Bert had the sweet satisfaction of beating Amundsen to the prize by just four days. Lt. Comdr. Richard E. Byrd and Lt. Floyd Bennett, flying with support from the navy and the National Geographic Society, took their triple-engine plane, the *Josephine Ford,* over the top of the world on May 9 to claim the aviation record;* Byrd, flying again for National Geographic, became first over the South Pole, on November 29, 1929. "Other than the flag of my country," Byrd said, "I know of no greater privilege than to carry the emblem of the National Geographic Society."[34]

Such talk was music to Gilbert H. Grosvenor, who had maneuvered the National Geographic Society into prominence and authority by the late 1920s, in part by his adept sponsorship of explorers such as Byrd, Bingham, and Peary, and by his later promotion of those he had not been farsighted enough to sponsor before their fame—Ernest Shackleton, Charles Lindbergh, Roald Amundsen, and others. Bert also made the most of his connections with presidents, military leaders, congressmen, conservationists, and powerful political figures, most of whom were happy to lend their names, and often their presence, for National Geographic projects. By his midforties, Bert could look with pride at the phenomenally successful organization he had built upon the foundations laid down for him by Gardiner Hubbard and Alexander Graham Bell.

---

*Byrd's claim was challenged later. But the weight of the evidence, while not conclusive, seems to support Byrd.

Who would inherit this painstakingly constructed enterprise? The most likely successor—indeed the only one—appeared to be Melville Bell Grosvenor. After a rough start at the U.S. Naval Academy, Melville had graduated in the upper half of his class in June 1923, when he received his commission as an ensign and joined the fleet. At age twenty-two, he had grown from an uncertain, awkward youth into a handsome young man, straight-backed, physically hard, and jumping with nervous energy. With his dark good looks and strong nose, he appeared the mirror image of his mother, who remained close to him throughout her life. They gravitated toward each other at gatherings of the clan, where they can be seen standing close and unashamedly holding hands in family albums. The attraction was natural since it was often said that Melville was really more of a Bell than a Grosvenor.[35] He had inherited all of Grandfather Bell's ebullience, along with the old man's habit of thinking big. An able writer and gifted photographer, Melville had been intimately trained by the two men who had contributed most to the character of the National Geographic Society. But before Melville could return to Washington to join his father at work, he was destined to serve out his obligatory four-year commitment to the navy.

The first order of business was for Ensign Grosvenor to get properly outfitted in some $600 worth of uniforms—winter and summer, formal and informal—which the navy required him to buy. "I will have more suits than my five sisters put together!" he told his father.[36] Over the next months Melville would be assigned routine duties, touring powder plants for inspection, organizing a navy contingent for President Warren G. Harding's funeral parade, steaming south to join the newly commissioned battleship *West Virginia* in Newport News. By all accounts he was a good officer, but the mind-numbing routine of the peacetime navy must have exerted a drag on Melville's buoyant disposition, and now that exuberant spirit had been boosted by a new development—Melville was in love.

The object of his affection was Helen North Rowland, a friend and New York roommate of his sister Gertrude, who had introduced the two in 1922 and watched the sparks catch. The dashing young officer and the delicate debutante became inseparable, stoking their romance with frequent visits and nonstop correspondence through Melville's senior year. Judging from the white-hot quality of Melville's letters to his *mother* at this time, it is probably merciful that his correspondence with Helen has vanished from family files. "Oh dear Mum," Melville wrote Elsie Grosvenor in 1923, "Helen has been so wonderful to me, about writing. Honestly. Oh she is developing so, and is so wonderful. . . . Oh Mum she is wonderful. Her letters are so good. She just grows and grows every day. I never saw anyone like her. . . . I just love that girl more every minute. Oh Mum. I just know she is the one for me. If you could see her letters. Oh gee. Never saw anyone like her. Never. Never. If only I may be worthy of her."[37]

No man's love letters should be called in evidence against him without cause, but Melville's reveal the salient features of his character. In love as in all else, he proved to be impetuous, trusting, and fizzing with enthusiasm. Like his grandfather, Melville did nothing timidly. The Grosvenor clan converged on Waterbury in January 1924 for the wedding of Helen and Melville. Seeing Bert step off the train surrounded by his wife, an aunt, and his five daughters, one of those present thought he looked like a sultan with his harem emerging in a cold New England drizzle.[38] Despite the adverse weather, both families plunged undeterred into the weekend round of toasts and parties and country club dinners, at which the numerically superior Grosvenors threatened to overwhelm Helen's smaller family. "It will be hard on her to have so many in-laws," one of the Grosvenors predicted.[39]

On this weekend, though, the prospective bride and groom appeared happy and strong as they stood before the Reverend Dr. Charles Allen Dinsmore in the First Congregational Church. Melville spoke his vows in a loud, firm voice that could have overpowered a gale. "Everyone could hear him and see what a fine boy he was and is," his sister Lilian wrote. When the ceremony was done, bride and groom hitched arms and made for the door, got most of the way there, and "so forgot themselves that they grabbed each other in a wild embrace in full view of the whole church!"[40] Their strong attachment made their brief separation difficult that winter, as Melville returned to military duty and Helen took a long-planned trip to Chicago with her mother. "I never never will go away from him again," Helen said. "I was so lonesome. I nearly went wild."[41] When they reunited, Melville began searching for a way out of the navy. In the aftermath of World War I, the service had finally begun to reduce its ranks and accepted his resignation without argument in March. He and Helen took a long vacation to Europe, returning to settle in Washington that autumn.[42] Melville began work at the National Geographic Society in September 1924, representing the fourth generation of his family there.

Bert pronounced Melville "tickled" with his new salary of $2,500 as assistant illustrations editor.[43] He would be working with the taciturn Franklin L. Fisher, illustrations chief, to select all photographs for each month's magazine. That the president and editor chose to place his son in that department suggests that he believed Fisher would act as a steadying influence on Melville; the move also shows the central importance photography had assumed at *National Geographic* magazine, where pictures filled more than 60 percent of each issue.

By the mid-1920s rich material poured in from all points of the compass— stories documenting the floating lives of boat people in China; tribal practices in Papua New Guinea; Mayan discoveries at Chichén Itzá, Mexico; advances in astronomy and domestic poultry; the first natural-color pictures from underwater; and new color coverage from the Andes, Ireland, and the villages of Alsace. Melville got a first look at all of these, as well as at the vivid work of Joseph Rock, the prickly adventurer who wandered China's southwestern frontier and shipped home illustrations of demon-exorcising dancers, sword-wielding princes, and lonely alpine meadows framed by snow peaks. Melville also sifted through the voluminous photographic harvest of Maynard Owen Williams, who bounced from the Inuit settlements of Greenland to the temples and churches of Jerusalem to the valleys of Liechtenstein with hardly time between assignments to mutter, "Oh, Thunder!"

For a young man of Melville's restless temperament, the daily rush of images and rapid change of subject provided an ideal pace. He enjoyed the work and attacked it diligently. He had an eye for pictures, relished his exposure to the often raffish photographers, and paid proper deference to Fisher and to Jack LaGorce, the loyal elders Bert had assigned to watch over him. Melville's natural modesty, cheerful personality, and eagerness to please made him a hit with colleagues, who looked upon him as a loveable sort of Prince Hal, so far removed from the throne that he posed no threat to the crown or to its senior ministers. Melville would remain in that subservient role for most of his adult life, never showing a hint of bitterness throughout an apprenticeship that stretched over more than three decades. The reason for Melville's long wait was Bert. He always intended for Melville to succeed him at National Geographic and put his son in position to do so. At the same time, Bert doubted Melville's ability to handle the top job. He did not trust his son.[44]

Ever the perfectionist, Bert was determined to be a good father. "The reason the children of great men don't measure up to their ancestors," he told his mother-in-law, "is not the children's fault so much as the fact that great men are so busy looking after their great ideas that they neglect their own offspring."[45] Melville, as Bert's first child and only surviving son, received the dubious benefit of perhaps *too* much attention from his father. Expecting so much of Melville, Bert was destined to be disappointed in him, perhaps because the two were so utterly different in their approach to life. Bert loved detail work, Melville hated it; Bert was focused, Melville scattered; Bert was rational, Melville intuitive; Bert was deliberative, Melville impulsive; Bert was a planner, Melville a dreamer; Bert was reticent, Melville effusive; Bert was precise, Melville vague; Bert was practical, Melville romantic; Bert was adult, Melville the eternal boy. From one generation to the next, the genetic cards had fallen in such a way as to make father and son almost unrecognizable. The main trait they had in common was an extreme sensitivity to criticism, which only exaggerated one's effect on the other. Melville loved his father but found him impossible to please. Bert's love for Melville was no less absolute but his son was flawed and erratic, a rough draft in need of an editor.

"Melville would try the patience of Job," Bert told Elsie when their son was thirteen. "Now he is very good at his work & then like today goes all to pieces."

"Melville is very clever when he exerts himself, but he is apt to be mentally sluggish and lazy. . . ."

"I often feel my own inability to handle Melville. He is very elusive in his work. He will work for a certain time and up to a certain point and then let everything go by the boards."[46]

In each of these instances, Bert spoke of the adolescent Melville, but his judgment changed little with time. Even when Melville was in his fifties, Bert still treated him as a child, which had the predictable effect of making Melville more excitable and childlike.[47] Thus, being born a Grosvenor was a mixed blessing for Melville, as it would be for his own children. The birthright gave him his start at National Geographic, but it also put intense pressure on Melville to prove himself,[48] a process that began just after his marriage to Helen. As the newlyweds prepared to travel abroad, Melville equipped himself with cameras and film from National Geographic and feverishly shot pictures of the European sights in hopes of impressing his new employer. Helen wondered why the son of Gilbert H. Grosvenor had to work so hard. Her offhand question, which would become a stressful refrain in the years ahead, went straight to the heart of Melville's struggle.[49]

Watching Melville from the sidelines, his uncle Ed Grosvenor missed no opportunity to boost the young man's standing with Bert. "You are lucky to have a fine, clean boy like that to go into your business," Ed said. "Would give anything to have one myself. . . . He's a lovely, adorable boy, enthusiastic about going into the NAG [National Geographic]. . . . He impresses me as having an excellent business head which I should think would prove very valuable in the course of time in handling the enormous business of the Society. I am very proud of my nephew."[50]

But neither an uncle's pride nor a mother's love nor a father's unremitting concern could shield Melville from the crisis that engulfed him on October 29, 1929. With hundreds of thousands of others on that Black Tuesday, Melville watched in horror, and then in rising panic, as the first wave of the Great Depression crashed ashore on Wall Street. It washed away more than

$10 billion—worth almost $100 billion today—during five hours of trading.[51] Melville's $9,000 loss would have been minuscule had it not represented three times his annual salary and a good deal more in symbolic terms for this high-strung young man who had two children to provide for, while upholding the family pride at the office. This latest failure would only confirm the lack of faith in him. The gathering sense of doom bore down upon Melville.

Just after midnight on October 30, Elsie and Bert received a frantic call from their daughter Gertrude, informing them that Melville, visiting New York when the crash came, had suffered a serious nervous breakdown.[52] He had been admitted to Presbyterian Hospital, where he would remain, sedated and resting quietly, for a week. Bert and Elsie rushed from Washington to show their support for Melville, who seemed to rebound quickly. After a few days he was embarrassed and restless to return to work. Doctors allowed him to leave the hospital but ordered further rest—away from the office and Washington. With Helen, Melville decamped to the Marlboro Blenheim Hotel in Atlantic City where he remained for most of November.[53]

Meanwhile, the family settled down to assess the aftershocks of the crash, financial as well as psychic. On the financial side, one of the hardest hit was Bert, who had staked $60,000 for investment by his son-in-law Paxton Blair, a young lawyer in Ed Grosvenor's Wall Street firm. Paxton had recently married Gertrude Grosvenor. She entrusted him with $15,000, which also went up in smoke;[54] how much of Paxton's own money was lost is unknown, but the sum was probably considerable. Despite this, Gertrude displayed no sign of her previous nervous distress, remaining strong to help her husband and brother through the crisis.

"It is a strange thing that a man of mature age in Paxton's position could have played with fire," said Bert's father, sounding like the New England preacher he had once been. "If Melville has lost only $9,000, the lesson may be cheaply won. Paxton's situation seems to me much more serious. He has lost much more than Melville, has imperiled more than money and his comeback will be much more difficult. I can well understand how the mere knowledge of Paxton's experience must have shocked and unnerved Melville."[55]

Bert's brother Ed had also been hammered. As a partner in one of Wall Street's most influential firms, he kept a sharp eye on the market, peppering his letters to Bert with advice—Sell AT&T! Buy Cuban Sugar! Buy United Fruit! Buy Singer! Never one to wallow in his own troubles, Ed did not mention the extent of his losses that fall, but one may guess their impact: "Hasn't the last month been hell . . ." he asked Bert that autumn. "Like everyone else have lost a lot and not all of it 'paper' either."[56] The strain took its toll on Ed's health; three months later, he died of pneumonia at age fifty-four.

For his part, Bert seemed to absorb the financial shocks with a calmness unusual in one so fidgety, perhaps because he knew that the Blair family would eventually restore his losses, or because he realized how much worse the disaster might have been had his investments been larger. "The past two weeks have wrought havoc in many prosperous homes," he told his father. "Thank God we personally have come through intact."[57] But the main reason Bert did not dwell upon money matters was probably because he was wrestling with a heavier concern—tormenting himself over Melville's breakdown, for which he took the blame. Had he tried harder, had he been more watchful, had he been a better father, Melville might have avoided this illness.

Bert's own father got wind of his son's self-reproach, sat down, and composed a soothing letter, reprising the role he played best when Bert's confidence flagged. "From the time Melville was a baby and you let no one disturb his sleep, you have constantly watched over him and shielded him," Professor Grosvenor wrote.

There is nothing you could do for him that you have left undone. . . . You have a rich reward in his devotion to you and his fine development. Lilian tells me that you torture yourself with imagining that you have let him work too hard and carry too many responsibilities and that therein is to be found the cause of his present collapse. I do not believe hard work ever injured anyone unless indiscretion of some sort or prior physical condition rendered him liable to disease. I can conceive of no indiscretion more annoying, more injurious and more perilous than speculation in the stock market. A "sure tip" or "inside information" is a sort of quicksand on which one ventures. When this heart-rending experience is past and Melville is thoroughly restored to health and happiness, you may thank heaven that he had this bitter lesson at the beginning of his career.[58]

By the middle of December, Melville had recovered enough to return to work in Washington, where his father pronounced him "stronger and better than ever."[59] But try as he might, Bert could not dispel the memory of his son lying helpless in a hospital bed, an image that now added to his doubts about Melville. How could a person of such fragile character withstand the pressures of running the National Geographic Society?

CHAPTER TEN
A GENTLEMANLY STANDARD

MAYNARD OWEN WILLIAMS IN XINJIANG, CHINA,
WITH CITROËN-HAARDT EXPEDITION, 1932.

As moonlight faded outside of Beirut, the sands of Bir Hassen began to glow with the first red of April 4, 1931. A camel's bell broke the stillness. This wake-up call was answered by a chorus of six or seven gasoline engines coughing to life, announcing the start of a most audacious road trip, an eight-thousand-mile overland drive from the Mediterranean to the Yellow Sea, the

Citroën-Haardt Trans-Asiatic Expedition. Maynard Owen Williams, stripped to his waist in the morning chill, quickly finished shaving, pulled on a flannel shirt, and did some last-minute repacking. He hurried to his seat in one of the ugly gray vehicles—half tank, half touring car—that would take him on the adventure of his life.

The cars, equipped with massive rubber half-track treads, effortlessly climbed the ocean of dunes in good military order, slipped through the Lebanon range, and disappeared in the direction of China. Over the next ten months the expedition, sponsored by the French automaker André Citroën and by the National Geographic Society, would crawl across Asia, retracing Marco Polo's travels from one end of the continent to the other. Williams, the lone American and sole Baptist among a hard-charging throng of French adventurers, would document the journey, along with the ebb of traditional culture from Syria to Persia, from Afghanistan to Kashmir. Even in the remoteness of Xinjiang, China, Williams would find modern ways creeping in, soon to overpower the old canzonet of braying donkeys and camel bells.

"That old camel bell," Williams wrote Bert Grosvenor as he set out that spring, "waked us from our slumbers on the morning of April 4 and has since called us to our meals. Perhaps, like the life it represents, it will fail to last forever. . . . The 'Unchanging East' is in a state of rapid flux. . . . The price of having been unchanging for centuries is now being paid, not in slow installments but almost at once."[1]

As Williams wrote those lines to Washington, the old certainties were falling away at home. The Great Depression had entered its third year and millions of innately restless Americans became even more so, taking to the boxcars and highways to find work in the greatest migration in U.S. history. For this "vast, homeless horde,"[2] as for those who would travel vicariously with Williams and the Frenchmen, the open road promised hope, the prospect of a fresh start, or at the very least an escape from current troubles. Among those set in motion by America's economic distress was Douglas Chandler, a sociable young stockbroker ruined in the 1929 crash. Deeply in debt, unable to find work, and weakened by a nervous collapse, he rallied long enough to borrow money and flee New York in 1931, at almost precisely the time that Williams set out for Asia.

Chandler roamed Europe with his wife and two daughters for several months. The change of scene helped him to shake off his recent disgrace, and he gradually began to build a new life. He bought a camera, took photography lessons, brushed off his old typewriter, and settled in Germany as a journalist, just as that country began to re-arm itself and attract the world's attention. A former newspaper columnist in his native Baltimore,[3] Chandler seemed to be in the right place at the right time; a good story was breaking around him, and he recalled a gregarious, round-headed man he had met socially in the old days—Jack LaGorce. Perhaps LaGorce would be interested in buying some European stories for *National Geographic*. Chandler renewed the acquaintance, struck up a correspondence, and impressed LaGorce with his work. The magazine bought some of Chandler's pictures for its archive and began suggesting assignments for him.[4] He would produce seven articles for *National Geographic* in the 1930s and see five of them published. Luck finally seemed to be running Chandler's way again.

Luck also favored the National Geographic Society in the early years of the Depression. With its broad base of subscribers and hefty reserves, the organization weathered 1929 and 1930 with net gains in membership. Then the pinch began. Subscriptions fell by almost 45,000 in 1931, from 1.25 million to 1.2 million; by 1933, memberships dropped below the psychologically significant 1 million mark, to 882,498. For an inveterate worrier like Gilbert H. Grosvenor, this decline was disturbing enough, but then the Depression assumed an ominous human form in the spring of 1932, when as many as twenty thousand out-of-work World War I veterans flooded the capital to demand bonus payments promised them for their military service. This self-styled Bonus Expeditionary Force (BEF), hoping to embarrass President Hoover and Congress into action by their presence, camped out in tents, packing crates, and cardboard shelters along the Anacostia River through the spring and into the summer.[5] Although the crowd was generally well behaved, and more pathetic than threatening, for Bert and others skittish about a Bolshevik uprising, the ragged, mud-spattered

BEF presence represented the first intimation of class warfare in the United States, a specter already menacing European capitals.

"The increasing number of veterans daily arriving in Washington may result in a very difficult situation for the government to handle," Bert told his father that June. "A large number of the so-called veterans never saw service, or if they did, did not cross the seas.* The promoters of the gathering here were Communists. The entire program was carefully prepared at Communists' headquarters in New York."[6] This was true to some degree—the Communist Party had planned such a march, but a grassroots collection of farmworkers, factory hands, and ordinary American families spontaneously began converging on Washington that spring, first by the hundreds, then by the thousands, before the Reds could organize their own gathering; a smattering of party members joined in, but they did not control the group.[7]

Aside from upsetting the decorum of life in the capital, the veterans also threatened to spoil Bert's plans for a grand reception for Amelia Earhart, who had recently flown solo from Newfoundland to Ireland, becoming the first woman to cross the Atlantic alone. Although the National Geographic Society had not sponsored her flight, Bert quickly arranged a special gold medal honoring her achievement, and he took over plans for her visit to the capital on June 21, 1932. While the BEF forces steamed like dumplings in the Washington heat, Bert met Earhart upon her arrival across town, escorted her to the White House to meet President Hoover, hosted a lunch for her at National Geographic headquarters, took her to meet the Speaker of the House and the president of the Senate, introduced her to three cabinet secretaries, stood by her at an afternoon press conference, and accompanied her to a White House dinner in her honor. He also arranged an award ceremony that night at Constitution Hall, the ornate new auditorium of the Daughters of the American Revolution. There before a standing-room-only crowd of three thousand, President Hoover pre-

*Under a "bonus bill" passed by Congress in 1924, those veterans who served at home during the war would also receive payment of $1 for each day served; overseas veterans would get $1.25. T. H. Watkins, *The Hungry Years* (New York: Henry Holt and Company, 1999), 132.

sented the new American heroine with the National Geographic medal and gave a speech that was broadcast across the country.[8] "Mr. Hoover received tremendous applause," Bert reported, "even more than Miss Earhart. . . . I believe Mr. Hoover's prospects for reelection are improving daily."[9]

Bert was relieved that the seedy crowd camped along the Anacostia had done nothing to mar the evening, which represented a moment of hope in an otherwise gloomy landscape. The BEF contingent remained in Washington a few more weeks until Hoover lost patience with them. Then, concerned about the prospect of civil disorder, he called in troops under the command of Gen. Douglas MacArthur, then chief of staff of the army, to rout the demonstrators. Vowing to "break the back of the BEF," MacArthur took a detachment of tanks, infantry, and cavalry down Pennsylvania Avenue on the night of July 28 and plowed into the crowd. Maj. George S. Patton led the cavalry, which cleared the capital of spectators and veterans alike. In the scuffle and confusion, at least two veterans died from gunshot wounds. Another lost his ear to a cavalryman's saber. Many were injured by tear gas. The Anacostia encampment went up in flames that night, and its stunned occupants straggled away, some heading for home and some for New York, where they set up camp in Central Park. For the moment, it seemed, the threat of revolution had passed—but so would Hoover's hope for reelection, thanks to the worsening economic picture and to his heartless image.[10]

By the autumn of that year, Bert could see that both the country and the president were in deep trouble. "Everyone here is very pessimistic," he told his father in October of 1932. "Four weeks ago things looked fairly cheerful, but . . . the general view among my associates is that the depression this coming winter will become increasingly severe. . . . The income of the National Geographic Society has suffered terribly this year; my salary is less; Elsie's income is one third what it was. Thanks to the prospects of a Roosevelt coming into the White House, my income and Elsie's will diminish still further."[11]

Despite Bert's sense of financial apprehension, the record indicates that he passed through the Depression virtually unscathed, retaining an annual salary of $85,000, equal to well over $1 million in today's money. John Oliver LaGorce, the next highest paid employee of National Geographic, earned $35,737 at the time.[12] Perhaps because of this cushion, combined with Bert's unshakable faith in the long-term health of National Geographic, he showed remarkable resilience through these difficult years, during which he conducted the organization's affairs as if prosperity was just around the corner. He not only kept a far-flung network of correspondents in the field—Douglas Chandler in Europe, Maynard Williams in Asia, and the sturdy Joseph Rock in China—he hired new staff through the 1930s and went boldly forward with a $600,000 expansion of National Geographic's administrative and editorial offices in Washington. He paid in cash for the new building, adding another section onto Hubbard Hall, which had once before been extended in 1913. The new quarters, anchored by a neoclassical façade with two matching wings, constituted a gleaming lesson in geography, with varieties of marble from Massachusetts, Minnesota, Tennessee, Belgium, Italy, Spain, and France, limestone from Indiana, and granite from North Carolina. Outfitted with a modern photographic lab, locust-paneled suites, a spacious auditorium, and a museum, the expansion impressed all comers with the organization's success and staying power when the building opened in 1932.[13] But as declining memberships reduced the National Geographic Society's cash flow the following year, Bert was forced to cut salaries (including his own) by 15 percent.* He avoided layoffs, for which the staff was grateful. "It was the depth of the Depression," one of them recalled. "We all accepted it—a job was the most precious thing in the world."[14]

Even with businesses falling around him, Bert determined to maintain appearances at the National Geographic Society. For its magazine, he insisted on the high production values—heavy paper, quality printing, and substantial page numbers—readers had come to expect each month. And throughout the Depression he made sure that the National Geographic Society remained involved in high-profile expeditions, which would keep its name before the public. The organization committed $11,000 to archaeologist Matthew Sterling's studies of Olmec culture in Mexico, and $20,000 to the Citroën-Haardt Expedition. In

---

*After two years of salary reductions, paychecks returned to normal.

1934, it gave $10,000 for naturalist William Beebe's bathysphere dive, jointly sponsored by the New York Zoological Society; Beebe, cramped in his steel diving ball with Otis Barton, descended to an ocean depth of 3,028 feet, a new world record.[15] That same year, National Geographic reached for the stars, sponsoring a manned stratospheric balloon flight with the U.S. Army Air Corps; the balloon, *Explorer,* climbed to 60,613 feet before it ruptured and forced the three aeronauts to parachute to safety.[16] Undeterred by that near disaster, the army and National Geographic joined forces again the next year, taking a new balloon, *Explorer II,* almost fourteen miles above Earth's surface—into the black of space. Their ascension to 72,395 feet established a new record for manned flight, which stood until 1956.[17] Such stories of pluck and courage received ample coverage in Bert's magazine, which must have offered some comfort to readers suffering through hard times.

In faraway Asia, economic worries seemed a million miles astern to Maynard Owen Williams, lumbering out of Persia and into the mountain kingdom of Afghanistan with the Citroën-Haardt Expedition in May of 1931. Supplied with Evian water, champagne, and crates of the best food from Fortnum & Mason of London, the motorcade crossed Asia in style, stopping for receptions with an endless succession of mayors, princes, and warlords. Williams wrote one letter home from a governor's garden, where he sat in the shade of a mulberry tree, watched a breeze stirring the pomegranate blooms, and listened to the music of a little stream as it cooled his canteen of tea for the long afternoon ahead. King Nadir Shah of Afghanistan received the group in Kabul. Children swarmed out of mud-walled villages from Islamkala to Ghazni to see the strange-looking vehicles passing through in clouds of dust, flags snapping on their hoods. Scores of Afghan men turned out to help wrestle the caravan across rivers, a scene that never failed to impress Williams.

"I continue to have a vast liking for these wild looking–men, who are direct, child-like in their curiosity and manly in their desire to help," Williams wrote Bert from Girishk, where the tribesmen cheerfully pitched in to help the cars across another river.

A knee-deep river requires several men to wade about and see how high they wet their legs. Then follows a long discussion as to which of several fords is the best, whether all the baggage shall be removed from the trailers, whether the beds can be safely left in place, etc., etc. A thigh-deep river requires all that in addition to the matter of dragging our auxiliary trucks across. If it should rain, a knee-deep river becomes a waist-deep river very soon. . . . The crossing of the Farahrud was a great show. At times there must have been five hundred Afghans on the two banks, and five hundred Afghans make a splendid sight, especially when one mixes in a few splendid horses and a handful of patient donkeys. . . . We have hardly seen a woman since we crossed the frontier. So most of us bathed *tout nu*. I have the reputation (here) of being a Puritan and to some of my friends it seemed strange that I should not wear a suit. . . . Being brought up in a Y.M.C.A. may not make you a voluptuary but it does make nudity in a swimming pool seem natural.[18]

A few days later Williams was dressed for a side trip to Bamian, the valley northwest of Kabul where giant Buddhist statues had watched over pilgrims and travelers since the year one. The statues, among the most important cultural sites of the region, had been carved in sandstone cliffs along the track that Buddhism followed to China. Williams, scrambling over the crags with a heavy camera outfitted with glass plates, took what were probably the first color photographs of the famous figures, whose enigmatic presence guarded the ancient patchwork of village and green fields until Taliban fundamentalists pulverized the site in 2001.[19]

Halfway through Afghanistan, Williams reported that he was still having a grand time on the road, but it was clear that the months of traveling had cooled his ardor. For one thing, he was surprised at the inordinate time he and other members of the expedition had to spend at official lunches, dinners, and receptions. "This afternoon," Williams wrote from Girishk, "I have to change my clothes, make myself more uncomfortable than I am in a flannel shirt, and waste endless time all to do honor to the Mayor. . . . One can't accept hospitality and then treat one's host like a bell-boy."[20]

Those diplomatic chores irked Williams because they ate into his time for real work. He tolerated the social diversions for the good of the story. But his relations with the group's leader, Georges-Marie Haardt, soon began to fray. Haardt, a Belgian who had helped build the Citroën Company into an automotive force before turning to a life of adventure, seemed to have had little patience for the strange needs of the photographers, artists, and scientists attached to his expedition. If Williams stopped to take pictures or lingered in a village for interviews, Haardt complained that he was sabotaging the itinerary. If Williams asked that more time be devoted to the mission's scientific goals, Haardt laughed at him. If Williams asked to peel off for a morning of shooting, his request was refused. To make matters worse, Haardt always established camp far away from settlements, which made it impossible for Williams to find photographic opportunities in the off hours. Somewhere between Afghanistan and India, it dawned on Williams that the imperious Monsieur Haardt cared nothing about his photographic work. The Belgian was more interested in his cars and his schedule than in his people—a clear triumph of industrial values over human ones.[21]

"There lies the real trouble," Williams blurted in a twenty-five-page letter to Bert Grosvenor, "that we will be prisoners of the cars themselves, and that a so-called scientific and artistic expedition will be subordinated to mere mechanical progress."[22] Williams could stand the fierce heat and freezing nights, the millions of flies chewing his ankles, the grinding sixteen-hour days to make just sixty miles. Indeed, the former missionary relished such privations, feeling that "a soft, easy life is a menace to man's development."[23] He could drink muddy water when the Evian ran out, endure the salt-encrusted underwear that creaked when he moved, tolerate the fitful sleeping on a rickety cot while his French comrades passed around lewd pictures. But the hardworking Baptist could not abide when Georges-Marie Haardt kept him from his mission. "Give me a chance to work and I'm happy," he said. "But pen me up and make it impossible for me to write or take photographs and I get sore—and show it."[24]

He showed it most dramatically that September, when a detachment of the Citroën-Haardt Expedition finally straggled over the Pamir Mountains and entered China through its back door in Xinjiang. Having determined that the vehicles could not surmount the rugged passes from Kashmir, the expedition had abandoned its cars at the frontier and covered the last few hundred miles the old fashioned way—on ponies, on yaks, and on foot. A reserve contingent of cars, stationed in Beijing, would drive westward to collect their colleagues in Xinjiang, then reverse course and travel east across the breadth of China to finish at Beijing and complete the grand design. Yet when it came time to record the Kashmir group's triumphal entry into China, Gaelic joie de vivre crashed headlong into Puritan rectitude on the old Silk Road.

Williams described the collision: "After taking my picture of M. Haardt and his followers, I tried to get them to stop in front of a glacier. . . . But they did not." They drove away, leaving Williams to catch up. When he did, he discovered his colleagues drinking champagne, celebrating for the expedition's movie cameras. Haardt then made the mistake of asking Williams to join the toasts. "I thought of America's [Prohibition] laws, of our influence with school children and of our close connection with the Government and I politely refused. He then dogmatically *ordered* me to join in drinking a toast." The Baptist exploded. "Enough!" he cried. "I am not going to drink for the movies and I am not going to lend myself to propaganda for champagne and against the laws of my country."[25] He stomped off.

Months later, as the caravan rolled into Beijing under grim February skies, relations between Haardt and Williams remained cool. But the troubles of the past year were soon forgotten when Williams learned that he had been named chevalier of the Legion of Honor, one of the highest tributes France could bestow.[26] For the rest of his life he wore the red ribbon in his lapel, proud to have been along for that long, dusty journey. He settled into a Beijing hotel and got down to work on his story, pounding out reams of manuscript pages and developing hundreds of glass photographic plates, which had miraculously survived the months of rattling, freezing, and broiling their way across Asia. He bundled off enough material to Washington for several stories and awaited the reaction from headquarters.

It was not long in coming: "Congratulations remaining with party under trying conditions," read the cable signed jointly by Gilbert Grosvenor and Jack

LaGorce. "Your photographs wonderful and you should feel no disappoint-ment. We are all much pleased with your work."[27] France's newest chevalier heaved a sigh of relief and began plotting his next assignment, which would take him to Poland and from there around the world several times more. "Never grieve for me if it is my good fortune to die with my boots on," he wrote Bert toward the end of their long association at *National Geographic*. "That's what I most hope for."[28]

Like others who distinguished themselves in the field, Williams knew he was lucky to be living away from office routine and the confinements of civi-lization. Even the dyspeptic Joseph Rock, who often complained about the hardships of China's backcountry, considered himself blessed for being there. "This is the greatest life and I am thankful to be able to enjoy it,"[29] he wrote. "Where I live we know nothing of depression. . . . There are no factories, no autos, nobody works for a living, that is in an industrial capacity, hence there are no hard times. Like everything else that hits Shanghai or the east and north of China, may it be war, depression, riots, etc. there is not the slightest repercus-sion here. We may as well live on the moon. . . ."[30]

Bert won the loyalty of such people by giving them ample time and gener-ous resources in the field. More important, he treated them with respect, no matter how much they whined about their troubles, clamored for attention, or behaved erratically. "Our job on the *National Geographic* magazine," Bert repri-manded an associate, "is to get the best material, regardless of whether we like or dislike the speech or manners of the man who has it."[31] If a job applicant or contributor had talent, Bert did not worry much about where he had gone to school or who his parents were, as long as he was loyal to the National Geo-graphic Society, did nothing to tarnish its reputation, and refrained from burn-ing down the building, which was often threatened when cigarettes tossed from the windows set the awnings on fire. "This morning we had another one which completely destroyed two awnings on the second floor and one on the first, as well as three lights of glass broken and resulting damage to paint," fumed an irate personnel manager whose inquiries produced numerous suspects but no confessions.[32]

For a straight arrow fixated on order, Bert betrayed a high tolerance for the eccentric individuals who gathered around him like mismatched moths drawn to the soft lights of the National Geographic Society. By the 1930s, that swarm included a Harvard graduate, a hard-drinking aviator, a radio announcer who secretly yearned to be a conductor, a publicist for an opera singer, a cigar-smoking diplomat, an Associated Press reporter, a writer from the West Coast who lived in his car, a science teacher from a Siamese missionary school, and an avid collector of records and pornography. They had little in common except for their sense of adventure, their love of writing and photography, and the confidence of Gilbert H. Grosvenor. "They're all a little queer," Bert admitted when one of his daughters asked why a visitor heaped mounds of sugar on his scrambled eggs at breakfast. "If they weren't queer," Bert explained, "they wouldn't be able to do the things they do."[33]

Bert kept the doors open for all supplicants, queer and otherwise, set adrift by the Depression, which brought some of the magazine's future stars to Washington. One of these, Howell Walker, was fresh out of Princeton when he applied for a job on the magazine in 1933. It was established that Walker had no experience, so he was sent away. "We need people who have seen the world," someone told him.[34] Two years later Walker was back. "I have just ridden a bicycle around the world," he quietly announced to Melville Bell Grosvenor. Melville was so impressed by Walker's initiative, and by his written account of the bicycle trip, that he steered the young man to a job in the photographic lab, an entry-level position for aspiring photographers. There Walker pulled on a smock and went to work alongside Volkmar "Kurt" Wentzel and Luis Marden, who spent their days mixing chemicals and developing pictures and their nights dreaming of shooting for *National Geographic*. All three eventually achieved that dream, securing much-coveted positions on the magazine's foreign editorial staff. Such things were possible in the meritocracy Bert established.

In the years following his breakdown of 1929, Melville had shown no danger of a relapse. He worked hard to establish a solid record at the magazine, assuming duties in addition to those he had as an illustrations editor. He took over

National Geographic's lecture program in 1933. An important aspect of the organization's public presence, the lecture series featured presentations by eighteen to twenty speakers who often contributed articles to the magazine. The program gave Melville an opportunity to meet key explorers and public figures—such as Richard Byrd, Charles Lindbergh, and William Beebe—and the reporters who turned out to cover them. Melville also got to know hundreds of Washington-area members who filled the auditorium for such events, which probably contributed to his sure sense of how readers responded to particular articles.

Throughout the 1930s Melville worked under the tutelage of Jack La-Gorce, who promoted the young man's interests, steered him toward good assignments, and watched over him like an uncle. One day in 1934 LaGorce summoned Melville to his office, a dimly lit room decorated with antique axes, knives, and other weapons, and began to rail about picture captions in a story about Rome. "Something's got to be done to improve these legends," LaGorce growled, slamming the manuscript on his desk. "Melville, I'm going to make you an assistant editor and put you in charge of the legends. You are to see that they're interesting, sprightly, and informative but above all accurate, and Lord help you if a mistake occurs. It is your responsibility. Go to it."[35]

Melville's retelling of that incident made it appear more spontaneous than it was in fact, since nothing happened at National Geographic without the consent of Gilbert H. Grosvenor. In this instance the Chief had already proposed that the responsibility for legends be taken from the overworked Jesse R. Hildebrand and given to Melville, along with the assistant editor's title. The shift, Bert said, would "save Mr. Hildebrand time and strength in searching out fresh authors, in writing articles himself and in other ways for which his splendid mind and talents can be more profitably utilized for the Magazine, thus saving much of his energy. . . . M.B.G. is the only one qualified in the Illustrations Division to tackle this tie-up."[36] For the first time in its history, the magazine known for its photographs now had a separate department to oversee the legends that complemented them.

Under Melville's guidance, legends developed into an editorial form unique to *National Geographic;* more than mere captions, these mini-essays attempted to lure reluctant readers into articles by means of exuberant writing that sometimes went overboard, a tone exactly mirroring Melville Grosvenor's approach to life. The personality of a new Grosvenor was finding its way into the magazine. At the same time, Melville and Helen were providing more candidates for the family business. Their third and last child, Gilbert Melville Grosvenor, was born in 1931, joining Alexander Graham Bell Grosvenor and Helen Rowland Grosvenor, known in the family as Helen, Jr., or Teeny.

These recruits were most welcome, as they filled in family ranks thinned by the depredations of time. Shortly after Bert's brother Ed died in 1930, their mother, Lilian Waters Grosvenor, followed, leaving a frail Professor Edwin Grosvenor to ride out the Depression alone. The professor, now in his mid-eighties, had demonstrated his brilliance in many ways, but he showed little skill at balancing a checkbook or running the household at 7 College Street in Amherst, which had always been the province of his wife. He was a soft touch who cheerfully gave money away to friends and limited his cigar expenses to one cent per smoke to make up the difference. Bert could see what was happening from afar and tried to help his father, just as the elder Grosvenor had come to Bert's aid when he had been a young editor battling for survival. He devised a budget for his father's weekly expenses, took over the professor's mortgage payments, and even told him how to rake leaves. He kept sharp watch, from a distance of several hundred miles, over housekeepers, gardeners, or others who might take advantage of the professor.[37]

As usual, Bert was overextended at home and at work. Not only had hard times reduced membership in the National Geographic Society; they were taking their toll on the family. In 1935, Gertrude's marriage to Paxton Blair ended in divorce after ten years and three children. And there were warnings of a rough passage ahead for Melville and Helen, who began spending their summers away from Baddeck because Helen felt like such an outsider among the Grosvenors.[38] Melville bought a vacation place on Gibson Island in Maryland, where he could indulge his love of sailing on the Chesapeake Bay and Helen found respite from the swarms of in-laws and cousins looking over her shoulder. On a happier note, Mabel Grosvenor was finally able, after long years of school and residency at Johns Hopkins, to open her pediatric practice in Washington—a development cheered by family friends like Jack LaGorce: "Dr. Mabel was in to see me yesterday looking very fit, and in the throes of painting and paper hanging in her new office," LaGorce reported to Bert in 1935. "I told her that as I am rapidly nearing my second childhood she would probably have me as an early patient."[39]

From this time forward, Bert's second daughter would be known in the family as Dr. Mabel. Bert, too, acquired his first honorific in the 1930s, beginning with a doctorate of law degree conferred by the College of William and Mary in Williamsburg, Virginia.[40] Having grown up in academia, Bert so relished the string of honorary degrees that eventually came his way that he listed them (Litt.D., LLD, Sc.D.) with his byline in National Geographic. He did not discourage colleagues from addressing him as Dr. Grosvenor in memos and in person. Perhaps in the belief that such titles lifted the National Geographic Society's learned image, he also arranged an honorary degree from George Washington University for Jack LaGorce, who was soon known to one and all as Dr. LaGorce. Apart from these small consolations, Bert found little to cheer about between the wars. Now in his fifties, he began to suffer from prostate trouble, which made him understandably irritable and finally required removal of the gland in 1933; this only added to Bert's testiness. "Father became quite dictatorial about that time," Dr. Mabel recalled. "He wouldn't carry on an argument. I wonder if he was dictatorial at the office too."[41]

He was. Bert began snapping at colleagues when he thought they had failed National Geographic in some way. If an employee short-circuited the chain of command or tried to engage him in argument, the gentlemanly mask came off, and a petty tyrant appeared. Never one for direct confrontation, he preferred to discipline subordinates through a trusted lieutenant, such as LaGorce, or by issuing a skewering memo, which would forbid further discussion. When Bert thought that Frederick G. "Ted" Vosburgh had stepped out of line, bypassing Melville Grosvenor by directing editorial queries to the ultimate authority, Bert crushed the young editor by complaining to his immediate supervisor. "I have come to the conclusion that Mr. Vosburgh is actuated primarily with thoughts for his own advancement, rather than for what is best for the Geographic. . . . He is not a good team worker."[42] Vosburgh resigned on the spot, but he was coaxed back after Bert came up with the peace offering of a paid vacation.

As the twentieth century grew more complex, it proved difficult to uphold the Chief's idiosyncratic editorial standards, which had been adapted from Alexander Graham Bell's idealistic view of the world. How could one make an article absolutely accurate (rule 1) without also making it controversial, or unkind to the subjects covered (rules 5 and 6)? Even the old campaigner Maynard Owen Williams could not achieve this balancing act to the Chief's satisfaction. Williams suggested an article on the new Soviet Union in 1933, pointing to the rapid changes there. "The Russia I knew seven and a half years ago no longer exists," he told Bert. "This year's splendid harvests, the new feeling arising from American recognition, the greater objectivity possible since friendly relations are established—all these factors will work in favor of outstanding material when we send a staff man to cover this, the most important assignment of the immediate future."[43]

Bert, who detested and feared the Soviets, made his views clear. "As long as I'm editor, The Geographic wont [sic] send you to Russia for an article. . . . You are too prejudiced to tell the Russia story to a Geographic audience. G.H.G."[44] Williams went on to other assignments and the Soviet Union passed much of the century without coverage in *National Geographic*. Since the magazine could not report honestly on conditions there without arousing controversy, Bert's solution was to declare the world's largest country a nonentity.

His anti-Soviet views must have been colored by personal experience, based on visits to Russia before and after the Bolshevik Revolution. As much as Bert deplored that rebellion, his family benefited from it in a small way. Visiting St.

Petersburg in 1928, Bert and Elsie bought cases of the Imperial Family's crystal and china at bargain prices.[45] Perhaps, eating from the old ruler's tableware, Bert was reminded too often of the turmoil, not to say the dispossession, the Bolsheviks had visited upon the established order. Throughout the 1930s, he groused about the "Stalin-Russian methods" that President Franklin Roosevelt brought to bear upon American industry through the National Recovery Administration.[46] And, like many others, Bert feared that a Red tide might sweep Europe at any moment unless a strong force resisted it. This may have been one reason that he viewed the rise of Adolf Hitler and the re-arming of Germany as positive developments.

"I hope I shall live to see Hitler unite all Germans of Germany and Austria into one powerful country," he told Jack LaGorce in the months just after Hitler had taken power in 1933. Bert pressed for a story on the new fascist regime, which Douglas Chandler provided.[47] Published in February 1937, his article "Changing Berlin" proved to be perhaps the biggest embarrassment in the history of the National Geographic Society.

Chandler, who had become increasingly anti-Semitic and sympathetic toward the Nazis during his long residence in Europe, painted a happy portrait of Berlin, a glittering city where red banners with bold black swastikas rippled along *Unter Den Linden;* where New Age architecture displaced tired old buildings; where coal was given to the needy in winter; where industry generously provided proper housing for thousands of employees; where "flaxen-haired, sun-crisped youths" splashed in the river and grew strong; where cattle were humanely slaughtered by electrocution rather than by crude kosher methods. Even the degenerate drinking habits of old Berlin moderated under the benevolent Nazi influence, according to Chandler: Beer consumption was down by 40 percent. "Now . . . the youth of Germany, striving for physical efficiency, scorn anything but the most moderate beer drinking," Chandler wrote. "They predict that paunchy waistlines and bulging necks will be unknown in the next generation."[48]

Even when Chandler went searching for flaws, he discovered few of them in Hitler's clean, well-lighted city. "Determined to hunt up the worst of Berlin's tenement areas, I plotted out a two-day walk through the quarters where the 'other half' lives," Chandler wrote. "Nowhere did I find a spot which measured up to my conception of a 'slum.' Many unfit dwelling houses on narrow streets have been torn down. In their place stand settlements—groups of apartments offering decent, moderately priced quarters for workers' families. Nearly 3,000 have been constructed. Some have small gardens attached."[49]

He was particularly impressed by the German version of the Boy Scouts known as the Hitler Youth, an organization "enormously popular with all classes." Preceded by white clouds of gulls, a cheerful assemblage of these boys, dressed in short pants and brown shirts, came tromping through the last paragraphs of Chandler's story, symbolizing the virility and strength he saw in Germany's future. "They are singing in accurately pitched, youthful treble that moving modern national song, the *'Horst Wessel Lied,'*" he wrote. Throughout Chandler's forty-seven-page report, readers found no foretaste of the horrors Hitler was about to visit upon the world. The month following Chandler's piece, the magazine published "Imperial Rome Reborn," an article by John Patric on Mussolini's Italy, which was only slightly less admiring of Il Duce than Chandler's ode to Berlin.[50]

Both stories met Bert's gentlemanly standards of accuracy without being truthful. As Europe deteriorated, and Hitler and Mussolini's true intentions were widely revealed, it would become increasingly difficult for *National Geographic* magazine to remain above the fray, a position from which it had advanced the view that the fascist regimes of the late 1930s were efficient, benign, even admirable. Coming from a respected magazine like *National Geographic,* this had been peerless propaganda.

In the meantime, Chandler continued to get regular assignments. With Bert's assent, he would be given stories on Belgium, the Baltics, Turkey, Yugoslavia, and Albania between 1937 and 1939. Despite the ample evidence of Chandler's reporting, his reverent references to Hitler in correspondence, and the complaints of readers, Jack LaGorce resisted any suggestion that Chandler held pro-Nazi views. When those views were finally confirmed in 1938, both Bert and LaGorce continued to support their star European correspondent and to give him assignments.[51] LaGorce maintained that Chandler's anti-Semitism had not affected his reporting.[52] By 1939, neither of Chandler's *National Geographic* patrons could ignore the evidence of events such as *Kristallnacht,* on which Nazis ransacked Jewish businesses all over Germany,* or the complaints of readers like William H. Danforth, a prominent American executive who had visited Chandler and come away exasperated. "The disturbing thing to me," wrote Danforth, then chairman of the Ralston Purina Company, "was that Mr. Chandler was more Pro-Nazi and Anti-Jewish than any man of any nationality that I met anywhere on our trip. His wife was just as vehement as he was. . . ."[53]

Against this tide of criticism, *National Geographic* finally severed ties with Chandler. LaGorce pulled him off the Albania assignment in April 1939, paid him in full, and never explained the real reason for their break. By that time it was likely that Chandler was on the payroll of Paul Josef Goebbels, Hitler's minister of propaganda, and probably had been since 1935, long before LaGorce and Bert had launched Chandler's *National Geographic* career.[54] Upset by the abrupt withdrawal of his American sponsors, Chandler grew increasingly strident and paranoid. As war approached, he blamed Jews for the loss of his *National Geographic* work, for the collapse of his career as a stockbroker, for the disappearance of his wife's fortune, and generally for all the ills of modern life. He began broadcasting for Nazi radio as Paul Revere in April 1941, at which time he urged the United States to stay out of the war between Germany and Britain. He became an eccentric figure around wartime Berlin, a tall, spare man with arching eyebrows, close-clipped gray hair, and a swastika pin in his neatly tailored lapel. He drove to work in a maroon Mercedes with American flags painted on its flanks, and made his broadcasts six times a week, spewing propaganda on behalf of Hitler. "Personally," he said in one such program, "I find many Jews most beguiling. I have no grudge against any individual Jew, but

---

*The Hitler Youth, so recently celebrated by Chandler in *National Geographic,* played a key role in "Crystal Night," so called because of the shattered glass their rampage of November 9, 1938, left in the streets.

collectively they are bringing ruin wherever they get a foothold. . . . Roosevelt is surrounded by Jews and is carrying out the Jewish plan enunciated by the Jews long ago for the world's domination."[55] After the United States entered World War II, he continued his broadcasts, mentioning his American associates Gilbert Grosvenor and John Oliver LaGorce from time to time.

This prompted concerned letters from *National Geographic* readers who made LaGorce squirm a good deal as he tried to rationalize Chandler's presence in so many issues of the magazine. "He is a free lance writer and photographer from whom we purchased several illustrated articles," LaGorce told Virginius Dabney, editor of the *Richmond Times-Dispatch,* "but he has never been at any time a member of The Geographic's staff or in its permanent employ."[56] LaGorce assured Charles O. Collett of Surrey, England, that it had been *National Geographic*'s "long-time policy to investigate the background and status of any free lance writer who submits material and we could find nothing against his record when he first submitted material for purchase."[57] To Alan C. Collins, a literary agent in New York, LaGorce suggested that Chandler's articles had never been solicited, but instead had appeared on the doorstep without encouragement from Washington. "From time to time we received, found available and purchased several articles from him in none of which was there a word or a line that smacked of propaganda," he wrote.[58] In no instance did LaGorce reveal that he had suggested the Berlin piece to Chandler, or that he had kept him busy with other commissions through the prewar years. LaGorce's handling of the Chandler affair struck some of his colleagues as an object lesson in why it was dangerous to have amateurs editing *National Geographic,* instead of editors with journalistic experience. "He wasn't an editor," said Ted Vosburgh, who believed that a seasoned journalist might have discerned Chandler's treachery. "He let himself be badly fooled. . . . The record shows that he ignored, or failed to see, several stoplights in the Chandler affair."[59]

The most glaring failure belongs to Gilbert H. Grosvenor—who overlooked no detail at the National Geographic Society, who read and approved Chandler's Berlin piece, who described it as a "fine paper."[60] He was able to do so because he admired Adolf Hitler, because he feared the Bolsheviks, and most of all because he was himself an anti-Semite. This allowed him to publish Nazi propaganda, assembled and placed by an enemy agent in the magazine he loved.

Douglas Chandler was seized by American officers in Germany on February 28, 1946, charged with treason, and returned for trial in Boston. Chandler, fifty-seven, looking worn and very frail, fidgeted through the testimony of sixteen former associates from the *Reichrundfunkgesellschaft,* the Reich Radio Corporation, in U.S. District Court. He was convicted that June, fined $10,000, and sentenced to life in prison.[61] None of his old colleagues from the National Geographic Society attended the trial.

CHAPTER ELEVEN
THE BIRTHRIGHT

JOHN OLIVER LAGORCE DISPLAYING WEAPONS COLLECTION
IN HIS NATIONAL GEOGRAPHIC OFFICE, 1936.

Hashime Murayama, a staff artist specializing in natural history for *National Geographic,* was a quiet, dignified man who might have been mistaken for a college professor. In his crisp lab coat and thick glasses, he could often be seen leaning over the microscope at his desk, close by the box of beetles, spiders, or other specimens he was about to immortalize for the magazine. Like his boss, Gilbert H. Grosvenor, the Japan-born Murayama was a stickler for detail. He devoured scholarly papers and interviewed experts when preparing for a story. He traveled cross-country to aquariums and museums to observe the subjects he did not have at hand; he would painstakingly count the number of scales on a sturgeon's flanks, study how a yellow perch turned in the water, scrutinize the geography of a bee's knee. The result of this elaborate research was to make Murayama's watercolors faithful to nature, but with no sacrifice of style. Even today his work breathes a delicacy and poise reflecting the understated character of its creator.[1]

Like many of his colleagues, Murayama had moved up the ladder during a long career at the National Geographic Society, advancing from his job hand-tinting black-and-white photographs in 1921 to become the magazine's first illustrator. His last work, published in the August 1941 issue, was characteristically meticulous, a portfolio of insect paintings entitled "A Rogues' Gallery of Imported Pests." Murayama's farewell piece accompanied an article by Ted Vosburgh entitled "Our Insect Fifth Column," which told how gypsy moths, corn borers, Japanese beetles, and other non-native vermin threatened American species. Both titles suggest the xenophobia the country felt at the onset of World War II, which would bring a humiliating end to the career of Hashime Murayama.

Although the timing and details of Murayama's departure remain murky— some recall that he was fired after the bombing of Pearl Harbor in December 1941; others that he left a year before that—there is little doubt about the reasons for it. Murayama was Japanese and, with the United States going to war against the Axis powers, Gilbert Grosvenor wanted no enemy sympathizers on his staff.[2] Except for the accident of Murayama's birthplace, there was no reason to suspect his loyalties; nonetheless, he was purged from the building so suddenly that he had no time to gather his personal effects or unfinished sketches. One colleague, hearing that Murayama had been fired, went to his office, rummaged around, and went home with several unfinished watercolors wrapped in brown paper.[3] With time the story of the artist's rude expulsion entered the lore at the National Geographic Society; one account of the affair had an irate Bert sputtering with rage after Pearl Harbor and ordering gardeners to uproot a Japanese cherry tree in the courtyard before Murayama's office chair was cold.[4]

"Even with all the prejudice against the Japanese at that time," recalled Andrew H. Brown, a former *National Geographic* magazine editor and an admirer of Murayama, "it is inconceivable to me that he would have done anything or even thought anything that would have been detrimental to our country. He was a wonderful, wonderful fellow. I never heard of any incriminating evidence against him."[5]

What happened to Murayama after leaving National Geographic remains a mystery. Was he deported to Japan with his wife and two children? Did he sit out the war in New York, where he had worked as an artist before moving to Washington? Did he stop painting? Did he send secret messages, encoded in his portraits of ants and moths, to his masters in Tokyo? No answers appear in the magazine's files, which must have been purged or lost after his exodus. He simply evaporates with the war, then appears for the last time in February 1955, with a brief notice of his death in the *New York Herald Tribune*.

Bert had been slow to perceive the threat from Hitler and his allies, but as the case of Hashime Murayama illustrates, the editor promptly—and quite ruthlessly—revised his views to suit the new reality, just as others were doing. Throughout the country anti-Japanese feelings ran high after the Pearl Harbor attack—higher than the bias against Germans. The Metropolitan Opera suspended performances of *Madama Butterfly* for the duration of the war but continued with most of its German repertory.[6] Bert followed suit at the National Geographic Society, showing Murayama the door while retaining Kurt Wentzel, a valued photographic technician and a native of Dresden who had emigrated to New York as a boy. Bert also cracked down on "enemy nationals" on the membership rolls, suspending delivery of the magazine to thousands of readers. Most of these were Japanese Americans who had been declared suspect by the Treasury Department.[7] His patriotic fervor also led Bert to ban certain subjects from the magazine; when in 1943 Jack LaGorce suggested an article on Sweden, Bert rejected his lieutenant's proposal. "Sweden is in the same class as Argentina," the editor proclaimed, "very pro-Nazi from the beginning of the war. Now Sweden is desperately trying to curry favor with the Allies. . . . I don't think we should play into her hands by giving space to her in the NAT

GEOG. MAG and thereby help her to build up goodwill."[8] Twenty years would pass before a Sweden article appeared.

Bert plunged into the war effort wholeheartedly, providing maps to the army's *Yank* magazine and to the British War Office. The Society gave five hundred thousand maps to the U.S. War Department and threw open the magazine's photographic files and map library to the army, navy, and to other government bureaus planning military operations.* Few reliable maps and charts existed in the government, which was unprepared for what was to be history's first conflict of truly global proportions. The National Geographic Society helped fill the information gap, especially for the Army Air Corps, which needed fresh cartographic data to cover vast stretches of new territory. With fighting in two hemispheres, intelligence gathering became urgent and increasingly sophisticated. A frequent wartime visitor to the magazine's Washington headquarters was Col. William J. "Wild Bill" Donovan, the first chief of the Office of Strategic Services, predecessor to the Central Intelligence Agency, which would retain its National Geographic ties after the war.[9] Military brass were also well represented on the National Geographic Society's board of trustees, among them Gen. John J. Pershing, Maj. Gen. H. H. "Hap" Arnold, and Rear Adm. L. O. Colbert. As in National Geographic's early days, the organization began to look less like a journalistic enterprise than an extension of the government.

This showed most noticeably in the magazine's war coverage, which was sweeping, upbeat, and solidly patriotic. The journal focused on the positive, showing happy troops, well-scrubbed warriors, and very little of combat's bloody reality. The wounded in *National Geographic*'s war wore pressed pajamas and big smiles as they displayed Purple Heart medals in their hospital beds, while a few pages later marines did battle with pie dough, turning out desserts in a battleship's galley: "Pie Like Mother Used to Make, and Maybe Even Better!" proclaimed the legend. Articles were often illustrated by official army or

---

*By National Geographic's estimate, the organization provided some 35,000 photographs to the government, which used them to plan bombing raids and landings around the globe.

navy photographs, to which the magazine seemed to have better access than other publications. Many pieces were signed by high-ranking officers, which lent a certain authority to the magazine's coverage. Several of these officers—such as Maj. Frederick G. Vosburgh and Maj. Fred Simpich—happened to be *National Geographic* staffers on leave for the war, which drained off the best of Bert's writing and editing staff. Melville Grosvenor, feeling that he should do his part, tried to rejoin the navy in 1943 but was rejected for health reasons. Now in his early forties, he had developed a heart murmur, which nonetheless allowed him to serve as an air raid warden in the Citizens Defense Corps.

While much of the magazine's war reporting was unmemorable, the same cannot be said of its cartographic effort, which reached a peak of timeliness, artistic achievement, and thoroughness during these years. Using his extensive network of government contacts, Bert anticipated strategic action and kept his cartographers cranking out newsworthy maps—on the Indian Ocean, the Philippines, the Low Countries of Europe, the Pacific Northwest, China, Germany, Japan, and other hot spots. These maps, known for their clarity and accuracy, were much in demand by civilians and military alike. Millions of radio listeners tracked the war news on National Geographic's maps, as did President Roosevelt, who outfitted his White House Map Room with them.

In this first-floor chamber—the very existence of which was kept secret for the duration of the war—FDR scanned new messages from Stalin and Churchill and wheeled himself slowly around the room to study maps of the terrain that would be forever changed by his decisions. Here the final plans for the European campaign were plotted on National Geographic maps, which also served as the basis for the postwar partition of Germany and Austria. Scrutinizing the proposed American sector marked up on one such map, FDR expressed concern about the geography: "I don't like it," he complained, recalling how he had pedaled a bicycle across that landscape when he was young. "Too hilly, too far removed from good ports," he protested, sending his military aides away for another round of cartographic strategizing.[10]

As the war ground toward its finish in 1945, military aides drew up plans for a five-stage invasion of Japan, sketching out their objectives on a National

Geographic map entitled "Japan and Adjacent Regions of Asia and the Pacific Ocean." By this time Roosevelt was dead, but President Harry S Truman continued to work long hours in the Map Room, where a young naval aide named George M. Elsey watched the president brooding over the invasion.

"Truman was appalled at the magnitude of Japanese forces," Elsey said. "I often saw him standing before the maps in deep reflection, profoundly disturbed by the estimates of American and Japanese casualties. This reinforced his decision to press Stalin to declare war on Japan and persuaded him to accept the recommendation of his military to use atomic bombs and bring the war to a close."[11]

When the war finally ended in August 1945, Bert and Elsie were in Baddeck, where they received the news from Jack LaGorce. "The celebration here . . . was terrific and lasted all night," LaGorce reported from Washington. "Mobs of people, by the thousands, glad of an excuse to raise hell, went all out to demonstrate their happiness. A news story printed was of a man and woman who stepped out of an automobile on the sidewalk in front of the *Post* building and stripped to their skins before everyone, then exchanged and dressed in each other's clothes. Sorry I missed that one!"[12]

For the National Geographic Society, as for the nation, there was much to celebrate. The magazine's circulation, having recovered from its Depression-era decline, now approached its earlier peak of 1.2 million. Advertising revenues were rising accordingly. Restrictions on paper use had been lifted, allowing *National Geographic* to resume printing covers on its sturdy seventy-five-pound stock. Senior staff members like Ted Vosburgh, Kurt Wentzel, and Andrew Brown were returning from overseas military duty to rejoin the magazine, which would be strengthened by their experience and travel. For the time being, Bert was happy to steer his journal back to the familiar terrain it covered best—archaeology, cultural geography, adventure, and natural history. Wars came and went, but there was something eternal in the depths of history and the cycles of nature—including the lives of worms.

How many other editors would get excited, as Bert did, over an impending story on worm farms? "This is *new* art—a most unusual natural history story,"

Bert informed a squeamish Jack LaGorce, who had expressed second thoughts about the worm article. "To me, the photographs are very unpleasant," LaGorce admitted.[13] Nonsense, Bert replied: "Darwin wrote a whole book on the earthworm, which is quoted today by naturalists. . . . Every fisherman, every country boy, has dug for worms and will be as astonished as I was by this extraordinary culture. . . . I predict a fine reaction from our readers for this is NEWS."[14]

The other news, which almost went unnoticed in the excitement of war and its aftermath, was that National Geographic had undergone a quiet revolution in the way it depicted subjects. Before the war, the magazine had just begun to make the shift from the large-format cameras and slow film that had ushered in the age of color.[15] Instead of film, the old Linhof and Graflex cameras had employed heavy glass plates coated with color emulsion. They had served the journal well, producing crisp, detailed color photography, which was among the best of the early twentieth century. But as Maynard Owen Williams and other pioneers of color could testify, the old color process proved something of a burden to the man or woman in the field. Not only were the glass plates fragile and the emulsions on them vulnerable to extremes of heat and cold; the plates also required inordinate care of handling when they were loaded into the camera, so that they lined up precisely behind a screen for the proper color registration. To avoid premature exposure, light-sensitive Finlay or Dufay plates had to be loaded in very dim conditions, which could be managed easily enough in a studio but spelled trouble for the field photographer, who often had to set up a portable darkroom in the middle of nowhere before composing the first picture (some added a black tent to their already considerable traveling impedimenta). Others, like Maynard Owen Williams, improvised: Photographing a dramatic performance in Greece, he borrowed two harp cases, threw a blanket over them, and disappeared inside to change plates while performers milled about onstage.[16] Early color specialists also faced the inconvenience of long exposures, which required them to open the camera's shutter for an interval sixty times that of black-and-white film. This required tripods to steady the cameras, and it meant that subjects, from emperors to camel herders, had to remain absolutely motionless for a full second or more during a photographic session; otherwise the resulting image would be blurred. This limitation explains why much of the prewar color photography in *National Geographic* and elsewhere looked stilted and posed. "It couldn't be helped," said Luis Marden. "The first thing you learned to say in any language was 'Hold still!'"[17]

As the technology of cameras and film improved with the 1930s, it gave rise to a new breed of innovative photographers—among them Alfred Eisenstadt, Henri Cartier-Bresson, and Margaret Bourke-White—all of whom embraced the 35-millimeter camera, particularly the Leica, and thereby changed the way people saw the world.[18] The new cameras, lightweight and relatively inconspicuous, allowed photographers to blend in, move fast, and capture imagery much more dynamic and spontaneous than anything that had come before. A newly devised high-speed film called Kodachrome, initially invented for the movies, took the 35-millimeter revolution a step further, making it possible to record vivid color pictures in a fraction of the time required by the old process. The new film and camera would eventually transform the National Geographic Society's approach to photography.

Before that could happen, however, traditionalists at the magazine had to be convinced that the new technology was more than a passing amusement. Old-timers dismissed the miniature camera as a toy when Luis Marden arrived at the National Geographic Society in 1934 with a Leica around his neck.[19] Ironically the Grosvenor family, which had done so much to popularize photography, was initially skeptical of the small cameras, perhaps in the belief that anything so much easier could not be as good. Faced with thousands of postage-stamp images from Tunisia and France, Bert Grosvenor rebuked Maynard Owen Williams, the man who had produced them, for wasting time with the little cameras. None of the diminutive pictures could be used, Bert complained. "The National Geographic Society wants deep-cut and detailed photographs," Bert wrote Williams in 1936. "I want you to stop this flood of postage stamps and send us some real pictures like you used to."[20] This sentiment was echoed by the magazine's illustrations chief, Franklin L. Fisher. When an illustrations editor presented him with a stack of 35-millimeter prints, Fisher tore them into small pieces and threw them to the floor. "Now pick those pieces of shit up and get out of my office!" he thundered.[21] Such displays did little to encourage use of the miniature cameras—but only temporarily.

The truth was that the little cameras recorded beautifully, but that *National Geographic*'s engravers had not yet developed a method for making acceptable printing plates from the 35-millimeter film. As soon as the engravers solved this problem, making it possible to reproduce images at any size for the magazine, Melville Grosvenor, whose influence was growing, led the charge for miniature cameras. He began traveling with a Leica, urged his father to master the sturdy little camera, and encouraged the magazine's photographers to study two fledgling picture journals, *Life* and *Look,* for inspiration. "From them we will gain some splendid ideas on the modern type of photography which is rapidly displacing the older kind made with larger plates in stand cameras with stiff, posed subjects," he wrote.[22] In an organization where photographic progress mattered above all else, Melville spoke for the new generation.

But he was not deemed fit to lead that generation. Even though all in the family had assumed that Melville would inherit the crown, Bert had always found reasons to delay passing it down, even after World War II ended and Bert turned seventy. Melville, now in his midforties and perhaps sensing that his time was slipping away, was clearly poised for the promotion. He wrote about it openly to his father.

"It's going to be my aim in the next year to get you to lay your mantle aside for awhile and take it easy," Melville wrote in 1945. "The Lord knows I'll miss you around that office, but I'll be darned if I won't be happier knowing that you are taking it easy and *enjoying* yourself. After all, Pop, you have been working and worrying about me and a lot of other people just about long enough. . . . It's time we took you in charge and did a little worrying and helping for *you.*"[23]

When Bert was away from the office, Melville took on many of his duties, which he shared with Jack LaGorce. "I feel that I have had a little bit of a taste of what you had to do when you first joined the Geographic," Melville wrote to

Bert in 1945. "I am editing manuscripts, correcting and cutting galleys . . . editing legends, approving and selecting illustrations, making up the Magazine, and, of course, doing the necessary editing of proofs. It has been a wonderful experience but hard work."[24]

About this time, Bert engineered his son's election as a lifetime member of the board of trustees, edging him closer to the center of power and ensuring the family's involvement in the future of the National Geographic Society. Yet Bert withheld the ultimate promotion: He still did not trust Melville, who had proven himself more than capable, but who remained emotionally erratic. His father worried about the psychic strain the top job might place upon Melville, who would perform diligently for a long spell and then make a small mistake that would erode his position. For instance, when Bert and Elsie returned from traveling abroad in October 1944 they expected Melville to greet them at the train station; instead they found this note: "I am sorry the big Cadillac is not there to take you home," Melville wrote. "It will be ready in a few days. In passing a parked truck the other day, testing the car, I went a little too close and bent the door handle on the right side. Damage was negligible—but doors won't close and new handles have not arrived."[25] Such antics, along with Melville's good-hearted enthusiasms, often ran afoul of Bert's taut sense of propriety. While his father was away one summer, Melville was on the point of allowing outside researchers access to the Bell family papers. Bert got wind of the arrangements and halted them. "I must caution you in your enthusiasm to be careful," Bert warned Melville, reminding him that the papers needed screening to avoid revealing family secrets.[26] This was just the sort of indiscretion that made Bert worry about Melville's judgment. Even when he doled out bits of privileged information to Melville, Bert felt it necessary to provide a caveat: "I have directed that the daily sheet be given you but please remember always to keep this report confidential," he said. "That means you must not show it to anyone, except Mr. Fisher."[27]

In each of these instances, the old clash—between the freewheeling Bells and the order-loving Grosvenors—was repeating itself down the generations. As Bert slid past normal retirement age, it became clear that his fatherly instincts for Melville conflicted with those he felt for his other child, the one made out of ink on paper. Having raised the magazine from obscurity, shaped its distinctive personality, and built it into a unique American institution, with the largest direct-subscription circulation of any monthly or weekly, Bert was not eager to give it up.[28] He still found a deep creative satisfaction in the magazine and its related operations; he never tired of poring over maps, combing through manuscripts, scrutinizing budgets, and marshaling the right balance of articles for each issue. Being compulsive and proprietary by nature, he must have worried whether Melville—or any mortal—could run the organization as effectively as he had done. He had also become more cautious with age, according to his grandson and namesake, Gilbert M. Grosvenor. "He saw the Geographic was finally a success and he didn't want to rock the boat. Circulation was growing and he was getting older and he just got very conservative."[29]

By this time the prestige of Bert's position had also grown to an enviable level—in the string of honorary degrees, the ready access to presidents and big-name explorers, the association with leading scientists, the white-tie dinners, the private dining room, the chauffeured Cadillac; any executive would have found it difficult to walk away from such perquisites, especially if these were burnished by the organization's reputation for altruism. He hoped that this sprawling family enterprise, and all that it stood for, would be passed along to Melville. But faced with the prospect of leaving, Bert found it easier to stay the course than to change it. And so Melville's apprenticeship continued long after it should have, as did Bert's tendency to second-guess his son's every move.

"Please instruct legends writers to use the full name National Geographic Society in legends," he warned Melville. "The abbreviation to Society deprives the organization of the advertising value."

"It is now incredible to me that I put my OK on this terrible, bumptious piece," Bert wrote Melville after rereading an aviation story by Gill Robb Wilson. "I think Wilson thinks we are a lot of dodo birds and has sold us some truck that he could not dispose of elsewhere. . . . Best wait until I return."

"I don't think you are posted on the map requirements of our members," he told Melville, "and I strongly recommend that you study the two articles in our map monograph. If I had let the NGS continue making picture maps . . . NGS maps would not have been worth a damn cent in World War II. . . . You Navy men don't have the slightest conception of what a good map should include. . . . Your enthusiasm for that South African map astonishes me."[30]

When Melville raved about a new portfolio of insect photographs, Bert rained on his parade, pointing out that the bugs appeared to be dead. "I would like to get a report from an entomologist about this," Bert wrote, suggesting a delay in the article. No detail escaped his notice. The lunch counter planned for a new cafeteria was too short. "There will be great congestion and delay unless this counter is lengthened," he warned Melville, "because when the new building is erected, we will be feeding probably at least five times as many people as we now feed here. . . ." The office linoleum was all wrong. "In all my experience," he wrote Melville, "I have never seen linoleum put down so clumsily as the recent placing of linoleum in the hall between my office and Mr. Hildebrand's office. . . . I hope no payment has been made. . . ."[31] When the linoleum crisis abated, Bert went looking for something else to supervise.

Melville seemed to take the second-guessing in his stride, perhaps because he so admired his father and knew that this occasional scolding was the price of training under a perfectionist. "You have the all-seeing eye," Melville told the Chief, deferring to his father as all others at the National Geographic Society had learned to do.[32] The generous spirit of Grandfather Bell burned bright in Melville, who approached office matters with the same gusto he did the other parts of life. The same could not be said of Bert, who brooded over office or family troubles even when he appeared to be engaged in something else— onstage at the White House, watching birds on his farm, sailing across Bras d'Or Lake, or flying off to Hong Kong.[33]

Now Bert worried over perhaps the most difficult problem of his long career, which appeared with the hard winter of 1947 and 1948. Having reached age seventy-two, he finally seemed ready to step aside in favor of Melville. Then his son threw the transition into confusion. A few days after Christmas in 1947, just after the wedding of his only daughter, Melville took off for a planned sailing trip in the Caribbean. While there, he mailed a letter home to Helen Grosvenor, his wife of twenty-three years, informing her that he was unhappy in their marriage and wanted a separation, perhaps even a divorce. She expressed shock, believing that their marriage had been happy. Melville thought otherwise, charging in subsequent divorce proceedings that Helen had abandoned their bedroom, had treated him cruelly and inhumanely, had ridiculed him incessantly, and had interfered with his work at the National Geographic Society. She had even demanded that he step down as Commodore of the Gibson Island Yacht Club, a high honor in the sailing community so important to Melville. She denied each of these claims, as well as the most serious one— that her abusive treatment had triggered a nervous collapse in Melville, in February 1948.[34]

According to Helen, the final crisis began with Melville's return from his Caribbean vacation in January 1948. Having already notified her of his intention to separate, he moved into a son's vacant bedroom in their Washington home. His later court filings contend that he arrived to find that she had abandoned their bedroom and "failed to conduct herself as a wife should." Helen countered that he began to behave erratically. He disappeared for days at a stretch, came and went at all hours, grew increasingly restless, and began muttering to himself. The collapse came on February 16, when Melville returned home about 11 P.M. in an agitated state.[35]

"I can't go on this way," he said over and over. "I don't know what is going to happen. I can't go on this way." When Helen moved to help him, Melville recoiled, barking, "Don't you dare do anything! Don't you dare touch me!" He announced that he was going to take a shower. While he did, a concerned Helen called Dr. Walter Myers, an internist married to Melville's sister Carol, and asked for help. Myers, who also happened to be a trusted friend of Melville's, appeared shortly, examined his brother-in-law, and ordered him to Washington's Emergency Hospital, where he remained for three days. Then he was transferred to a psychiatric facility, the Sheppard and Enoch Pratt Hospital in Towson, Maryland, where he would stay until his release on April 25, 1948.[36]

During Melville's hospital confinement, Helen wrote to him three times a week and sent him presents four or five times. Melville had one of his sisters return each letter and each gift "with the message that he had no love for sale," according to Helen. When Melville was released from the hospital that spring, he sent a chauffeur to gather some suits and personal possessions from their home, to which he never returned.[37] The marriage had unraveled.

According to members of the family, a major source of friction between Melville and Helen had been her feeling of not fitting in with the Grosvenors. "She was jealous of his attachment to the family," said Mabel Grosvenor.[38] Others who married into the clan had similar problems of acceptance—either you fit in and enjoyed it, or you were the perpetual outsider. "The family was self-contained," said Grosvenor Blair, one of Gertrude Grosvenor's two sons. Blair described the closeness of this large family as a mixed blessing: On the one hand, those born into or embraced by the family found comfort in it; on the other, this insularity afforded the insiders little experience of the outside world, which made it difficult for some of the Grosvenors to adapt there. And like Helen Grosvenor, Blair's father, Paxton, always felt like an outsider. "He deeply resented the family," Grosvenor Blair recalled. "My father couldn't understand why people would want to go out on a boat and get wet and seasick. He hated Baddeck."[39]

Of course Bert and Elsie felt dismay not only at Melville's emotional troubles but also at the prospect of divorce in a family that held strong views against it. (Alexander Graham Bell had once suggested that men who violated the marriage vows should be sent to prison.) Bert, who aspired to the same lofty objectives in family life as he did at the office, considered divorce embarrassing and shameful—a position he would soften as three of his five married children eventually divorced.[40] In Melville's case, Bert soon rallied to support his son. "There is nothing to be ashamed about your present perplexity," he told Melville. "There is no reason that you should apologize to your conscience for your present uncertainty about your future. You have taken the honorable position by withdrawing from Washington in order that you may get yourself into shape to decide what is the wise course for you to adopt. Take plenty of time."[41] Both Bert and Elsie hoped for a reconciliation. "I can't get my mind off Melville and his difficulties," she wrote Bert from Florida in 1948. "If only they could reach a new understanding, or if not that Helen would be accommodating. It's a mess any way! It's only a wonder he's stood the situation as long as he has. He's such a loveable man . . . I know how upset you must be."[42]

Bert's upset deepened when the family troubles turned into news. Because of the Grosvenors' prominence in Washington, local papers covered the divorce closely. The *Washington Post* gave it front-page treatment when Melville sued for divorce in August 1949. "Son of Editor: M. B. Grosvenor Seeks Divorce; Calls Wife 'Cruel, Inhuman,'" proclaimed the headline; beneath it, an article described their separation, Melville's breakdown, and his hospitalization. The *Evening Star,* whose owners and editors had long served on National Geographic's board, covered the story more discreetly, placing it on page A-3.[43] Helen provided an opportunity for follow-up coverage when she counter-sued Melville on grounds of desertion. To make matters worse, this unseemly tale played out smack in the middle of what was to have been a festive year celebrating Bert's golden anniversary at the National Geographic Society. At the end of that year, a county judge found that Melville had deserted his family and granted a divorce in Helen's favor. Their two adult children were by now on their own, Teeny having married in 1947 and Alec having enrolled at the Naval Academy, where he was a midshipman when the divorce became final. Helen was given custody of their remaining child, Gilbert M. "Gibby" Grosvenor, eighteen, then a student at Deerfield Academy.[44] He remained loyal to his mother for the rest of her life, taking care of her as other Grosvenors closed ranks behind Melville.[45]

After his well-publicized illness and marital crisis, Melville slowly began to rebuild his life. He sailed with friends, played golf at the Chevy Chase Club, and threw himself into seasonal chores at Wild Acres, where he arranged for the alfalfa harvest, supervised the laying of a new sewer line, and kept the fences mended. He commandeered a garage apartment at the farm as his bachelor quarters, which he described as "cozy and nice." His sunny outlook was returning. "I certainly do love it out there," he reported to his father. "Your little wren was chipping and flitting in and out of the lamp, just as it or its forebears have for many years," Melville wrote.[46] He resumed a busy schedule at work, where he hoped to win back his father's confidence and esteem.

Bert did his best to encourage Melville. He ordered a promotion for his son in October 1948, shortly after his return to work. "From now on I would like to have you assume the title of Chairman of the Lecture Committee," Bert wrote. "You have shown such excellent judgment in selection of the lectures that the membership attendance has been steadily maintained and increased."[47] Melville profusely thanked his father. "This promotion comes at a particularly appropriate time for me since I have just concluded my twenty-fifth year of assisting or arranging The Society's lecture course," wrote Melville, who had been previously known as the secretary of the lecture committee. In another man, this response might be read as ironic; in Melville, who was incapable of sarcasm, the appreciation was as grateful as it seemed.[48]

A few months after Melville's divorce, he married again, this time to Anne Elizabeth Revis, a spirited young woman from Charlottesville, Virginia, who had joined the magazine as Melville's secretary a few years before.[49] She and Melville had fallen in love while he was still married to Helen. Of course the close-knit and gossipy staff at headquarters knew. They had watched Melville's car purring off in the dusk, discreetly stopping a few blocks from the office to admit the tall dark-haired woman they knew from work. Since nothing occurred at National Geographic without Bert's knowledge, he must have known too. Perhaps by then the old puritan had grown so accustomed to Melville's impetuosity that this latest development did not surprise him. In any event, he not only welcomed Anne into the family but arranged a writing assignment for her. Her article, "Mr. Jefferson's Charlottesville," was published three months before the wedding.[50] Unlike her predecessor, the new Mrs. Grosvenor adored Melville's parents. She shared her husband's enthusiasm for sailing, for world travel, and for life at Baddeck. She thrived in the family and so, now, did Melville, with no recurrence of his previous mental distress.

He assumed more responsibility for the day-to-day operation of the magazine. By the early 1950s he was hiring new staff in consultation with his father, rushing maps for the outbreak of war on the Korean peninsula, prospecting for fresh photographic talent, and encouraging the work of an obscure new contributor named Jacques Cousteau. As his confidence bloomed, so did Melville's competitive spirit; he tracked coverage in *Life, Holiday, Collier's,* and other periodicals and juggled *National Geographic*'s schedule in an effort to scoop them.[51]

Bert, seventy-eight, must have felt that both Melville and National Geographic were strong enough for him finally to relinquish the title of editor and president on April 1, 1954, fifty-five years after he had gone to work for Alexander Graham Bell. In his letter of resignation, Bert singled out Melville's importance to the organization. "I have been favored 30 years by the presence of a strong son beside me," Bert told the trustees. "He has proved an alert, indefatigable and gifted editor, and in his 30 years' service has contributed greatly to the progress of The Society and its Magazine."[52]

Then Bert named John Oliver LaGorce, seventy-three, to succeed him. The trustees created a new position for Bert, who became chairman of National Geographic's board. Melville, fifty-two, was promoted to vice president and associate editor. He would have to wait awhile longer.

CHAPTER TWELVE
MELVILLE TAKES CHARGE

MELVILLE GROSVENOR JOINS PALEONTOLOGIST LOUIS LEAKEY
DIGGING FOR FOSSILS IN TANZANIA, C. 1965.

Sensing Melville Grosvenor's disappointment at being passed over for the editor's job, Maynard Owen Williams, the retired dean of *National Geographic*'s foreign staff, found the right words to ease this latest setback. "The son of a big man has a hard row to hoe," the old missionary wrote to Melville, "even if he is a big son of a big man. I have watched you develop for many years and I am proud of you. . . . Any alternative to your eventual Presidency and Editorship is unthinkable. The Society is NOT a family affair, but no one will ever make the same impress on it as the Grosvenor family has made, is making, and will make." Williams counseled Melville to be patient and to trust Jack

LaGorce's allegiance. "You can always be sure that although Dr. LaGorce's prime responsibility is to The Society, you will never lack his personal loyalty and support." Williams was right in that assessment, and in his belief that LaGorce's regency would be brief, like that of others who would bridge the gap between Grosvenors.[1]

LaGorce's appointment lasted just three years, serving as a reward for his long work at National Geographic while providing Melville's final polishing for the top job. During his tenure as caretaker the innately conservative LaGorce performed his role with little flourish, taking no risks and following the publishing formula the Chief had adhered to for more than half a century. LaGorce crowded the magazine with colorful pictures, avoided controversial subjects, cultivated influential friends in the capital, and watched circulation hover between 2 million and 2.5 million, a membership level then at the limits of the magazine's printing capacity.

"He wouldn't change anything," Melville recalled. "His whole editorial policy was: 'Steady as you go, don't upset the apple cart.' And I went along with it because that's what he wanted."[2] Confident that the momentum of the National Geographic Society would sweep it along until Melville was ready to take control, LaGorce did not worry about competition from upstart magazines. "Personally," he told Bert, "I do not feel that *Life, Look,* or any other picture magazine today competes *with* The Geographic in its field of *sustained reader interest,* which, in the final analysis, is the acid test. . . . The very reason why five times more volumes of The Geographic are kept and bound than any other magazine is because of their permanent value to be re-read and for reference. . . . The average person can look at the contents of [*Life*] . . . and read the legends of a single issue in from 12 to 15 minutes, and from inquiries I have made I cannot believe that the copies are frequently re-read or *saved.* . . . Please do not lose sight of the fact that the one and only object of the owners of *Time, Fortune,* and *Life* is to make money from advertising income and that both *Fortune* and *Life* have largely copied our formula . . . so should we try to copy them?"[3]

The unspoken answer was an emphatic *no,* and so the magazine remained essentially unchanged through the sleepy period the novelist and former

National Geographic executive Charles McCarry described as its "long after-noon."[4]

LaGorce followed his predecessor's policies in accepting no ads for liquor, tobacco, patent medicines, or real estate schemes, all of which were believed to lower the dignity of the organization and to conflict with the National Geographic Society's altruistic mission. "We have deliberately refused through the years the millions of dollars that were available in advertising liquor and cigarettes," Bert explained, "because we realize from personal observation the evil consequences on the health of these drinks and smokes."[5] (To this day, the magazine accepts no ads for liquor or tobacco in its North American editions, but it has succumbed to the allure of Viagra, Plavix, Zōcor, and Allegra.) Even though LaGorce had run the magazine's advertising department for many years before he became editor, he shared his mentor's view that commercial pitches had no place alongside editorial material; both men kept advertising scrupulously segregated from articles. LaGorce also continued Bert's policy of limiting advertising to no more than 10 percent of the magazine's total pages; then as now, ads represented a fraction of *National Geographic*'s revenue, most of which came from circulation.

Under LaGorce, the membership seemed to change as little as the magazine, its readership reflecting the look of mainstream America—solidly middle class, somewhat innocent, self-assured, and most definitely white. LaGorce, who had been the architect of a National Geographic Society policy to ban blacks from membership in the 1920s, continued to promote their exclusion well into the century. Even as President Harry S Truman was beginning to desegregate the armed forces in 1948, LaGorce and Melville Grosvenor made it clear that African Americans were still unwelcome as members—at least in Washington—where it was feared that they might show up at lectures.

Without mentioning race at all, Melville raised the issue in a note to his father on November 19, 1948. "I think if we should continue our policy of keeping the rolls officially closed by allowing the membership committee to fill vacancies of 100 to 150 a month by qualified new members, we can continue the high standards of our Washington membership and avoid unpleasant situations," wrote Melville, then in charge of the lectures program.[6] A few days later, LaGorce weighed in with his recommendation to Bert. "Re: Washington Members," he wrote, "For reasons which are sound in my opinion I am not in favor of taking, i.e. accepting a number of new Washington area members at this point. Let us see what the Administration will do as to the Negro question hereabouts before we find ourselves in a jam over the lectures. There will be plenty of time by next mid-summer to review the matter."[7] Bert's reaction is absent from the record—but there can be no doubt that the policy of excluding blacks could have persisted without his approval. Racism remained a fact of life in the National Geographic's upper echelon from one regime to the next, remaining that way until a new generation of leaders began to assert their control in the 1970s.

Throughout the LaGorce transition, Melville Grosvenor bided his time, repressed his natural tendency toward impatience, and learned to accept it when his initiatives for modernizing the National Geographic Society were rejected by his elders. Melville, now headed into his midfifties, had been well tempered by failure, by family strife, and by having his emotional struggles revealed to the public. Against this backdrop of embarrassment and disappointment, the temporary deflection of his dreams for National Geographic must have seemed minor. "Most of my ideas I'd presented before I became editor, but the older men turned them down," Melville recalled later. "So I put them in the back of my head and waited."[8]

Melville used the time profitably, building his self-confidence and absorbing the standards his father had so carefully put into place at the National Geographic Society. Now that Bert had relinquished day-to-day control of the magazine, he seemed eager to pass along tips and encouragement to Melville, who received an intense tutorial on how to run the enterprise. Bert encouraged Melville to trust his instincts, to continue his innovations with color photography, to participate in decisions of the research committee, to keep hiring young people, to assert his right to confidential data, and to worry less about stepping on toes. "I have in my time saved many a story turned down by a subordinate but which later proved a 'winner,'" Bert told Melville. "So *I caution you most earnestly,* do not allow yourself to accept always a subordinate's recommendation. You are over cautious of offending, or rather I should say, of hurting the feelings of a subordinate by disagreeing with his opinion."[9]

For one who had done so much to erode Melville's confidence and to hold him back, Bert now behaved like a man transformed. Reviewing a recent issue of the magazine, Bert raved about it. "Most beautiful, most newsy & interesting and most informative and educational also," he wrote Melville. "Your direction of this superlative Magazine is evident on every page. If you can match this issue every month we will soon get to three million. . . ."[10] A few months later, he paid his son the ultimate compliment by invoking the spirit of Alexander Graham Bell, who had encouraged Melville's exuberant approach to life. "I'm delighted with your competent manner of handling the many diverse problems of the Nat Geog Magazine," Bert wrote. "Your Grandpa Bell often said to me that he thought you had unusual executive talent. You are proving that he was right in his observation."[11] But Bert still fretted about his son's health. "You have a tremendous job," he told Melville. "You are worrying because you are tired. Try and relax. . . . You must watch yourself and get some relaxation every weekend. When a man gets tired, he begins to worry and then worry wears him out and handicaps his thinking and judgement. . . . I predict you will be a greater Editor than G.H.G. or J.O.L., provided you take care of your physical body." Bert concluded with a crucial bit of fatherly advice: "Wear a hat in cold weather and in hot weather."[12]

To avoid undermining Melville's authority, Bert kept any criticism of the magazine within the family. *"Please remember,"* he wrote Melville, *"I give no comments* about anything or any person except complimentary comments except to you. So if anyone says your Father said this or that, that person is misquoting me. . . . I am resolute not to criticize to anyone about anything, now that I have retired."[13] Thus, while Bert remained as chairman of the board, he established that position as a largely ceremonial one. Still, the old perfectionist could not resist the urge to let fly with an occasional comment, but he delivered such criticism softly—a break from past practice—to keep from upsetting Melville. "I would not be human if occasionally I did not frankly comment about a thing or two," he apologetically wrote on the eve of his son's fifty-fourth birthday. *"You are the only one that I can do this to. . . .* I beg you not to take too seriously when I deprecate something or other. Times & conditions *change.* I am the last one to deprecate a change. I want *everything* to improve and I rejoice that you are not fettered too much by precedent."[14]

During this period, Bert even took the extraordinary step of admitting a mistake that called attention to Melville's superior editorial judgment. Over the misgivings of both Bert and LaGorce, Melville had convinced the magazine to undertake extensive coverage of the coronation of Queen Elizabeth in 1953 and the British conquest of Mount Everest in 1954. Both articles, Bert now conceded, had been runaway hits. "Each *seemed* expensive," Bert recalled, "but judged by the gain and enthusiastic reception of those articles by our readers, the cost . . . was a bagatelle measured by results and when calculated on a distribution of the cost among 2,100,000 members. Neither of those wonderful articles cost as much as one cent per member. . . . *So go ahead and don't feel you have got to get the permission and approval of J.O.L.* for everything you want to accomplish," Bert wrote, suggesting that Melville avoid undue deference toward LaGorce. "J.O.L. is much pleased with your direction of the Magazine," Bert assured his son. "I've always given him a free hand when I've been away from the office and I'm sure he wants you to carry on with a free hand when he's away."[15]

Finally, Bert urged Melville to remember that the key to National Geographic's success had been its willingness to be different from other organizations—in its approach to stories, in its insistence on excellence, and in its way of treating members as a special class of readers. "It has been our constant plan to instill in members that being members of the N.G. Soc. they get special privileges. I know we get criticized because we don't do what others do but nonetheless our policy has built up a renewal rate that is, I believe, unique."[16]

Facing the biggest responsibility of his career, Melville grew stronger at the prospect, in large part because of his father's unwavering support; it was the same sort of steadying Bert's father had provided through his own difficult inaugural as *National Geographic* editor. Although Melville still had moments of doubt, they were no longer debilitating, and he faced the future hopefully, piling up plans for his administration while hiring the young editors and writers who would help him bring those plans to fruition. The young staffers provided a much needed jolt of energy for the magazine, which had grown tired and predictable with its success. Mary Griswold, who would doggedly coach Jane Goodall, Dian Fossey, and future stars through many articles, came on board as a picture editor in 1956; despite Miss Griswold's complete lack of publishing experience, Melville, acting on impulse, hired the young woman on the spot. "I need young people around me!" Melville gushed to his surprised recruit.[17] He also collared a brash young navy veteran named Wilbur E. "Bill" Garrett, who had caught Bert Grosvenor's eye during a visit to the University of Missouri in 1954. Melville installed Garrett, a recent graduate of Missouri's journalism school, as a picture editor alongside Mary Griswold. They were soon joined by other recent Missouri graduates, Thomas R. Smith and Robert L. Breeden, who would rise through the ranks with Garrett to assume key leadership positions. The Midwestern trio of Garrett, Smith, and Breeden would come to be known as the Missouri Mafia.[18]

Like many other staffers Melville would hire, these young people came from modest backgrounds, reflected the wholesome American values most *National Geographic* readers were looking for, and demonstrated the family's egalitarian attitude about recruiting and promoting talent. Although both Melville and his father had happily drafted employees with Ivy League credentials, the old school tie was no guarantee of a position at National Geographic. "He may have a fine connection with Boston's first families and be a St. Mark's and Harvard graduate," Melville wrote to his father regarding a well-recommended applicant, "but still I do not think he is anyone we would wish to employ on our staff. . . . In addition to this fellow not being a good writer, he has a very unattractive appearance and, we think, a bit 'peculiar.' "[19] The peculiar applicant was sent away.

In the first step toward broadening National Geographic's operations, Melville hired a talented pair to start the organization's book publishing program—these were Howard E. Paine, a brilliant designer and idea man from a Massachusetts ad agency, and Merle Severy, a Columbia graduate and former leader of the U.S. Army Band. Bill Graves, a newspaper reporter and former Foreign Service officer whose manic energy matched Melville's, soon joined the magazine staff, along with Robert E. Gilka, a photo editor from the *Milwaukee Journal,* then known for its innovative use of gravure printing and color pictures.[20] From this roster would emerge two future editors in chief, a legendary director of photography, and a surrogate son for Melville. Almost unnoticed in the flurry of hiring, Melville's younger son, Gilbert Melville "Gibby" Grosvenor, slipped through the marbled entrance at Sixteenth and M Streets to report for work as a picture editor on the first day of summer in 1954.

Gibby Grosvenor, twenty-one, a recent Yale graduate, was a shy young man with the dark features and earnest manner of Alexander Graham Bell, the orderly habits and cautious style of Bert Grosvenor, and the tenacity of both. The young man seemed to possess little of Melville's characteristic impulsiveness, explosive enthusiasm, or charming style. Nor had Gibby, who had remained distant from his father since the divorce a few years before, shown the slightest interest in joining the magazine until recently. Instead, he had remained close to his mother, gone off to Yale, and signed up for pre-med courses, an independent path that would have led him away from family entanglements in Washington. Then Gibby suddenly changed course as he approached his senior year, surprising all in the family—and perhaps himself—with a late-blooming love for photography and journalism. "I never planned on a career at National Geographic," said Gil (as Gibby would be known as an adult). "It just sort of happened."[21]

The first break in Gil's favor came when his elder brother, Alexander Graham Bell Grosvenor, chose a naval career over the National Geographic job for which he had been groomed all of his life. Melville, who was never very good at masking his feelings, made it clear that Alec had been his first choice. One is struck by how much Alec dominates Melville's family letters—he is the boldest sailor, the best photographer, the most promising writer, the toughest patient, the favorite son.

"Gil was sort of left out in the family," said his aunt Mabel Grosvenor. "Mel was very interested in Alec, the first boy. He got all of the attention."[22] But once

Alec graduated from the U.S. Naval Academy and began his career as a highly regarded aviator, Melville's favored successor was lost to the magazine. "It wasn't that Alec was negative on National Geographic," said his cousin Grosvenor Blair, "he was just positive for the navy."[23] With Alec out of the running, relatives encouraged Gil to carry on the family tradition. Melville was more than supportive—he was eager to have Gil on board, but also wary of pushing his son into a National Geographic career, given Gil's lingering bitterness from the divorce.[24] Indeed, Gil's mother had threatened never to speak to him again if he joined his father at National Geographic, an institution she still viewed with suspicion.[25] Countering this was the positive pull of a family figure whom Gil revered, his grandfather and namesake. Gilbert H. Grosvenor saw promise—and perhaps something of himself—in the hardworking grandson whom he occasionally met for lunch in a private dining room at headquarters, there to discuss sailing adventures, photography, and family news. "They were close," said Mabel Grosvenor, "very close. Gil worshiped my father."[26] In many ways, Bert's relationship with Gil was analogous to Bell's with Melville; in each instance, the aging patriarch, freed from the daily routine of raising a family, had the leisure and the standing uncomplicated by too much familiarity to instruct the younger generation in the ways of the older one.

Gil's opportunity came unexpectedly. Restless in the summer before his senior year of college, he joined friends in the Netherlands, where American students were helping with flood relief in 1953. Almost as an afterthought, Gil took a camera with him that summer, started using it, and found that he not only enjoyed photography but seemed to have inherited the family eye for pictures. Exceedingly modest by nature, he returned to Washington with a bundle of photographs and notes, unsure whether they were good enough to show to his father and to Uncle Jack LaGorce. To Gil's surprise, not only did his work receive favorable reviews; it was soon published under his byline with the title, "Helping Holland Rebuild Her Land."[27]

With that, the fifth generation officially took its place at the National Geographic Society. If young Gil had been a talented outsider instead of a Grosvenor, it is doubtful that he would have found his way so easily into print, much less into a berth at National Geographic. Family connections had given Gil his start at the magazine, just as they had his great-grandfather, his grandfather, and his father. Years after Gil was on board, Melville Grosvenor still bristled at the suggestion that nepotism had worked in his son's favor. "He came to us and he has written some marvelous articles and he has a talent for executive and administrative work," Melville testily told an inquisitor. "What am I going to do? Push him out because he is a Grosvenor?"[28]

The last Grosvenor stayed—but at a cost, striving to prove himself through a long career, even after the rest of the family had died or moved away. Of course it had been the same for his father, who had struggled in the shadow of Gilbert H. Grosvenor. Melville had known the daily pressure of coming into the office, where all eyes were upon him, "everybody looking because I was the son. . . . I was the son of the president. They were wondering what to do with me. . . . I had to carry on."[29] The long journey to the top finally ended for Melville on January 8, 1957, when National Geographic's board of trustees named him to replace John Oliver LaGorce as editor and president. Melville was fifty-five.

He began his new job by imposing a term limit on himself, intending to step down after ten years to make way for the next generation. Knowing his time was limited, Melville launched a blizzard of initiatives to transform, modernize, and invigorate the family institution that had become staid and stuffy with age. "It was the beginning of the dawn," said Bill Garrett, describing Melville's revitalizing effect on the magazine.[30]

Among other initiatives, Melville stepped up his drive to recruit young talent and made plans for an ambitious new headquarters building and a separate fulfillment center. The new headquarters, a white marble tower planned for completion in 1963, would be the National Geographic Society's fourth home in Washington, where Bert had reported to that shabby rented room on April's Fool's Day in 1899. With the construction of this new building, designed by Edward Durell Stone, the organization now occupied a full city block, bounded by Hubbard Hall and its old holdings on Sixteenth Street, the new tower on Seventeenth, with M Street forming the northern border.

In addition to moving the organization into a modern building, Melville was also taking the National Geographic Society into modern times. He allowed employees to take a coffee break, a concession that prompted the *Washington Post* to tweak its old-fashioned neighbor. "Up to now," the *Post* reported, "the Society has officially recognized Java only as an island in Southeast Asia."[31] Melville clipped the piece, laughed it off, and proceeded with his modernization campaign. He tripled the magazine's photographic budget, expanded the cartographic operation, and did away with the requirement (and the pretense) of memberships being dependent on nomination.[32] He unleashed a small army of writers and photographers to prepare the stories he knew would be popular with *National Geographic* readers. Informed that his lieutenants planned to send only ten or so people to record Winston Churchill's funeral in 1965, Melville exploded, "Oh, come on! This is going to be big!" He sent twenty-six staffers to record the event. For good measure, he commissioned a reminiscence by former president Eisenhower, "The Churchill I Knew," and bound a thin vinyl recording of the late prime minister's speeches into the August 1965 issue. Readers loved it. "They told me it couldn't be done," Melville recalled, beaming at the memory.[33]

Melville showered the magazine with attention and money, but he also put a new emphasis on its parent National Geographic Society. Seeking to reach new members, he began to diversify, taking the National Geographic Society into television production and other ventures. At the same time he arranged for the magazine to be printed on new high-speed presses in Chicago, a shift from the archaic sheet-fed process, which had hampered circulation and constrained the use of color in the magazine. With the new web-fed presses *National Geographic* magazine could spread color throughout its pages. A shift in printing contracts seems like the most routine of business updates, but this one had far-reaching consequences. The new presses, so much faster and more efficient than the old ones, enabled the magazine to meet demand triggered by the postwar economic boom. When Melville became editor in 1957, circulation stood stagnant at 2.1 million; when he stepped down after a decade, membership had rocketed past 5.5 million.[34] This rise in circulation filled the organization's treasury to overflowing, allowing its annual dues to remain at the bargain rate of $6.50 through Melville's administration, even as the organization spent freely for building, hiring, fieldwork, and scientific endeavor. By 1967, the Committee for Research and Exploration was handing out $1 million in annual grants, helping to popularize the work of Louis and Mary Leakey, Jacques Cousteau, Jane Goodall, and others National Geographic would help to bring into the spotlight.[35]

The magazine's new readers received a journal that looked decidedly like the old *National Geographic* magazine, filled with articles on faraway places and the wonders of nature. But the magazine was also becoming brighter, sharper, and less old-fashioned under Melville. Because of his printing innovations, color photography enlivened more of its pages, and while coverage remained staunchly optimistic, the first glimmers of reality began showing in its articles—on Israel's Six Day War of 1967, on Algeria's drive for independence, on Thailand's delicately balanced economic hopes. Supreme Court justice William O. Douglas contributed a timely piece on Inner Mongolia, a culture caught between China and the Soviet Union. The work of Magnum photographers—the best photojournalists in the business—added spice to the magazine's traditional fare. Melville even broke his father's long-standing moratorium on the Soviet Union, assigning a Russian-speaking history professor, Thomas T. Hammond, to produce an article for the September 1959 issue. His thoughtful, well-balanced piece, accompanied by handsome photographs and a new map of the western Soviet Union, gave readers an overdue glimpse of the superpower. More important, such coverage represented the first inroads on what came to be known as Bert Grosvenor's "three monkey" rule—by which writers and photographers had been expected to see no evil, hear no evil, and speak no evil.[36]

For Melville Grosvenor, who had led a sheltered and privileged life, the real world served as a source of occasional bafflement and constant surprise, as revealed by his reaction to a story on the Hungarian crisis of 1956. Reviewing a series of black-and-white photographs of displaced families dragging their few possessions through the train station in Vienna, the editor seemed on the verge of tears. "Those poor people!" Melville cried. "All they have is in those little bags—their clothes, their papers, their stocks and bonds!"[37] He was not joking. The editor of National Geographic retained this boyish guilelessness into old age; friends and colleagues found it to be one of his most engaging qualities. "He was an innocent man," said Luis Marden, "in the best sense of the phrase. He had the curiosity of a child. 'What is that? What makes it work?' That curiosity spread round the place."[38]

Under Melville's stimulus a flood of recruits filled the ranks left vacant by the death and retirement of Bert's old staff. Many of the new arrivals were drafted by James M. Godbold, a former marine fighter pilot Melville had hired away from the Minneapolis Star-Tribune and installed as director of photography, a newly created position that signaled the ascendancy of pictures at National Geographic. Godbold, a supremely confident Mississippian who was built like a fullback, became affectionately known as the Mississippi Raider for his ability to lure young talent away from the best newspapers in the United States. They came to Washington from Ohio, Wisconsin, Minnesota, and from throughout the heartland, eager to work with the well-regarded Godbold and the young picture editors Melville had engaged. In the process, it was inevitable that the new generation would clash with the old guard, and the youngsters gradually lifted the magazine out of the tired habits that had characterized the 1940s and 1950s.

Melville introduced expanses of white space around articles and gave the cover its first makeover in decades. One by one, he ordered all but a scattering of oak leaves removed from the front of National Geographic, where they had been lovingly maintained as a border since 1910. He listed the titles of articles on the spine of each issue so that they could be scanned on the shelf. He changed the name of the magazine from The National Geographic Magazine to The National Geographic, finally reducing it to National Geographic. And he began to publish color photographs on the cover, which had previously consisted entirely of type. This last innovation Melville instituted with characteristic flair. Knowing that many of his associates (including his son Gil) opposed cover photos as unnecessary in a magazine with no newsstand sales, Melville planned an object lesson that might have been devised by Grandfather Bell. Melville summoned the senior staff to his office and started tossing copies of *National Geographic* to the floor. Some of those present wondered if the boss had lost his grip. "Now," Melville demanded, "let's see if anybody can tell me which is the latest issue."[39] None of his associates could do so without crawling around and shuffling through the magazines. Next Melville threw down ten prototypes with a photograph on each cover. "OK," he said, "pick out the Japanese issue." Gil Grosvenor dived into the pile and came up with a geisha cover. "That's what we're going to do," Melville announced. Two months later, in September 1959, he ran the first color photograph on the magazine's cover, a picture of the American flag rippling in the breeze.[40]

Unlike his father, Melville seldom brooded over such decisions; instead, he embraced an idea, quite often on the basis of pure instinct, and promoted it with evangelical zeal. If opposition developed to one of his initiatives, it seldom seemed to bother Melville, who had absolute confidence in his judgment as editor. "He was a gambler," said his sister Mabel. "He didn't mind taking risks. If he turned out to be wrong, he was perfectly happy to live with the results."[41]

Carolyn Bennett Patterson, a writer and editor who began working with Melville in 1949, recalled how he swept colleagues along with his enthusiasm. "I remember once an article on spiders," she said. "Well, most of the editors weren't terribly thrilled, but MBG [as Melville was known at the office] went on about how *marvelous, marvelous* they were. He was so excited that we grew thrilled ourselves; it wasn't phony, just contagious excitement. Then, of course, we passed the thrills on to our readers." Some of those readers found Melville's initiatives disturbing, particularly when he introduced articles about physical anthropology and evolution in the magazine, prompting howls of protest and resignations from readers in the Bible Belt. But there was no stopping Melville, who was confident that new readers would replace the old. "I was 55," Melville explained. "I couldn't waste any time."[42]

Younger colleagues scrambled to keep pace. "You'd often find yourself almost running beside him," said Luis Marden. "You would catch fire from him."

The excitement worked in both directions. Once, when editors were discussing a prospective article on the French exploration of Canada, Melville blurted out his support for the idea: "By God, that's right!" he said. "Everybody wants to read about those Canadian *voyeurs!*" Regardless of what they were called, Melville knew that he liked the proposal and that it would be a popular subject, which proved to be the case. "He was Mr. Everyman," said Marden. "If Melville was interested in something, you could be sure the average and not-so-average reader would be."[43]

Melville's affinity for the membership seemed almost as finely tuned as his father's had been. Although Bert had brought photography into prominence at *National Geographic,* he was at heart an editor rooted in the literary tradition. He valued scholarship and good, clear writing as much as he did photography. His Yankee reticence made him uncomfortable expressing emotions in person; instead he most often entrusted such sentiments to paper. Melville, being more intuitive by temperament, strongly favored the visual side of *National Geographic,* which flourished in his time. He was an indifferent reader who, like many of his members, preferred to move fast and look at pictures. Because he grew quickly bored with manuscripts, he detailed colleagues like Ted Vosburgh and Franc Shor to provide him an assessment of articles.

Melville's father had been a distant and undemonstrative figure who earned the respect of associates; Melville, by contrast, was loved. Six years after his death, one of his editors could still say: "I miss Melville every day." They remember his administration as the golden age of National Geographic, a time when the staff's ebullience matched the country's and when technology was changing exploration, as it was the magazine and the Society—researchers, writers, and readers had new access to ocean depths and coral reefs, to the frontiers of space, and to discoveries in the soil of Africa, where paleontologists were unearthing the secrets of human origins. All would be documented by *National Geographic,* which had unprecedented resources at its command, backed by the unwavering support of its editor and president. "When you'd done something he liked, he'd put that big paw of a hand on your shoulder, and with a huge smile he'd let you know that he liked you and liked your work," said Tom Smith, one of the young editors Melville welcomed into the extended family he was gathering at the organization.[44]

Melville's paternal feelings came rushing to the surface when his colleagues faced any personal crisis. When a colleague's son struggled with severe psychiatric problems, Melville hurried in to offer advice and comfort, a gesture the recipient remembered with tenderness fifty years later.[45] When a respected associate suffered a mental breakdown and was hospitalized for treatment, he awoke one afternoon to find Melville sitting beside his bed, holding up a clock. "All you have to do is keep yourself under control from here to here," Melville advised, indicating a fifteen-minute increment. "Then do it for a little longer. Don't think beyond that. Before long you'll be out of here good as new."[46] When one of Melville's most promising young writers, Peter T. White, returned from a harrowing assignment in Laos, discovered extensive changes in an article he had written, and exploded at Melville, the editor soothed his visitor.

"Peter, you're not well," Melville interjected. "You have worms. Somebody else came back from a trip and talked to me like this and *he* had worms." Keeping an eye on the excitable author, Melville reached for the telephone. "I'm going to send you over to my own doctor right this minute!" he declared. White, whose anger had melted under this unexpected show of affection, retreated meekly, went to the doctor, and never shouted at Melville again.[47] (White did not have worms.)

As Melville's generous nature became known around the office, it was said that some unscrupulous employees waylaid him with sob stories on long elevator rides and emerged with fatter paychecks. "You'd see them lined up outside the men's room waiting for a shot at Melville," said Richard E. Pearson. "People tried to take advantage. He was an easy mark."[48]

But Melville also proved himself to be quite protective of the magazine's quality, which caused some outbursts still remembered by those who worked with him. Bill Graves, then a young writer just starting at *National Geographic,* recalled the day that he and others were summoned to their boss's office to find him on his feet, pacing around, and visibly upset. Melville closed the door and slammed the current magazine down on his desk.

"This is shit!" he said. "Shit!" He called their attention to the repetition of the word "honeybee" fourteen times in the legends for a story on that subject. "Don't I pay you good money, Bill?" he asked Graves, who had written the offending legends but dared not tell his outraged boss that synonyms for "honeybee" were very rare. "You could see the thunderclouds over his head," Graves said, recalling how Melville made each of those present feel personally responsible for degrading *National Geographic*'s standards. "But when we got up to leave, Melville parked himself by the door and didn't let anybody out without a big pat on the back and an apology: 'Now I don't want you guys to take this personally,' he told us."[49]

It was seldom difficult to gauge Melville's feelings. "He would have made a lousy poker player," said his son Gil.[50] Melville would invariably squirm and pout when he was disappointed or bored by story ideas or pictures. If he liked something, as was more often the case, he demonstrated it by slapping his knee, pounding the desk, or pulling his hair—sometimes all three. "Marvelous! Marvelous! That's marvelous!" he would shout, popping up from his seat to pace about, hands chopping the air. He darted around looking for maps, consulting old magazines, and spinning out the names of friends who might add something to the subject.[51] He thought nothing of crawling across the floor for a better view of a well-used globe that stood in the corner of his office. Indeed, it was from this position—on his hands and knees—that Melville was struck with one of his most successful ideas. Having been invited to visit Antarctica in 1960, Ted Vosburgh went to discuss the trip with Melville. "Oh, that's wonderful!" said Melville, leading Vosburgh across the room, where the two of them got down on the floor to study the globe and work out the route from Christchurch, New Zealand, to Antarctica. Because these places appeared on the underside of Melville's globe, he had to strain to make out the shadowy antipodal coastline. "That's not right," Melville said. "There should be a globe you can pick up in your hands and look at all sides." Melville called in the National Geographic's chief cartographer, explained the problem, and the organization soon produced a freestanding globe, which could be lifted from its stand like a beach ball. National Geographic immediately sold almost three hundred thousand of them.[52]

Such successes could be credited to Melville's unique combination of confidence and impulsiveness, which occasionally landed National Geographic in trouble. The *Wall Street Journal* chided the magazine in 1966 for doctoring pictures, a temptation that had become irresistible given the technical wizardry available in Melville's new photographic labs. There, in pursuit of photographic excellence, male genitalia were obscured by artfully placed foliage or loincloths; scratches and refractions were removed from aircraft windows through which pictures had been shot; rainbows were brightened for dramatic effect; and cigarette butts were erased from the floor. Technicians added color to black-and-white photographs on occasion.[53] Perhaps this tinkering represented a modern adaptation of Bert Grosvenor's impulse to prettify unsightly realities. After a few embarrassing episodes in which such alterations were made public, the magazine eventually adopted a policy of not tampering with photographs.

Melville's eagerness, combined with his strong sense of patriotism, also led him to cooperate with the U.S. intelligence services, just as *Time* and other publishers did during the Cold War. When the Federal Bureau of Investigation approached on a matter of national security in the 1960s, Melville found it difficult to say no, and the FBI was soon installed in a National Geographic office affording an unobstructed view, through a gap in the magnolias, of the Soviet embassy a few hundred feet down Sixteenth Street. From the mid-1960s through the early 1970s, gentlemen in dark suits came and went from a room on the southeast corner of National Geographic headquarters at all hours. One agent trained his attention on the embassy, while another guarded the door of room 411, which displayed a brass plaque identifying the tenants as the "Mid-Atlantic Research Committee." The room, occupied twenty-four hours a day, was outfitted with black curtains and black walls to prevent KGB agents from seeing inside the FBI outpost. National Geographic's maintenance staff was forbidden to vacuum the room or to empty its trash, which was handled by a private crew. When Joseph E. Steptoe, a thirty-five-year veteran of the National Geographic engineering staff, was asked about the FBI office, he said: "The FBI didn't have an office here. . . . But I can tell you all about the Mid-Atlantic Research Committee, whose job it was to spy on the Soviet embassy."[54]

At some point, the FBI quietly decamped. Whether the surveillance post yielded useful counterintelligence is unknown, as the bureau has never acknowledged the existence of its Sixteenth Street perch. Because the magazine's ties to intelligence services came to be common knowledge around the National Geographic Society, it is quite likely that those links were also familiar to unfriendly governments in the Soviet Union, China, and Eastern Europe, where the magazine's writers and photographers had just begun working again. This concerned young editors like Bill Garrett, who urged Melville to reconsider the organization's policy of cooperation. "Given the places we were going, it would not take much for the host countries to manufacture some evidence and throw our people in jail," Garrett said. "I told Melville we ought to cut it out."[55]

He did not cut it out. He loaned Barry C. Bishop, a member of *National Geographic*'s foreign editorial staff, to the CIA for an extended spy mission in Asia while continuing to list Bishop on the magazine's masthead.[56] Under Melville and his successors, the organization also made its personnel available for debriefings when they returned from foreign assignments, on which they collected maps, census reports, and other useful data.[57] They shared such information with intelligence experts; in turn, National Geographic got access to material, especially cartographic data, not generally available to the public.[58] These entanglements, while unacceptable by today's standards of journalism, continued the cozy arrangements the National Geographic Society had maintained with the government since its founding. As his predecessors had done, Melville invited government officials onto the board of trustees, published them regularly in the magazine, and generally treated the Society as a friendly extension of the government.

This tradition would be slow to change. But in most other ways National Geographic broke free of its Victorian roots under Melville, who fostered an unparalleled season of growth. Perhaps no observer was more delighted by

Melville's success than Bert Grosvenor, who had spent so much of his life harboring doubts about his son. By 1965, membership renewals hit 91 percent, representing the most loyal following in the industry. Revenues reached $48.3 million that year, chiefly from membership dues. Income from investments accounted for $1 million; net advertising revenues earned more than $1.5 million on sales of $6.4 million. Globes, maps, and books added $4.8 million.

In addition to these undeniable measures of achievement, the eldest Grosvenor could open each month's edition of *National Geographic* and admire the masthead, which now displayed multiple printings of the surname he had so long ago identified as the "most aristocratic and handsomest name there"— *Grosvenor, Grosvenor,* and *Grosvenor.*[59] For Bert, who had worked so hard and cared so much, that masthead constituted an uplifting sight, and Melville's unexpected success was a source of keen parental pride. Although he quietly—and presciently—disapproved of some of Melville's diversification schemes, believing that new ventures would divert attention from the Society's core business, Bert scrupulously followed his rule of deference to the new editor and president.[60] Once, as the trustees were discussing Melville's plans for the first National Geographic atlas, which Bert had always opposed, he feigned illness, excused himself, and left the boardroom to avoid undermining his son.

"*You are a genius,* and I am very proud of you, my dear son," Bert wrote in one of many effusive letters he sent to Melville.[61] Bert's handwriting, once firm and clear, had grown spidery. He was ninety. He had recently buried Elsie Grosvenor, his constant companion of more than sixty years, and had begun making plans to join her. He often visited her grave at Rock Creek Cemetery in Washington, walking among the hillside tombs of departed Bells and Hubbards with a letter in his wallet. The letter, from 1956, and deeply creased and battered from frequent study, was from Elsie: "Dearest Bert," it read. "You have been such a wonderful husband to me. I love you so very much! Your inarticulate wife, Ellie Girl."[62] Throughout their lives together, Elsie had remained strongly independent and every inch a Bell, testing Bert's patience, his need for quiet and control, his sense of order. Despite this—or perhaps because of it—

they had loved each other deeply, even as the animated Bell spirit clashed with the punctilious Grosvenor one.

When the scattered clan gathered in Baddeck for the summer of 1965, Bert, ever the detail man, called in the grandchildren to say goodbye. "I just wanted you to know how proud I am of you," he told Grosvenor Blair, whose legal career was flourishing. "My family has always meant more to me than National Geographic," Bert told his surprised grandson.[63] When that summer ended, Bert was too weak to leave Canada, where he lay in an upstairs bedroom at Point House, surrounded by boxes of correspondence and newspaper clippings. The old editor was fussing over his own epitaph, writing and rewriting, when Melville burst in on him one morning. "For heaven's sake, Father, put that away," Melville said. "It's depressing!"[64] Bert complied. He slipped away in February before finishing this last writing assignment.

His most lasting monument, more permanent than stone, would be one made out of paper, in which his writers and photographers reached out to millions of readers around the globe each month.

CHAPTER THIRTEEN
THE BEST OF TIMES

LUIS MARDEN AND MELVILLE GROSVENOR IN TONGA FOR THE CORONATION
OF KING TAUFA' AHAU TUPOU IV (*IN BACKGROUND*), 1967.

Halfway across the Pacific, equidistant from New Zealand and Panama, an open boat pulled away from the steamer *Rangitoto* and made for an island fringed by thudding surf. Luis Marden, perched on a thwart in the little boat, savored his first glimpse of Pitcairn Island, its ragged black volcanic peaks gradually enlarging out of the swells in the evening light. One of Marden's seatmates, noticing the piles of diving gear on deck, broke the spell of the moment, addressing him in the island's patois of English and Polynesian: "I heardsay you gwen dive in Bounty Bay," the man said, thumping the crate containing Marden's air tanks for emphasis.

"I am," Marden admitted.

"Man, you gwen be dead as hatchet!"

Marden hardly had time to consider this prediction before the three rocks guarding the entrance to Bounty Bay—Flattie, Mummy, and Duddy—flashed into view. Just then the boat's captain maneuvered to catch one of the big Pacific combers, and urging his oarsmen to pull away, raced his boat uphill on the back of a wave, threaded among the guardian rocks, and sluiced down to shore in a gratifying sizzle of white foam. One of the islanders came out through waist-deep water to meet the boat, for there was no beach in Bounty Bay, and invited Marden onto his back. He carried *National Geographic*'s representative ashore like a minor potentate.[1]

Thus began the two-month assignment Marden considered the most satisfying of his long career, which reached its peak during Melville Grosvenor's term as editor and president. For Marden as for others, the Melville Era, from the mid-1950s to the late 1960s, was a time of unparalleled possibility, reflecting the national confidence of those postwar years. Staffers traveled first-class, often using the "Gifts to Natives" line in their expense books; they dined at the best clubs in London and Hong Kong and enjoyed access to kings, princes, and tribal chieftains from Hunza to Borneo. One discerning editor, who chatted as easily in Mandarin as he did in French, maintained a private wine cellar at the Ritz in Paris; another fished in Generalissimo Francisco Franco's private trout river outside of Madrid. Such were the occasional perquisites of traveling for the magazine, but more often discomfort and loneliness were the field man's most frequent companions.* Those who withstood malaria were ready targets for hepatitis, dysentery, frostbite, divorce, bad rum, worse food, and eccentric medical care. One writer, confined to a seedy hospital in the Peruvian highlands, had his eye disorder treated with a poultice of pig excrement wrapped in a leaf; another, crossing the Saudi Arabian desert, patched his furiously leaking radiator with a mixture of camel dung and barley paste.[2] Writers and photographers were expected to be inventive, to absorb local customs,

---

*Most were, and still are, men.

and to come back with the deeply reported, luminously photographed articles that only National Geographic's lavish budget and expansive schedule could produce.

With Melville at the helm and money in the bank, the National Geographic Society seemed to be all over the globe and all at once. Not only did the Society contribute $64,000 toward the establishment of Redwoods National Park in the 1960s, the magazine also documented the birth of the U.S. space program, the first decade of Jacques Cousteau's underwater discoveries, and the pioneering studies of human origins by Louis and Mary Leakey, the Kenyan paleontologists. All benefited from the enthusiastic support—financial and emotional—of Melville Grosvenor and National Geographic. So too did Jane Goodall, a young Englishwoman who had no college degree, no scientific training, and no experience abroad when she first applied to the National Geographic Society for a grant in 1960. Her research project, which focused on the wild chimpanzees of Tanzania, had been dreamed up by Louis Leakey, who thought the study might offer valuable insight into the antecedents of human behavior; of course this seemed like madness to the credentialed members of National Geographic's research committee, who reluctantly agreed to finance the project under relentless bombardment from Melville Grosvenor and Louis Leakey. "Louis was a great salesman and he talked MBG into giving Jane a chance," one of the skeptics, Gilbert M. Grosvenor, recalled years later. "I thought they were out of their minds. . . . But in point of fact, MBG and Louis were absolutely right and the research scientists were absolutely wrong. . . . My father and Louis were two peas in a pod."[3] Both men, impulsive operators who shared the gift of open-mindedness, had seen some unspoken promise in Jane Goodall, whose decades of primate research would transform understanding of our closest relatives and form the model for subsequent studies by Dian Fossey and Biruté Galdikas, two other Leakey recruits sponsored by the National Geographic Society.

"Melville knew how to say yes," said Luis Marden, who found his boss to be a more than eager sponsor for his own adventures at the National Geographic Society—whether that meant diving with Jacques Cousteau, combing Madagascar for the fossilized egg of an extinct bird called aepyornis, making sound recordings of Tongan music, or probing for the remains of the HMS *Bounty*. After letting Marden's Pitcairn Island proposal ferment for several years, Melville suddenly approved his plan for a story on the *Bounty,* the armed British vessel that Fletcher Christian and other shipmates had wrested from Lt. William Bligh in the mutiny of 1789. Retreating to Pitcairn with a handful of Tahitian men and women, the mutineers had burned and sunk the ship, settled down, and sought to avoid detection, which they eluded for almost two decades. They were less successful, however, at finding happiness on their island. Their community gradually degenerated in a cycle of jealousy and violence, leaving but two of the mutineers to die natural deaths. Marden planned to document the lives of the surviving Pitcairners and to search Bounty Bay for the ill-fated ship. It was an ambitious plan to begin with, but Melville made it more so on the eve of Marden's departure in 1956.

"Oh, by the way, you're going to make a film, aren't you?" Melville asked.

"Nobody said anything about it," replied Marden, who until that moment thought that he was packed and ready to travel.

"Oh, you *must* make a film," Melville said. "You *must!*" Springing from his chair and pacing the office, Melville explained how such a film would be a natural for the National Geographic Society's lecture series. "People expect that sort of thing!" He was also beginning to think about television, a new medium that offered fresh ways for National Geographic to disseminate its message. At the very least, the networks might welcome footage if Marden made some discovery during his Pacific adventure. So with only two or three days remaining before he sailed, Marden rounded up standard movie cameras, specialized underwater movie cameras, several miles of 16-millimeter film, and other equipment, and added these to the mountain of crates he had already assembled for Pitcairn. The reciprocal of Melville's saying yes was that his subordinates were expected to do so as well.[4]

Marden was greeted warmly on Pitcairn Island, a community of 153 souls who earned their living from fishing, subsistence farming, and selling fruit and crafts to passing ships. He was shown up a long hill, past a sheltering banyan tree, and into the tidy home of Fred and Flora Christian, who would provide room and board during his weeks on the island. Despite the settlement's violent origins, it had developed into a quiet, sober community, thanks in part to Seventh-Day Adventists who had converted the whole island a few generations back. In a place where gossip was the worst crime, the island's jail stood empty until Marden arrived; he was invited to fill the two cells with his diving gear, camera equipment, and random baggage; having already researched the abstemious habits of the Adventists, he had hidden a bottle of whiskey in one of his bags. He left it there and settled into the community, which fattened him on feasts of mangoes, watermelons, coconuts, pineapples, homemade bread, and roasted goat. He roamed Pitcairn gathering notes, sailed out on fishing excursions, and took measures to ensure that he did not end up "dead as hatchet" while diving. Recruiting two young islanders, one a direct descendant of Fletcher Christian, he trained them to scuba dive in the tricky currents of Bounty Bay, where the Pacific Ocean, unobstructed for thousands of miles, made a boisterous landfall. Day after day, Marden and his assistants shrugged into scuba tanks, pulled on flippers, and tumbled in the strong currents as they scoured the bay's coral-encrusted floor for some trace of the *Bounty*. They found nothing except for a bronze rudder pintle coated with coral, a heap of iron ballast bars piled up near shore, and widespread discouragement at every inquiry about the ship. "It-sa gone," islanders told Marden time and again. "Nothing left."[5]

As New Year's Eve of 1957 settled on the island, Marden began to worry. He had a month of fieldwork behind him, two weeks before him, and little to show for having come so far. The *Bounty* might have vanished, but its curse still seemed very much alive as he trudged up the hill alone that night, entered the jail, furtively extracted his bottle, and poured himself a consoling tot. He sat in the dark on a wall, dangling his legs over the edge of a cliff, admiring a New Year's display of stars from horizon to horizon, and muttering to himself about Fletcher Christian.[6] Perhaps his grumbling found its mark. Within days, Marden and islander Len Brown began diving in a new part of the bay fifty yards offshore, far from where they had found the rudder pintle. Carefully searching the ocean floor with his face mask scraping the bottom, Marden saw what seemed to be a trail of worms. In fact, they were curved sheathing nails caked with coral. He reached above him, grabbed his colleague's hand, and pumped vigorously. The trail of nails led to a field of copper fittings, chain links, oar locks, and other bits of metal now green with age and cemented to the ocean floor by coral. Using a chisel, Marden painstakingly freed each of the objects, cleaned them up, and saw that many bore the unmistakable mark of British ownership, a broad arrow stamped into the metal. Marden had discovered the resting place of the *Bounty*.[7]

"Have found the curlew's nest," he relayed in a coded message to Washington. Given the magazine's long lead time, he wanted to protect his scoop, and he knew that Melville would understand the enigmatic cable, which harked back to the hottest ornithological news of 1948, when a National Geographic researcher had discovered the nesting site of the bristle-thighed curlew in Alaska.[8]

With his mission accomplished and story secured, Marden began the long journey home, shipping ahead a cache of movie film and photographs and beginning a memorable fourteen-thousand-word account of the *Bounty* adventure. The article, which led the December 1957 issue of *National Geographic* under the title, "I Found the Bones of the *Bounty*," attracted attention around the world, boosted the magazine's profile, and landed Marden on *The Today Show*. There, thanks to Melville's foresight, Marden was able to show film clips from Pitcairn. The initial wave of interest also produced a fateful invitation for Marden to appear with Melville on *Omnibus*, a highly regarded Sunday afternoon television program on NBC moderated by Alistair Cooke. That venue was particularly appealing because the network agreed to broadcast Marden's film, *The Bones of the* Bounty, in color. It also offered an opportunity for Melville to gauge public reaction to the programming he wanted National Geographic to produce on a regular basis.

So it was with a keen sense of anticipation that Melville joined Marden in New York on February 9, 1958, for their *Omnibus* appearance. They found the studio in the old Broadway Theater, and reported to the makeup department, where they received a thorough brushing and heavy coatings of powder to make them presentable for the broadcast that afternoon. With time to kill before air-time, Marden suggested lunch at a nearby deli and made for the door. Melville, looking apprehensive, grabbed him by the arm.

"Do you think it's all right to go out in the street all made up like this?" asked Melville, glancing over his shoulder.

"Skipper," said Marden, using the affectionate term by which Melville's colleagues addressed him, "this is probably the only place in the country where nobody would notice—it's Broadway!"

Millions tuned in that afternoon to see the National Geographic Society's first television film, with commentary by Melville and Marden, along with that of Mendel L. Peterson, curator of naval history for the Smithsonian Institution, and Harvard University's Douglas L. Oliver. All went smoothly until Cooke asked Melville to talk about the history and purpose of the National Geographic Society. Melville began with the National Geographic's founding in 1888, warmed to his subject, and was soon off on a monologue. He hardly paused for breath, having been advised by the program's director to take all the time he needed. When Cooke tried to break in, Melville ignored him. Forgetting that he was on camera, a desperate Cooke pointed to his watch and helplessly gestured to his director. Finally Cooke interrupted Melville in full flight, prompting the Skipper to hold up his hand like a traffic cop. "Now just a minute, young man," he told his surprised host. "You asked me to talk about the Society and I'm trying to tell you about it." Cooke apologized, Melville wrapped things up, and the show ended with a spectacular Nielsen rating of 19.3, which meant that the *Bounty* film, and the melting of Cooke's famous sangfroid, had been seen by 7.8 million viewers, roughly one-fifth of those owning television sets. Columnists from the *New York Times* and the *Washington Post* gleefully needled Cooke for his loss of composure, but this did little to dampen their enthusiasm for National Geographic's television debut, which the *Times* termed "compelling" and the *Post* "a commendable study of a fascinating topic."

Of course the most enthusiastic review came from Melville himself. The

Omnibus experience crystallized his intention to take National Geographic into the brave new world of television. By 1961, the board of trustees had authorized the creation of a television department, which would be headed by Robert Doyle, a Washington producer for NBC's *Huntley-Brinkley Report*.[9] Before this moment, the National Geographic Society had one main purpose—to produce the magazine. Branching out into this new venture represented a great risk for the organization, which had no broadcasting experience and was unlikely to see any financial return for many decades to come.

Melville never betrayed a moment's doubt that the plunge into television, or any other initiative on his growing to-do list, would succeed. The late-blooming executive who had spent so long in the shadows now struck friends and family as the picture of confidence. "He never really looked over his shoulder," said Gilbert M. Grosvenor, describing his father's operating style as editor and president. "I never heard him speak about what other people might think of what the magazine looked like. . . . When it all came down to the bottom line it was what his gut told him." Melville largely shunned the market testing and readership surveys just then coming into vogue, believing that his long experience of meeting with and listening to member readers provided all the audience data he needed.[10]

His instincts were seldom wrong. An impatient reader himself, Melville encouraged his writers to follow a homespun formula that became the trademark of his editorial era. This style included heavy use of the first person, a certain naïveté of viewpoint, and a reliance on vivid description occasionally bordering on the excessive. In Melville's day, a yellow warbler was not merely a bird, but a "capricious sunbeam" brightening the garden.[11] A cliff became a "frowning brow of rock."[12] A sky full of parachuting soldiers sprouted "nylon mushrooms."[13] Lakes were quite often jewels. Melville made no apologies for this prescription; his belief was that readers would be lured by flashy writing, which had to compete with the magazine's rich photographic offerings. His journal also encouraged an affable persona in its authors, who seemed quite neighborly and approachable in print, betraying no airs as they roamed the world making friends. A reviewer from *Newsweek* derided the magazine's style as the "Dear

Aunt Sally" approach to journalism—the sort of criticism that stung the younger generation at National Geographic, where Bill Garrett, Gil Grosvenor, and articles editor Allan C. Fisher, Jr., hoped to make the journal writing less clichéd.[14]

Fisher, a gifted editor who often found himself recasting articles to suit the perceived tastes of National Geographic's mass membership, became convinced that readers would support a more literary, less formulaic approach to stories. He enlisted Gil Grosvenor in his reform plan, and the pair of them proposed a readership survey to Melville. "I can tell you right now how the survey will turn out," Melville told his visitors. "Eighty-two percent of those polled will say they like the writing and do not think the style should be changed. But I see neither of you believes me. So go ahead and run the survey. It's worth $12,000 for the education of you two." When the results came in, Melville was off by two percentage points; 84 percent of those surveyed said that they liked the magazine's writing style and wanted to keep it that way. It would be years before Aunt Sally was forced to retire. Meanwhile Fisher abandoned his plans for elevating National Geographic's text, while he continued to marvel at Melville's mystic bond with his readership. "How could he predict so uncannily what the result would be?" Fisher wondered. "He could only say that he had known the Geographic and its readership so long and so intimately that he just knew the result."[15]

Melville also knew, as some people never do, that he had been favored with a second act in life. Now in his sixties, he sat at the top of a respected national institution at a time of unprecedented growth. Even more, on the heels of his disastrous first marriage he had embarked upon a richly satisfying one with Anne Grosvenor, who shared his passion for all matters pertaining to the National Geographic Society. She joined in the embassy dinners and lecture evenings so important to Melville's work and accompanied him on foreign trips ranging from Tonga (where Melville represented President Lyndon B. Johnson at the coronation of King Taufa'Ahau Tupou IV) to the Serengeti (where Louis Leakey and Anne had to restrain Melville from jumping out of their Land Rover to photograph lions).[16]

She helped entertain successive waves of visiting adventurers. To honor Sir Vivian Fuchs, the British explorer who first traversed Antarctica, Anne arranged an elaborate dinner at Wild Acres, where guests were greeted by a mannequin in polar gear, and dined on pheasant from Wisconsin. Dessert consisted of a cartographically correct cake depicting the southern continent set off by a blue sea of spun sugar. Red icing marked Sir Vivian's route across the cake. "We hated to eat it," Anne explained, "but all of us could only consume part of the Weddell Sea."[17] Anne was there for the small occasions as well, watching over her often distracted husband, whom she helped to find missing notebooks, to track down lost addresses, to police his diet, and to remember his appointments—always with a smile and a kind word, in marked contrast to Melville's first wife, who seemed to delight in his mistakes.[18]

Shortly after Anne and Melville's marriage in 1950, they built a handsome new house at Wild Acres, the sprawling Maryland farm that had served as the center of family life for most of the century. There among the flowering magnolias, forsythia, and crab apples, they welcomed a second family for Melville, a first for Anne—Edwin Stuart Grosvenor was born in 1951, and Sara Anne Grosvenor, in 1956. Like many fathers who start new families late in life, Melville took a special comfort in these youngsters. Their presence provided not only welcome stimulation ("I need young people around me!") but also the heartening thought that the Bell-Grosvenor line would endure, perhaps even into some future iteration of the National Geographic Society.

"He is a husky lad and has the Bell nose," Eddie's father wrote to a friend in 1951. "You can't mistake him. We are very pleased with his arrival on the scene."[19] As Eddie developed, it became clear that he had also inherited the driven spirit of the Grosvenors, along with the Bell tendency toward stubbornness. At age seven, he escaped the dentist's chair, sprinted up a flight of stairs, and held his mother and dentist at bay with two fistfuls of sand from the stairwell ashtrays.[20] When a nursery school classmate called Eddie a crocodile, the editor's son wrestled his accuser to the ground and beat him up, an incident that seemed to please Melville greatly.[21] "He is so full of ideas and independence it takes patience to handle him," Melville reported to his own father in 1958. "I guess he is like me in many respects!! A little of your child psychology will help me in handling Eddie."[22] Sara Grosvenor developed her own methods of humbling her brother, whose pajamas she hid in a garden urn.[23]

As Melville's new family grew, so did the fortunes of the National Geographic Society. By the mid-1960s the revenues piled up faster than even Melville could spend them. The Society paid its employees decently, maintained private dining rooms with liveried waiters for its senior staff, and kept a fleet of chauffeured cars for its officers. At Melville's urging, the organization had spent $11 million on the headquarters building designed by Edward Durell Stone and opened in 1964 to favorable reviews by Ada Louise Huxtable, then the architecture critic for the *New York Times*.[24] Less charitably but no less accurately the journalist Tom Kelly called the structure a "monument to the fact that there is no money in the twentieth century which can be so lavish as non-profit money."[25]

Successful nonprofits like the National Geographic Society had little reason to hold on to their funds, which had to be invested or spent each year. Thus there was every incentive for Melville and others planning the new building to add elaborate touches—French walnut for executive offices on the ninth floor; a four-ton camera for copying maps; an exquisite ground-floor museum with a black granite reflecting pool and a giant globe an operator could tilt and spin with electronic controls. Even the trees guarding Melville's parking lot were noteworthy—a stand of mature magnolias forty feet tall, which staff members dubbed the Grosvenor National Forest.[26]

Such displays did not go unnoticed by other publishers, who complained that National Geographic's status gave it an unfair advantage in the marketplace, where the organization's exemption from taxes, its high sales volume, and its freedom from dividend payouts allowed it to offer its magazine, books, and atlases at deeply reduced prices. "We're jealous," said Caleb D. Hammond, president of C. S. Hammond & Company, publishers of maps and atlases in Englewood, New Jersey. "Everyone in the ball game should play under the same rules." These complaints, and protests from other publishers, led the Internal Revenue Service to reassess National Geographic's tax status in the late 1960s.

Under new rules the agency put into effect at the time, the organization was required to pay tax on income from advertising sales while retaining its exemption from taxes on income and real estate. The bite might have been worse had Melville discontinued his father's practice of cultivating patrons in high places. During Melville's time, he arranged life memberships for Sen. Stuart Symington of Missouri, Lyndon B. Johnson, Jacqueline Kennedy, and Gen. John D. Ryan, commander of the Strategic Air Command, among others. At the same time, Melville assembled a well-connected board of trustees who helped steer National Geographic through this period of rapid growth, among them Laurance S. Rockefeller; Crawford H. Greenewalt, chairman of Du Pont Company; Chief Justice Earl Warren; William McChesney Martin, Jr., chairman of the Federal Reserve; Benjamin M. McKelway, editorial chairman of the *Washington Star;* Gen. Curtis E. LeMay, air force chief of staff; Juan T. Trippe, chairman of Pan American World Airways; and Conrad L. Wirth, former director of the National Park Service. Officials from the National Geographic groused about the IRS ruling and began paying the new taxes, which made hardly a dent in the organization's prosperity.[27]

The positive mood at National Geographic reflected Melville's ebullience, just as earlier decades of the magazine had been strongly colored by Bert Grosvenor's stuffiness and exactitude. In the early 1960s, the institution also absorbed the energy and optimism of a nation just stretching awake under the stimulus of the Kennedy administration. Brimming with confidence and a belief in their destiny, Americans were going out into the world to share their knowledge as Peace Corps workers, to explore the frontiers of science, and to install new governments in Africa, in Asia, and in other regions deemed important to U.S. interests.

In the midst of this surge of national pride the American Mount Everest Expedition aimed to put the first Americans on the 29,035-foot summit in 1963. Although assaults on Everest have become commonplace, there was nothing ordinary about the expedition Norman G. Dyhrenfurth had in mind for

National Geographic that year. The proposal from Dyhrenfurth, a native Swiss with extensive climbing credentials, came at a time when only six humans had stood at the top of Everest and lived to tell about it—Sir Edmund Hillary and sherpa Tenzing Norgay in 1953, followed by four members of a Swiss party in 1956. Dyhrenfurth's American team would try to match that record, while also sending detachments to the neighboring peaks of Lhotse (27,890 feet) and Nuptse (25,850 feet) to complete an unprecedented Himalayan triple slam. As if this were not enough, Dyhrenfurth also developed an ambitious scientific program in consultation with National Geographic's Committee for Research and Exploration; expedition scientists would conduct solar radiation studies on Khumbu Glacier, assess the physiological impact of high-altitude work, and examine the psychological dynamics of teamwork on the mountain.[28] This research plan attracted the interest of NASA, the National Science Foundation, and the Office of Naval Research, all of whom agreed to support the expedition. So did the National Geographic Society, which had been lured by the prospect of adventure and by the solid scientific enterprise. But it also helped that the Everest story would have high entertainment value off the printed page, a new consideration that emerged during Melville's era; this would become increasingly important to his successors.[29]

Dyhrenfurth dangled two particular inducements before Melville Grosvenor. First, he reserved a spot on the climbing team for Barry C. Bishop, a capable young photographer with mountaineering experience who had recently joined the staff of National Geographic; second, Dyhrenfurth proposed to make a documentary film about the expedition, which he was uniquely qualified to do, for in addition to his climbing background, Dyhrenfurth was a producer who taught filmmaking at UCLA. Given the positive reaction to National Geographic's Everest article of 1953, Melville knew that the story of this expedition would find an enthusiastic reception. And since the Society had just established a television department with the express purpose of creating a regular documentary series, Everest offered the perfect opportunity to launch it. Thus, under Melville's prodding, National Geographic weighed in as the primary sponsor in May of 1962, which helped Dyhrenfurth to attract other contributors. Melville's commitment, which eventually ran to $175,000, paved the way for the most ambitious, best-equipped, and biggest group yet to lay siege to the mountain.

When the expedition set out from Kathmandu in the spring of 1963, it took a thousand porters and sherpas to carry their equipment to base camp. There, almost thirty tons of gear was marshaled for the final assault, which would be supplied with new tents, sleeping pads, Eiger boots, reindeer fur boots, 12,000 cigarettes, 216 bottles of oxygen, portable laboratory equipment for blood sampling and urinalysis, drilling tools for collecting ice, and waterproof diaries for recording moments of psychological stress. The party carried seven movie cameras, 28,000 feet of movie film, and the first freeze-dried food on Everest, including crab imperial, pork chops, and chicken tetrazzini. Because *National Geographic* was involved, each of the twenty team members carried a new Nikon-F camera, winterized to withstand temperatures of minus twenty-two degrees Fahrenheit; each was also issued an ice axe with a screw-in camera attachment for steady shooting.[30] One could see that this group was different— dressed in pastel blues, in bright saffron, and in burnt orange, not primarily for safety but for the visual boost it would bring to television and magazine coverage.

This last touch was Dyhrenfurth's idea. Cool under pressure, steady in his optimism, and possessed of high energy, Dyhrenfurth combined a Hollywood promoter's charm with a Teutonic eye for detail. His carefully selected Everest team was the highest-educated one yet, with four Ph.D.s, three M.D.s, and four master's laureates in its ranks, mainly to satisfy the scientific objectives essential for selling the expedition to National Geographic. But he also made sure there were hard-core climbers in the team who stood a good chance of reaching the top, among them Jim Whittaker, a seemingly indestructible Mount Rainier guide with many seasons under his belt; William F. "Willi" Unsoeld, a former Tetons guide and a graduate of several Himalayan expeditions; Thomas F. Hornbein, a Seattle anaesthesiologist with mountaineering credentials from the unforgiving heights of Pakistan and Alaska; and John E. "Jake" Breitenbach, a tough young Wyoming guide and veteran of Mount McKinley.

But even the best-staffed and equipped of expeditions remained vulnerable to the imperatives of nature, as members of the American Everest team learned that spring. While laying a route through the jagged terrain of Khumbu Glacier, a contingent of five men was surprised by an avalanche that came roaring down the mountain, scattering the party, injuring a sherpa, and leaving young Jake Breitenbach buried under tons of ice, rocks, and snow. Although Breitenbach's colleagues dug furiously to find him, his body could not be recovered.* At twenty-seven, he was one of the youngest members of the group.

The expedition shook off this tragedy of March 23 and pressed on, edging up toward the high camps through harsh weather, through arguments over their objectives, and through growing misgivings about the advisability of the grandiose plan for a triple slam.[31] More than a month would pass before Jim Whittaker and sherpa Nawang Gombu trudged through a ground blizzard to the top of Everest, where they became first to plant the Stars and Stripes, along with the National Geographic flag, on May 1, 1963. Dyhrenfurth, accompanied by Sherpa Ang Dawa, had attempted to follow the summit team that day. But, burdened by a movie camera, tripod, two still cameras, and film, and faced with a dwindling oxygen supply, Dyhrenfurth made the decision to turn back just eight hundred feet short of the top, a courageous call that bewildered his climbing partner.

"Up go, Bara Sahib?" asked Ang Dawa.

"No, Ang Dawa," Dyhrenfurth said, shaking his head. "Down go."[32]

As the spring weather grew warmer, increasing the dangers of avalanche, the expedition finally abandoned its three-summit plan and sought to deliver as many team members to the crown of Everest as time allowed. Tom Hornbein and Willi Unsoeld would pioneer a difficult new climbing route via the West Ridge, reaching the summit on May 22. On that same day, teammates Lute Jerstad and Barry Bishop climbed to the top by the traditional South Col course Hillary and Whittaker had taken. Hornbein and Unsoeld came down by the South Col, which established the first traverse of Everest. This was but one of

---

*Jake Breitenbach's body was found, heaved up by the ice, by Japanese climbers six years later (the *Washington Post*, Oct. 8, 1969).

the mountaineering records the Americans set that spring. They got the most people to the top in a single expedition (six) and the most in a single day (four). But what remains most notable is that the four men who reached the summit on May 22 lived to tell about it.

Traveling in two teams, they arrived on top very late in the day, so that all four climbers—Barry Bishop, Lute Jerstad, Willi Unsoeld, and Tom Hornbein— left the upper reaches as darkness settled on the mountain. Exhausted from weeks of living at high altitude, wrung out from the day's exertions, and running low on oxygen, the four could not make it back to the safety of Camp VI a few hundred feet below them. They were forced to spend the night at 28,000 feet with no food, no tents, and no sleeping bags as temperatures reached eighteen degrees below Fahrenheit.

Settled on a narrow rock ledge with his colleagues, Lute Jerstad did not recall sleeping at all, but made a stubborn effort to stay awake. He stuffed his mittened hands into his armpits and knocked his feet together to keep the blood flowing. "As grimly as I had vowed to reach the summit," he said, "I now felt an indescribable drive to live through that night."[33]

His climbing partner Barry Bishop remembered the creeping cold. "My feet, which while I was still moving, had been growing colder to the point of agony, had now lost all feeling, and the tips of my fingers were following them. We curled up in our down jackets as best we could. Then after a while I was lying dazedly on my back, with my feet propped up like two antennae, wondering, almost too weary to care, how badly they were damaged. I tried to wiggle my toes but felt nothing."[34] Frostbite was setting in.

By pure chance, it was a rare night of calm weather on the mountain, which rattles with high winds through most of the year. Perhaps to their own surprise, all four explorers were there to watch the sun edge over the snow peaks the next morning, which spilled a most welcome light up South Col and up the long white slopes to the crown of Everest itself, which glowed benignly in the clear morning air. Bishop thought to get his camera, remembered his numbed hands, and simply admired the new day, which would prove to be an agonizing one for him and for his colleagues. Beginning at first light, the four men stumbled down the mountain, met a rescue party that had turned out to intercept them, and worked their way painfully down from the airless heights, finally straggling into Advance Base Camp in the dark—at about 10:30 P.M. on May 23.

In the course of that long day, the severity of the frostbite suffered by Bishop, Jerstad, and Unsoeld revealed itself. Working down through the high, frozen country, the wounded climbers walked unsteadily but without pain. With blood now rushing back to their damaged extremities, their feet began to thaw and feeling returned, and with it came an excruciating pain that seared up through the soles and touched something deep within the brain. (For some reason, Hornbein alone was uninjured.) The worst pain came the next day, May 24, as they made their way down through the Khumbu Glacier, now shifting with the warmth of spring. The jagged trail disappeared and reappeared, heaving up blocks of ice and rock. The wounded party hobbled along, stubbing their feet on the broken trail and singing out in agony. Watching his friends from behind, Hornbein wept. When they emerged from the glacier at base camp, it was clear that Bishop, Unsoeld, and Jerstad could make it no farther on their own. They were carried down to Nanche Bazaar, eighteen miles distant, by relays of sherpas working in twenty-minute shifts; first in baskets, with the wounded riding backward, affording them a jostling view of Everest diminishing behind them; then piggyback style, looking forward to the brown hills of Nepal.[35]

Neither the mountain nor the men who had conquered it would ever be the same. Everest was littered with the remains of tents, oxygen bottles, broken boxes, old socks, bits of tin foil, beer cans, and other jetsam from the heavily supported, thousand-strong army the National Geographic Society had unleashed on the mountain. An unintended legacy of this amply reported expedition—which produced miles of movie film, more than twenty thousand color photographs, and voluminous text—was to make the mountain seem less special, more approachable, somehow smaller. Before ten years had passed, larger expeditions would swarm over Everest, straining to find yet another first for the cameras and record books. A thirty-four-member Japanese team landed in Nepal in 1970 to film Yuichiro Miura rocketing down South Col on skis at a speed of 150 kilometers per hour and braking his descent with parachutes.[36] From there it was a short jump to the tragedy of 1996, when a combination of harsh weather, ill-prepared climbers, and crowding on the summit led to twelve deaths in one month, as chronicled by Jon Krakauer's *Into Thin Air*.[37] With the passing years, respect for Everest had degraded to a point succinctly described by Sir Edmund Hillary: "It's all bullshit on Everest these days."[38]

But in 1963 the achievement was real, and National Geographic made the most of it. Melville Grosvenor arranged a Hubbard Medal for the American Everest team. President John Kennedy agreed to present the medal in a rose garden ceremony on July 8. Barry Bishop, who eventually lost all of his toes and the tips of his little fingers to frostbite, was wheeled in for the ceremony, which was attended by most members of the team, including five sherpas who had been flown in for the big day. Only Willi Unsoeld, still hospitalized for frostbite and hepatitis in Kathmandu, was unable to attend; Jake Breitenbach's widow came to see Norman Dyhrenfurth receive the gold medal from President Kennedy. In turn, Dyhrenfurth presented Kennedy with an American flag Bishop and Jerstad had carried to the summit. Someone helped Bishop out of his wheelchair to stand for a photograph, which shows Melville Grosvenor beaming at the president, Bishop squinting in the bright July sun and looking painfully thin, with fellow members of the Everest team arrayed behind him. Kennedy, under the usual strains of presidential pressure, must have been grateful for that day's diversion, which gave him the opportunity to celebrate mountaineering as "a special form of the vigorous life" and to praise the Everest party for reaching the "far horizon of experience."[39]

That summer Bishop entered a cheerless rhythm of visiting doctors, working hard, and drinking deeply. "Liquor sure helps frostbite," he wrote a friend. "So far I have not found a more enjoyable anaesthesia for my painful feet."[40] His magazine deadlines kept him busy sorting through rafts of notes and thousands of photographs for his article, "How We Climbed Everest," which appeared as the cover story for the 75th anniversary issue in October 1963, along with pieces by Dyhrenfurth ("Six to the Summit") and by the Hornbein-Unsoeld duo ("The First Traverse").[41] In writing to friends, Bishop put a brave face on his trauma, which he dismissed as a "minor inconvenience. . . . I know that I have had my moment of truth in climbing the old girl," he wrote. "Had somebody told me prior to making the attempt . . . that I would suffer this injury, I would have still proceeded as I did. Yet one doesn't repeat it."[42]

Within months, however, Bishop was ready not only to repeat the adventure but to enlarge upon it. Over drinks in Washington with Gen. Curtis LeMay, a trustee of the National Geographic Society and U.S. Air Force chief of staff, Bishop cooked up a high-risk, high-altitude spy mission to India, which might have been devised by Ian Fleming: Bishop proposed to recruit and train a team of American and Indian mountaineers to carry the pieces of a lightweight nuclear-powered surveillance device to the top of Kanchenjunga, India's tallest mountain (28,146 feet) to monitor nuclear testing across the border in China. Once on the mountain, the four pieces of the tracking system would be assembled and a small nuclear generator known as a SNAP unit (System for Nuclear Auxiliary Power) would be inserted in the device to collect signals from China.[43] At a time when satellite monitoring was not yet effective, such ground-based surveillance was considered essential to national security, for China had just completed its first successful nuclear test on October 16, 1964. This had set off alarms at the highest levels of the U.S. government. The National Security Council approved the Bishop-LeMay surveillance plan, which the Central Intelligence Agency took charge of planning and implementing. Barry Bishop stepped in to recruit and train members of the U.S. climbing team, including his old Everest partner Lute Jerstad, who had recovered from his frostbite with all toes and fingers intact.[44] Preparation for the mission, known as Operation HAT (High Altitude Test), began late in 1964 with visits to CIA headquarters at Langley, Virginia.[45] Both Melville Grosvenor and Melvin M. Payne, the executive vice president of the National Geographic Society, agreed to loan Bishop to the CIA for eighteen months, beginning in January 1965.[46]

When this initial phase of planning and recruitment was done, Bishop led his handpicked team of six U.S. climbers to Elmendorf Air Force Base outside of Anchorage, where the group stayed in the officers' club, learned to assemble the monitoring device, played furious games of volleyball while waiting for the weather to clear, and finally honed their mountaineering skills on nearby Mount McKinley—all under intense cover of secrecy. "It was all very cloak-and-dagger," recalled Rob Schaller, a Seattle surgeon who signed on for the mission, was assigned the pseudonym Norris P. Vizcaino, and was paid $1,000 a month by the agency.[47] By the time the expedition actually began in September 1965, the objective had shifted from Kanchenjunga to Nanda Devi, a somewhat less formidable but no less strategically situated Indian peak at 25,645 feet. Bishop's role had become advisory rather than active, perhaps because his injured feet restricted his ability to climb. Although he remained on loan as a consultant to the CIA, the mission went forward without him—and it proved to be a disaster.[48]

A small contingent of American and Indian climbers quietly began their ascent of Nanda Devi that autumn, carrying the monitoring equipment and generator in five cardboard boxes; it was easy to identify the box containing the generator, which felt warm to the touch.[49] The expedition went without incident until mid-October. As the party approached the summit, a severe storm swept down on the group, forcing them off the mountain. During this retreat, expedition members stashed all five boxes on a ledge at 23,750 feet, secured the cargo with heavy ropes and pitons, and felt certain that the cache would remain safe until conditions permitted a return in the spring. Yet when the CIA fielded a follow-up expedition in June 1966, team members were shocked to find that an avalanche had sheared off a corner of the mountain, sweeping the stash of surveillance equipment down a glacier feeding the Ganges River.* All that remained on the mountain were a few wires and one battered box containing a receiver.[50]

Indian authorities shut off access to the area and the CIA sent technicians to continue the search, which proved futile. Nervous Indian officials began to monitor levels of radioactivity in the river, a source of drinking water for much

---

*The CIA mounted a third expedition to the region in 1967, when it successfully planted a monitoring device near the summit of Nanda Kot, a sister peak to Nanda Devi. The device was in service for a year.

of the subcontinent and a waterway sacred to Hindus throughout the country. No appreciable surge in radioactivity was reported; none of the seven plutonium capsules from the fuel pack was ever found; and the strange adventure of Nanda Devi lapsed into silence for more than a decade.[51]

It sprang back to life in 1978, with an investigative report by Howard Kohn in *Outside* magazine. Kohn's article, "The Nanda Devi Caper," sparked a political furor in India and a ripple of follow-up articles in the *New York Times*, in the *Washington Post*, and in the lively Indian media, which had a field day with the story. The U.S. ambassador was summoned to India's Foreign Ministry for an explanation, which did little to ease strained relations between the countries.[52] Under sharp questioning in the Indian parliament, Prime Minister Morarji Desai fended off attacks from opposition parties but admitted that the Nanda Devi project had been approved at the highest levels of government—although the CIA still refuses to confirm that it ever took place. Desai promised a full investigation and assured citizens that the missing nuclear material posed no threat to public safety.[53] This was most likely the case, according to a commission of scientists who investigated at the government's request. Even if all 1,734 grams of plutonium from the SNAP device had been dumped directly into the river—which is very unlikely to have happened—the radioactive material would have been rendered harmless by the Ganges, which produces a flow far in excess of the five thousand gallons a day needed to dilute such a pollutant, according to the panel: Furthermore, the commission noted, the generator's plutonium 238 was not fissile, and was thus incapable of producing the sort of explosive chain reaction that occurred at Chernobyl.[54] With time, the Nanda Devi controversy faded away.

Meanwhile, the man who started it all returned to civilian life. Barry Bishop earned a Ph.D. and settled into a long, uneventful tenure that stretched over thirty years at National Geographic, where his Everest boots were enshrined in a glass case and he served as secretary of the Committee for Research and Exploration, eventually assuming the chairmanship of that panel. On Melville Grosvenor's order, the sparse evidence of Bishop's espionage career was locked away in a company safe, so that few of Bishop's colleagues ever knew about his secret life.[55] Nonetheless his earlier achievement in the Himalayas was enough to make him a figure of admiration at the office; he was the burly man with the elaborate Old World manners who wandered the halls with an odd, jaunty limp.

Melville Grosvenor's gamble on Everest not only launched Bishop's brief espionage career, but also provided the National Geographic Society with a new way to tell its stories. Dyhrenfurth's treatment of the 1963 expedition exceeded expectations. Despite the harsh conditions of that spring on the mountain, he returned to Washington with masterful footage of struggling climbers, of snapping flags and howling winds, of the shock of a young colleague's death, of triumph on the summit. This summit footage, wobbly and unfocused though it was, constituted the first motion pictures from the top of Everest. Orson Welles narrated the hour-long film. Welles's hiring was controversial—because of the expense involved (a $90,000 fee) and because the prospective narrator had exiled himself to Europe, where he was ducking the Internal Revenue Service, which was eager to collect back taxes from him. Moreover, he was traveling with Rita Hayworth. Welles eventually agreed that National Geographic could pay his fee directly to the IRS. This allowed him to return home with no fear of arrest.[56] The great actor rose to the occasion, delivering just the right blend of drama, elegance, and understatement for "Americans on Everest," first broadcast by CBS in 1965. The film, seen by 26 million viewers, won favorable reviews and inaugurated the National Geographic Society's first documentary series, which won a George Foster Peabody Award that year. That series, consisting of four hour-long specials per year, would move to the Public Broadcasting Service and continue for more than thirty years, earning regular Emmys and wide recognition.

Melville's willingness to strike out in these new directions had been inspired by his Grandfather Bell, who had often encouraged his understudy to "leave the beaten track occasionally and dive into the woods."[57] Melville had done that with his campaign of rapid diversification and growth, which cer-

tainly broadened the reach of National Geographic. But his modernization program also had the unintended effect of diluting his family's influence. By the
time he was ready to retire in 1967, the organization had developed into such a
complex operation, with multimillion-dollar budgets, a staff of nearly two
thousand, and projects around the world, that managing it effectively seemed
beyond the reach of a single individual. This, at least, is the argument Melvin
M. Payne put forward when he moved to take control of National Geographic.

CHAPTER FOURTEEN
FACING LIFE

GILBERT M. GROSVENOR STANDING IN CONTROL CENTER, APPLAUDED BY
MELVILLE GROSVENOR WITH TED VOSBURGH (*EXTREME RIGHT*), 1967.

An atmosphere of determined calm, encouraged by soft gray carpet, burnished walnut paneling, and dim lighting, permeated the complex of executive suites on the ninth floor of the National Geographic building, where the ambient noise dropped a few decibels and a discreet notice on the stairway admonished visitors from below to wipe their feet before entering. So it was that the soberly dressed men and women emerging from the stairwells and elevators on August 1, 1967, behaved with all due reverence, like worshipers filing into a sanctuary. They spoke softly, made their way across the marble lobby, and found their seats in the Control Center, a narrow room where the beige curtains were

drawn tight against the sunlight. This ominously named chamber, tucked behind a heavy bronze door at one end of the hall, existed for a benign purpose:
It was here that the Executive Editorial Council, a group of twenty or so senior
magazine editors, met regularly to plan future issues. On this wilting summer
day, however, the editors gathered not to discuss schedules but to witness a rare
tribal ritual, the orderly transfer of power from the Grosvenor family to their
designated successors.[1]

After an exhilarating decade as editor and president, Melville Bell Grosvenor,
sixty-six, was stepping down, true to his word, to make way for a new generation of leadership, although it could hardly be called a younger one. Effective
immediately, his two jobs—one editorial, one administrative—would be split
between Ted Vosburgh, sixty-three, who would replace Melville as editor, and
Melvin M. Payne, fifty-six, who would take over as president and chief executive officer. Despite Melville's reservations about splitting the duties of editor
and president, he reluctantly acceded to the change, which Mel Payne, the chief
beneficiary of the new arrangement, convinced him was not only necessary but
unavoidable. No single individual had the skills to do both jobs, Payne had argued: Vosburgh could handle the editorial duties, but he had no administrative
experience; Payne had proven himself an able executive, but he lacked an editorial background. Melville's son Gilbert M. Grosvenor, thirty-six, was considered too young and green to step into either of the top jobs. So for the first time
since 1920, when Gilbert H. Grosvenor had claimed the joint duties of editor
and president, the jobs were split. The family gave them up with the hope, but
with no assurance, that Gil or another member of the clan might recover them.[2]

Under new arrangements approved by the trustees, Melville would become
chairman of the board and editor in chief; the former was an honorary title created for his father, the latter an honorary one brought back for Melville, with
neither position entailing operational responsibilities. "Harry Luce in his later
years operated much the same way as I expect to . . . horizontally rather than
vertically in the echelon," Melville wrote his friend William O. Douglas, the
Supreme Court justice, that year.[3] The last piece in the realignment involved

Gil Grosvenor, now in charge of long-range planning in the Control Center. He had spent his thirteen-year career training for greater responsibilities—writing and shooting on foreign assignments and doing brief stints in various departments, from picture editing to lectures, from the business office to fulfillment operations, from production editing to the Control Center—all to give him the broadest possible understanding of an increasingly complex organization. Under a deal worked out between Melville Grosvenor and Mel Payne, Gil would be seconded to Vosburgh as associate editor and vice president, placing him in position for the editor's job when Vosburgh's interregnum ended in two or three years.[4]

For Gil and other magazine staffers gathering in the Control Center on that August morning, their new editor was a known quantity. Vosburgh, a veteran of thirty-four years at National Geographic, was the steady, decent man who scrutinized every headline, paragraph, phrase, and semicolon written for the magazine; he made sure that no mistakes or vagaries of expression appeared in the journal. He once stopped the presses—at a cost of $30,000—to replace a restrictive comma inadvertently dropped from an article.[5] "When he edited something," a colleague recalled of Vosburgh, "he really edited it right down to the last comma, period. There used to be a saying around here, if it got by Ted, it was all right."[6] Throughout Melville's frenetic and occasionally disordered reign at National Geographic, it was the detail-oriented Vosburgh who stayed late, pored over proofs, and grilled writers so remorselessly that one of them complained that Vosburgh had "an instinct for the capillary."[7] Nonetheless, he gathered up loose ends, got the magazine to press, and provided occasional reality checks for the high-flying Melville. "Ted was the man who held onto the tether of Melville's magnificent airship," said Bill Garrett, who worked closely with both men.[8]

Like Vosburgh, Melville's replacement as president was a cool customer who appeared to be the antithesis of his predecessor. Although Mel Payne had worked at National Geographic since 1932, he remained something of an enigma to colleagues. By turns charming and icy, he was the sort of man who might stare through you instead of answering a question, just to make you repeat it.[9]

He was known for his exceedingly cautious style, his business acumen, his devotion to hard work, and his unshakable love for the National Geographic Society. Of humble origins, he was orphaned at age ten and he had been scrambling ever since, trying to advance in a world where, he told friends, "every hand was raised against me."[10] Upon graduating from high school in Washington, D.C., he had begun work at National Geographic as a secretary, taking shorthand and typing letters for Thomas W. McKnew and other executives in the business office. At the same time he earned a law degree by attending night school and early morning classes. As McKnew moved up the corporate ladder, Payne rode his coattails, eventually becoming executive vice president, the number two administrative position. As a young man, Payne had been approachable and friendly with colleagues, but as he advanced through the ranks, he gradually withdrew from such relationships, perhaps in the belief that they might compromise his position.[11] Even those who admired Payne found it hard to get close to this terse man, who grew more taciturn with age, wasting neither money nor words. What few words he did speak could be quite blunt, as the magazine's senior editors learned in their initial meeting with the new president on August 1, 1967.

"As of today," he told the editorial group in the Control Center, "the ten golden years are over."[12] He let the chill sink in. Then he announced that the National Geographic Society could no longer afford to be so freehanded. Payne foresaw the remarkable growth known under Melville Grosvenor coming to an end. To prepare for that decline, Payne was going to insist upon a more businesslike culture within the organization. He imposed a cost-accounting system for the first time in National Geographic's history. He scrutinized printing contracts, kept track of paper clips, and told the staff to cut back on newspaper subscriptions. He served notice that he would crack down on expense accounts, which had been generously given and lightly policed during Melville's regime.[13] Payne added a page of advice in the pocket-size expense books writers and photographers carried on assignment; these tips bear repeating not only for the interest they may hold for frugal travelers today, but also for the insight they offer into the flinty mind of Melvin M. Payne:

Although most employees returning from field assignments submit their expense accounts promptly, others procrastinate to a degree that it is sometimes necessary to intervene officially. This is never done when the delay is due to the pressure of publication dates. However, such delay is caused by the failure to maintain the necessary data while in the field and the reconstruction of expenditures is not only arduous and time-consuming but it is also inaccurate. This sometimes results in a personal financial loss.

I have found that the simplest way to avoid this situation is to review and enter all my expenditures at the end of each day, matching receipts to the items entered on the daily record by parallel numbering, and then keeping them safely in a clasp-type envelope.

If I am reasonably satisfied that I have entered all my expenses, I do not run a total for the day in every case. However, upon returning to the office, because everything is current it is a matter of something less than an hour to total the account and balance expenses against the advance.

By keeping this record up to date on a daily basis, not only will you be able to turn in your account properly upon your return to the office, but you will protect yourself against a personal "deficit" payment. . . .

    MMP

After the initial shock of Payne's pronouncements, his austerity program proved to be more symbolic than substantive; except for the minor annoyance over expense accounts and a new budgeting process, the magazine's resources continued to grow, and so did memberships, despite the new president's warnings.[14] Perhaps Payne's comments were meant to establish dominance over his new charges, and to draw a contrast between his style and Melville Grosvenor's seat-of-the-pants management approach, which had been a constant worry for Mel Payne.

"Payne thought that Melville Grosvenor would drive us into bankruptcy," recalled Bill Graves, an editor who knew both men. "And Melville tried to do that every day. But guess what? The more we spent, the better it got! We couldn't get rid of the money fast enough."[15] Twenty years after Payne's gloomy forecast, the magazine's circulation had doubled again, reaching almost 11 million and providing most of the National Geographic Society's revenue. "The magazine was growing at a great rate," Gil Grosvenor recalled of the late 1960s. "Books were selling. Everything was coming up roses. But Mel Payne felt it was going to stop. . . . To an extent he was right—but way ahead of his time. He wanted to go down in the history of National Geographic as the person who saved the Geographic from itself."[16]

Saving the National Geographic Society meant saving it from the Grosvenors, in Payne's view. "Mel Payne *hated* the Grosvenors," said Richard E. Pearson, a veteran of three Grosvenor administrations at National Geographic. "I never understood it. The Grosvenors put him where he was. Maybe it was out of resentment. . . . The family was like royalty here."[17] It was true that Payne, the orphan who eventually made good, seemed to harbor an antipathy toward the founding clan, which had been blessed by advantages that he would never know. From his youth Payne had learned how unkind the world could be, whereas Melville Grosvenor's innocence of that world could be nothing short of breathtaking. Once, when he encountered Ralph Gray, one of his editors, at a summer picnic, Melville noticed that he was licking a piece of colored ice.

"Ralph," Melville asked. "What's that thing in your mouth?"

"That's a Popsicle," Gray replied.

"What's a Popsicle?" Melville asked.

"Well," Gray said, "I guess you just freeze some Kool-Aid."

"What's Kool-Aid?" Melville asked.

This circular conversation continued until the object of Melville's curiosity melted, leaving Ralph Gray with an indelible memory of an otherwise forgettable afternoon. "You have to realize you're on a different plane . . . when you talk to Melville," Gray explained later. "He doesn't know the world."[18]

Despite their contradictory perceptions of the world, Melville Grosvenor and Mel Payne had established a workable—if not always smooth—relationship at the office. In part, this was because Payne handled a myriad of administrative details for Melville, which left him free to live large and dream big. Payne also kept his resentments well concealed as long as he worked for Melville. And he grudgingly admired Melville's editorial flair and his capacity for inspiring loyalty in those who worked for him. For his part, Melville must have known of his colleague's antagonism, but he was not threatened by it; indeed, Melville often came blustering into Payne's office to tell him how much he treasured his friendship. One day he pulled up a chair right next to Payne's and confided: "You are just like a brother I never had."[19]

Such displays might have narrowed the distance between Melville the Romantic and Melvin the Pragmatic, but they did nothing to ease the way for Gilbert M. Grosvenor, who seemed to attract Mel Payne's enmity from the day he began work at National Geographic in 1954. Given Payne's background and distaste for nepotism, his wariness of Gil, while not commendable, was understandable: Gil had come straight to the magazine from Deerfield and Yale with no experience; he had instant access to the organization's top brass; he received one good assignment after another; and he raced through the hierarchy, replacing his grandfather and namesake on the board of trustees in 1966. Thus at age thirty-five, he found himself sitting across the table from Payne and other graybeards in the boardroom. It was too much for Payne.

"He just did not believe that Gil should be working at National Geographic," said Bill Graves. "It was the silver spoon problem."[20]

Gil understood Payne's resentment better than anyone. "He was self-made," Gil recalled. "I think it disturbed him that someone could come into the organization and start with a leg up . . . someone who had a direct link to the decision makers. That always bugged him. The [Grosvenor] name bugged him."[21]

It bugged Gil too. He told friends that his name was a burden at work, where he felt that colleagues judged him on the basis of family connections instead of ability.[22] From the other direction, he agonized over the weight of expectations placed upon him by his family. As their fifth-generation representative at National Geographic, he knew that the family was measuring him against the brilliant Alexander Graham Bell, the respected Gilbert H. Grosvenor, and the lovable Melville Bell Grosvenor. How would this youngest Grosvenor fit? The family was watching to see, just as they had gathered on the shore at Baddeck with binoculars one summer to gauge how Gil handled the *Lady Anne*, his father's custom-built yawl and something of a family icon. "They're all wait-ing for me to screw up," Gil told friends as he took the helm, glanced nervously toward land, and hurried the boat out of sight as fast as he could.[23]

"Gil was a very unhappy young man," said Gifford D. Hampshire, a fellow picture editor who befriended the young Grosvenor in his early days at National Geographic. Still bruised from his parents' divorce and somewhat estranged from his father, Gil wrestled with self-doubt and conflicting loyalties, wonder-ing whether to stay at the magazine or to strike out on his own. "He knew very well that he had accidentally become the inheritor of the magazine—well, really the whole National Geographic Society—if he wanted it," Hampshire said. "He was troubled about whether to take on that responsibility."[24]

Hampshire, who assumed the role of elder brother and sympathetic lis-tener, gave Gil some advice: "If you stay," he told his young friend, "you will spend the rest of your career . . . surrounded by a bunch of sycophants who tell you what they think you want to hear and a minority of others who are skepti-cal of you or disloyal to you just because you are a Grosvenor." Hampshire ad-vised Gil to leave National Geographic, to get experience at a newspaper or another magazine, and to make a name for himself independent of the family. Then he could return if he chose. Unbeknownst to Hampshire, this was the same advice Gil got from his brother Alec, who had turned down opportunities at National Geographic in favor of a military career. Gil ignored them both—in part because he loved the brand of photojournalism then emerging with the new generation Melville had recruited, in part because he did not wish to dis-appoint his grandfather, who was keen on Gil's entering the family business.[25] And so Gilbert M. Grosvenor stayed for the next fifty years.

In that time, he would watch the organization broaden its reach and fill its bank account to unprecedented levels; plunge into a circulation nosedive; add millions to its foreign-language readership; endure the convulsions of budget cuts and large-scale downsizing; suffer through a string of failed succession plans; raise a new headquarters building; put geography back into American classrooms; launch and abandon new publishing and broadcasting ventures; form partnerships with Ted Turner and Rupert Murdoch; farm out its mapmaking functions to a commercial publisher; invest in a string of struggling IMAX movie theaters; and put its name on a line of snow globes with little plastic tigers inside. Through all of this, Gil would be drawn into a succession of leadership struggles that threatened his family's legacy and tested the depths of his resolve. Had he taken the advice of Alexander Graham Bell Grosvenor and Gifford Hampshire, he might have had an easier life of it, but not a more stimulating one.

This self-conscious young man was making the rounds of his training regimen in 1960 when he drew an assignment in Eckington, the Washington neighborhood where the National Geographic Society housed its membership service and fulfillment operations. There, surrounded by occult files of mailing lists and Address-O-Graph machines, he met Donna Kerkam, a recent graduate of Sweet Briar College working as a clerk.[26] From a prominent Washington family, she was supremely self-assured, attractive, and a provisional member of the Junior League. She caught Gil's eye, the two began dating, and engagement announcements soon appeared in the society pages of the *New York Times,* the *Washington Post,* and the *Washington Star.*[27] Married in 1961, Gil and Donna began a furious regimen of embassy dinners and social receptions; at the same time they formed lasting friendships in the tight photographic community associated with National Geographic. The couple received occasional notice in the social pages, which never failed to mention Gil's association with National Geographic and Alexander Graham Bell. Before long Gil and Donna added a new generation to the family line—Gilbert Hovey Grosvenor II, born in 1970, and Alexandra Rowland Grosvenor, born in 1973. Donna shared Gil's enthusiasm for globe-trotting, and accompanied him on magazine assignments to Bali, Ceylon, and Monaco, earning bylines with him. Outspoken and opinionated, Donna strongly identified with the environmental movement then in its nascent stages, telling one interviewer that she believed if God returned to Earth, he would come back as a blue whale.[28] She also made it clear that she did not expect to spend all of this life as an unpaid model for *National Geographic.* "I was tired of wearing an orange dress and standing in the ruins for the spot of color in Gil's photos," she told the *Washington Post.* She took crash courses in photography from National Geographic's professionals, furthered her studies at the University of Missouri School of Journalism, and before long she was receiving her own assignments from *National Geographic*.[29]

When Gil began work at the magazine, his father remained a distant, busy figure; he was ready to help his son in a pinch, but he could not provide the steady, almost daily, flood of advice and encouragement that Professor Edwin Grosvenor had lavished upon the young Bert Grosvenor, or much later in life, that Bert had given Melville. Gil was left to learn the ropes without his father's guidance, but with plenty of coaching from Tom McKnew of the business office, Ted Vosburgh in the manuscripts department, Gifford Hampshire in illustrations, and Robert E. Doyle in fulfillment—all trusted colleagues Melville encouraged to help his son.

Like others of the new generation at National Geographic, Gil had watched his father launch the organization in new directions, but most of the innovation had changed the look of the magazine rather than its substance. Photography had reached new heights under Melville, but it would take years for the organization to find its journalistic voice, one that would acknowledge the harsh realities of life without being overly pessimistic about them. For Gil Grosvenor, Bill Garrett, Gifford Hampshire, Bob Breeden, and others who proudly called themselves the Young Turks, the transformation of *National Geographic* could not come soon enough. "A bunch of us would go to lunch together and plot against the magazine," said Hampshire. "Our revolutionary thoughts were for more white space, modern picture displays, and more realistic coverage of the world."[30] Among themselves, the Young Turks grumbled about the headlines that Ted Vosburgh placed on their articles ("Finland: Plucky Neighbor of Soviet Russia; Fireflies: Torchbearers of the Twilight"). And they complained about the predictable pictures Melville chose for the magazine's cover, which emphasized bright colors, adventure, and friendly natives. "MBG had a sixth sense of knowing what the readers wanted," said Gil. "It wasn't always what the Young Turks thought . . . readers wanted. I can remember many, many instances where there were choices of putting in a very fine dramatic photograph versus a young woman in a bathing suit on a beach. MBG would . . . choose the girl in the bathing suit and say 'You've got to have something in this magazine for the tired businessman at night.' We all sort of rolled our eyes. . . ."[31]

After Melville's exit Vosburgh was inclined to continue the safe course set by his predecessors. His socially conscious young staff pressed for coverage of Vietnam and Southeast Asia (nine articles in seven years), and they managed to get previously forbidden words ("threatened," "polluted," "imperiled") into headlines on occasion. But the march toward reality was a slow one, as Vosburgh often reminded his colleagues. "The Geographic's way," he told them, "is to hold up the torch, not to apply it."[32] Thus as the turbulent decade of the 1960s played out, the magazine continued to do what it did best.

That meant short shrift to unpleasant subjects, such as the burning of Washington in April 1968. As the skyline glowed red with fire that spring in the aftermath of Dr. Martin Luther King, Jr.'s assassination, the editor of *National Geographic* sharpened his pencils, closed the door, and went to work on legends for a hyena article. He telephoned Carolyn Bennett Patterson, the legends editor, to help him.

"Did you know about the rioting, Ted?" she asked. "The city is on fire."

"Oh yes," said Vosburgh, "I did hear something about that. Are you on your way up here, Carolyn? I want to get hyenas off to the printer."

Somewhat concerned for her safety, she joined Vosburgh on the ninth floor, where they sat side by side on a couch, methodically reviewing each hyena caption. Vosburgh seemed unfazed by the trouble outside. He asked questions, made fixes, taking his time. Police blocked off surrounding streets. The office closed. Black smoke swirled overhead as the riots burned closer. "You're sure about this?" Vosburgh asked, tapping his pencil on a sentence. "Do we need this comma? Would you consider a semicolon here?" Their deliberations were interrupted by a sharp knock at the door. Vosburgh's secretary appeared, and he snapped.

"I told you I didn't wish to be interrupted," he said. "We are clearing hyenas!"

Hyenas got cleared that night and off to the printer. Patterson got safely home. Vosburgh, who had been a lieutenant colonel with Gen. George S. Patton's Third Army in World War II, shrugged off the distraction. "I had seen no reason to panic," he said. "Washington was not burning, just a small section of it."[33]

With Vosburgh and Mel Payne running National Geographic, the semi-retired Melville Grosvenor had time to enjoy his second family. He and Anne bundled their teenage son, Eddie, and their young daughter, Sara, on marathon voyages of the *White Mist,* a yawl that became a familiar fixture in *National Geographic.* Readers were treated to Chairman Melville's lengthy articles on the Greek isles, the Canadian coast, and other vacation spots, thoroughly and ably documented with photographs by Eddie, the clan's newest photographic talent. Vosburgh rearranged whole issues to accommodate these sea stories, but he balked at delaying an article on the solar system to make room for a *White Mist* voyage up the Hudson to the St. Lawrence River.

"I'll need at least 55 pages," Melville told Vosburgh.

"But Melville, that's more than we're giving the whole solar system," said Vosburgh.

"Yes," said Melville, "but there are no *people* out there."

Vosburgh moved the planets, making way for Melville's article, "North Through History Aboard *White Mist,*" which led the July 1970 issue.[34]

The family still had some influence at National Geographic. So as Vosburgh's tenure as editor neared its end in 1970, Melville quietly began to lobby key board members to ensure that a Grosvenor would have a leadership role in the future. After more than fifteen years in training, Gil, thirty-nine, was poised to take over as editor. Melville knew he could count on his friends among the trustees—Connie Wirth, Laurance Rockefeller, Ted Vosburgh—to support Gil. And Gil had developed close ties with two other board members, James E. Webb, the former NASA administrator, and William McChesney Martin, former chairman of the Federal Reserve Board.[35] But with Mel Payne's opposition to Gil well known, the votes of other trustees could not be certain, and even likely allies such as Bill Martin were sensitive to perceptions of nepotism at the National Geographic.

"I hate nepotism," Martin told the *New York Times.* "But I've interviewed forty-three members of the staff to find out who would be the best choice. I give up. It was unanimous in favor of young Gil."[36] Thus, faced with the prospect of breaking with tradition or continuing the Grosvenor line at the National Geographic Society, the board approved Gil's appointment as editor in June 1970, to take effect in October.

His moment of triumph was short-lived, however. Mel Payne, who had another five years to serve as president, summoned Gil to his office to make certain that he knew Payne had opposed the promotion. "You tend to listen to other people," Payne bluntly told Gil, whom Payne thought was less qualified for the editorship than other candidates, in part because of the young man's personality. Yes, Gil was earnest; yes, he had worked hard; yes, he was a gifted photographer who had demonstrated a knack for administration. But this self-effacing Grosvenor struck Payne and other detractors as unsure, indecisive, and averse to confrontation. More important, Payne worried that Gil might be too heavily influenced by Bill Garrett, who had emerged as the most aggressive of the Young Turks. If Garrett pushed things too far and too fast on the magazine, would Gil be able to resist him? Payne knew that Gil and Garrett had become fast friends in recent years, vacationing together, trading dinners with their families, and planning an exciting future for the magazine.

"You put too much faith in Bill Garrett," Payne told Gil. "I'm not sure that Bill Garrett is good for this magazine. He is more interested in news coverage, and that's not what we are all about. I think he is going to have too much influence on you."[37] Although unable to block Gil's appointment, Payne had nonetheless delivered a sobering message: If Gil or his friend Garrett made any mistakes, Payne would be there to pounce on them. With this dark cloud trailing him, Gil began his new job as editor of *National Geographic.*

Despite Payne's misgivings, Gil proved to be an excellent and innovative editor. He delegated responsibility to key deputies, cultivated support among trustees, introduced new substance to the magazine, and watched the membership grow by another 3 million during a decade as editor; circulation passed the 10 million mark in 1978, bringing *National Geographic* third in line behind *TV Guide* and *Reader's Digest*. To recruit young readers, Gil persuaded the board in 1975 to launch *World,* a children's magazine that achieved a circulation of 1.3 million in its first year.

Mel Payne could hardly argue with such success, but he occasionally felt it necessary to reprimand Gil when his associates strayed from the path of decorum. Photographers, adventurers, scientists, and other world travelers were expected to conform to local customs while on assignment for *National Geographic,* but this also applied when they visited Washington. There they were meant to dress like FBI agents, the men in dark suits and white shirts, the women in conservative dresses, especially if they appeared in the tenth-floor dining room known as the Masthead. Lunching there in 1973, Payne was shocked to see a member of Gil's editorial staff out of uniform. The perpetrator, Payne reported to Gil, was breezing about the room "in a Mexican outdoor barbecue shirt, i.e., neck open, short sleeves, shirt tails out, with red, yellow, and purple roses! This goes over great in Mexico and at summer picnics, but certainly is not suitable for our dining room." The barbecue shirt, Payne feared, was evidence of a disturbing trend toward "excessively informal dress" at the National Geographic Society, where open-necked, short-sleeved, tails-out, go-to-hell safari shirts had been sighted on more than one occasion. Standards were disintegrating.

"If the building were not completely and efficiently air conditioned which makes it entirely comfortable to work in a light jacket," Payne wrote Gil, "this sort of dress would be understandably acceptable, but it is not necessary and before the situation becomes contagious, I recommend that you speak to these people and reaffirm the Society's long-standing policy. Once this business gets out of hand, you and I both know there will be no limit."[38] Gil handled the matter quietly; the Masthead reverted to its safe monochromatic palette, and the Young Turks, who were rapidly becoming Gray Turks, quietly resumed their campaign of reform, which was transforming the magazine into a modern publication.

The look of the journal improved markedly, due in part to smarter design, and in part to faster printing and better color made possible by a change to gravure printing. The quality of the writing, photography, and editing improved as well, thanks largely to the recruiting efforts of Bill Garrett and Bob Gilka, who scoured other magazines and newspapers for professional journalistic talent, brought them to the magazine, gave them ample field time, and challenged them to do their best work.

Gil's appointment as editor came at a time of increased public sensitivity to environmental issues—with the advent of Earth Day in 1970, new legislation to safeguard the air and water, and the expansion of protected public lands. From the era when Gil's grandfather helped to establish the National Park Service, National Geographic had demonstrated a strong interest in conservation; Gil and his lieutenants took that tradition a step further, signaling a new approach to the magazine's coverage of the environment. A package of articles and maps entitled "Our Ecological Crisis" led the December 1970 issue, which confronted readers with oil-soaked birds, massive fish kills, strip mining, and belching smokestacks. This green-leaning style would become an important aspect of National Geographic's character for the rest of the century.[39]

By publishing stories from the real world, Gil faced a new balancing act. If he was to keep readers abreast of the times, he had to do so without alienating the magazine's core constituency. So in addition to articles on environmental trouble, oil shortages, North Korea, Siberia, East Berlin, Haiti, Chile, Cyprus, and the Middle East, he provided readers with an abundance of the familiar—liberally illustrated pieces on exotic cultures, paleontology, archaeology, ants, birds, chimps, sharks, rats, parks, science, and space exploration. While this standard menu made up most of National Geographic each month, the new editor's willful departure from his grandfather's doctrine of gentlemanly journalism did not go unnoticed.

Members of the American Jewish Committee picketed the headquarters building in April 1974 to protest an article on Damascus, which glossed over Syria's mistreatment of Jews in that country. Gil was forced to issue a rare apology in his editor's column that year, admitting that "occasionally events move too fast for us and we must pay the price. . . . Now, after months of carefully reviewing the evidence, we have concluded that our critics were right. We erred. . . . We realize that to report at all on the modern world is to risk such eventualities." The magazine continued reporting on the modern world, mixing the bad with the good, with Gil wondering when Mel Payne was going to squawk.[40]

At an age when many are eager for retirement, Payne showed no sign of flagging. When at sixty-five he was required to give up the presidency of National Geographic, Payne became chairman of the organization. This required Melville Grosvenor to get out of the way—a contingency Payne had foreseen and planned for. In his last year as president, he had appointed a management continuity committee to prepare for the succession. Among other things, the committee recommended that Payne should be chairman and that Melville Grosvenor should be given the title chairman emeritus; Melville would be allowed to retain his title as editor in chief, along with a salary, but for only three years.[41]

According to Robert C. Seamans, Jr., one of the committee members appointed by Payne, Melville had been doing little to earn his $40,000 salary, which appeared unseemly. "We decided that management of the society needed to be broadened out beyond the Grosvenor family," said Seamans, dean of the MIT engineering school. "Not only did this make common sense, but also the Geographic was getting involved in areas other than its publications." The committee also recommended that an outside trustee be named chairman at the end of Payne's term,* and that the titles of editor and president be kept separate in the future, given the growing complexities of the organization. Finally the com-

---

*Payne and others at National Geographic were outraged by this suggestion, which was not only rejected in 1976 but has been ignored in the quarter century since then. The chairman had always come from inside the National Geographic Society.

mittee proposed a caretaker president to replace Payne; he was Robert E. Doyle, a sixty-year-old veteran of the organization's fulfillment and membership operation. This last suggestion anticipated that Gil Grosvenor would be in line for the president's job by 1980.[42]

After barely a year in his new position as chairman, Payne moved to demonstrate the board's dominance over Gil Grosvenor and the magazine. The opportunity came in 1977 when the journal published a string of articles on Harlem, Cuba, South Africa, and Quebec. Although these pieces were quite mild by prevailing standards, the fact that National Geographic had taken notice of the real world prompted a flurry of protests from readers, from conservative critics, and from trustees concerned about the magazine's drift into new territory.

Cuban exiles in Miami complained that a January 1977 piece on Cuba had been too soft on Castro and communism; the apartheid regime in South Africa grumbled that a June 1977 article, "South Africa's Lonely Ordeal," had been too critical; Anglophones protested that the April 1977 coverage of Quebec gave undue prominence to French-speaking separatists. The syndicated columnist Kevin P. Phillips wondered if National Geographic was suffering from an infectious case of "radical chic," while the very conservative watchdog group Accuracy In Media denounced the magazine's "inaccurate and distorted" style of journalism and launched a letter-writing campaign to call trustees' attention to the problem. To make matters worse, the magazine's New York advertising office, sensing an opportunity to modernize National Geographic's stodgy reputation, placed full-page newspaper ads in the New York Times and the Wall Street Journal heralding the February 1977 Harlem article as a harbinger of the magazine's new style. "It's time you took another look," the ad proclaimed, inspiring Mel Payne to do just that.

Payne announced in May 1977 that he would form an ad hoc committee of the board to examine why the magazine was getting this sudden, unwelcome attention. He was especially concerned about reactions to articles on Cuba and South Africa. "There has been too much emphasis on disagreeable aspects," Payne explained. "In the future, there won't be so much emphasis of that nature."[43] While protesting that he had no intention of censoring the magazine, Payne appointed a trusted ally, board member Lloyd Elliott, president of George Washington University, to chair the committee. The group was directed to interview Gil and other editors to determine if the magazine had strayed from established policy, and to report its findings to the full board that fall. Payne's action, which came as the family's influence was beginning to fray, plunged the normally collegial organization into a power struggle unmatched since 1900, when Gilbert H. Grosvenor had wrestled John Hyde for control of the magazine. This time the atmosphere became so suspicious that Melville Grosvenor, believing his office telephone to be tapped, used pay phones to call his family.[44]

"I have been somewhat amazed to discover the level of animosity which currently exists within the Society," Elliot said as the board gathered to launch the ad hoc committee that June. "This is not a witch hunt," Payne announced. "This is simply an examination of the heavy criticism we have had. I am hopeful that out of it will come new harmony so that the situation will not arise again."[45]

Payne's comments brought howls of protest from the only editors still on the board. Both Ted Vosburgh and Melville Grosvenor saw the maneuver for what it was, a power grab, an insult to Gil Grosvenor, and perhaps an attempt to scuttle his chances for promotion to the president's job in 1980. No matter who was editor, such an oversight committee posed a threat to the independence of National Geographic, which had been free of interference from above since 1900, when Bert fought off a similar assault.

"It looks like a railroad job to me," said Melville Grosvenor, who objected that the ad hoc committee was stacked with older trustees, with none representing the younger generation. Besides, he said, the articles under scrutiny seemed well within acceptable bounds and the few letters of complaint were inconsequential, "like a flea on an elephant's back. . . . I have read the article on South Africa several times," he said. "I think it explains fairly both sides of the problem. . . . We should not have to come to the board every time there is an article like this. It is the re-spon-si-bil-i-ty of the editor," he said, slapping the table with each syllable. "If the editor does not do his job, the board can fire him. I don't believe I would serve as editor with a committee like this looking over my shoulder. . . ."[46]

Ted Vosburgh expressed complete confidence in Gil Grosvenor, who sat quietly through most of the meeting. "You have seen the record of growth of the Society under his editorship," Vosburgh said. "I don't think we need a committee breathing down his neck. You are always going to have criticism and that should not panic anybody."[47]

Payne icily replied that he had not panicked.

Perhaps too late, Melville's friends on the board, Connie Wirth and Laurance Rockefeller, tried to deflect the issue, suggesting that no committee investigation was necessary. But Crawford H. Greenewalt, a Payne ally and former chairman of the Du Pont Company, pressed the matter: "The point is that the magazine has been subjected to criticism," he explained. "It has to do with the reputation of the magazine. . . . If the proposition is that the board has nothing to do with editorial policy, then frankly I don't see how, as a trustee with responsibility on my shoulders, I could remain on the board. . . . We have the ultimate responsibility for the conduct of the Society's operations."[48]

Payne's forces carried the day and the board authorized Elliott's committee to investigate. Chief Justice Earl Warren, who had voted for the investigation, tried to reassure the Grosvenors and their supporters: "When the matter is looked into, it should lead to support of the editor. . . . I read the [South Africa] article and it would never have occurred to me that anyone except the most intense partisan could take exception to it. . . . If I were on the editorial operation, I would regard this as a means of making clear that editorial freedom, like academic freedom, is supported by the board."[49]

The jurist's soothing interpretation went up in smoke as soon as word of National Geographic's inner turmoil leaked out that summer. The press rightly gauged the story as a fight between the forces of change and hidebound tradition at an institution known for playing it safe. Writing in the *Washington Post*, Kathy Sawyer reported that the magazine under Gil Grosvenor had pursued "a less evasive approach to controversial topics," which had caused dissension be-tween "some of the old guard, led by chairman Melvin Payne, and those who fa-vor the policies of the young editor. . . ."[50] Her front-page piece produced a second wave of coverage, including a prominent *Newsweek* article, "The Geo-graphic Faces Life," and an extensive piece in the *Los Angeles Times,* "Geographic: From Upbeat to Realism" by A. Kent MacDougall. According to MacDougall, the magazine had "departed from its traditional policy of avoiding the unpleas-ant, the ugly and the controversial. Besides the standard Geographic fare . . . the magazine has been taking a franker look at pollution, poverty, hunger, dis-ease and despotism. Such unpleasantries are still not emphasized but neither are they glossed over."[51]

In an interview with MacDougall, Payne made it clear, even as the commit-tee's deliberations were taking place, where his sympathies lay—he sided with the white South Africans criticizing the magazine and staff writer Bill Ellis. "The writer's personal bias showed through," Payne told MacDougall. "You have to watch writers; every once in a while they'll slide one by you." For his part, Gil lamented that the squabble had become public. "I don't think it is proper for any board to wash its dirty linen in public," he told MacDougall, "particularly when we haven't established that the linen is dirty."[52]

Through several tense sessions that summer, Gil met with the ad hoc com-mittee, patiently answered their questions, and provided them with stacks of documentation to support the factual content of the articles under scrutiny. "The game plan," Gil recalled, "was to drown them in information. The stack was this high," he said, holding his hand two feet in the air. Gil invited commit-tee members to go through each heavily annotated article and urged them to question Margaret Bledsoe, the head of the magazine's formidable research de-partment. She and her associates had subjected the disputed articles to the usual fact-checking process, which required multiple sources for each statement of fact, routed the manuscript to a panel of consultants and to the State Depart-ment for comment, and gave all persons quoted a chance to review their state-ments before publication. Representatives of countries mentioned in the article were given the manuscript and had their objections considered. "We had every goddamn thing we could find from the researchers," Gil said. "The committee was totally ill-equipped to go through the articles the way a researcher and an editor would, and I reminded them that we did this for every article."[53]

The committee seemed impressed by the detailed proofs Gil had provided them, which removed any suggestion that the magazine had misrepresented the facts. But the tone of those articles was still a concern, and a larger question remained: Were real-world stories a proper subject for coverage in *National Geographic*? Gil maintained that they were. "We have always prided ourselves on giving background information on sensitive areas of the world," he told the committee. "The question is whether we can be prepared for some criticism. We publish 87 stories a year. One or two may be a problem. . . . If it is the determination of the board that we not get into politically active areas, that would be a change in policy," Gil said. The magazine had not changed, Gil said. The world had changed. The magazine was adapting to it.[54]

In this and other ways, Gil carefully walked the line between deference and independence in his dealings with the ad hoc committee. He also removed any opportunity for criticism by privately admitting that, faced again with the decision to cover Cuba or South Africa, he would have deferred both stories; but he held firm to the principle that responsibility for such actions rested with the editor, not with an oversight committee.

Thus, having established that the articles had been sound, if at the fringes of what was considered comfortable for the National Geographic Society, the committee neared the end of its investigation with very little to say against the editor or the magazine. There was no formal finding that Gil had made a mistake or an error in judgment, no conclusion that the articles had been a departure from policy, and no directive that the magazine should avoid such topics in the future. Nor were there threats about making the ad hoc committee a permanent panel for editorial oversight, a proposition broached by Lloyd Elliott and promptly shot down by other committee members.[55] The group was satisfied to leave responsibility for the magazine with the editor, who was encouraged to keep the board apprised of prospective articles without seeking advance approval for them. Since Gil had been doing this routinely, he happily agreed to continue that practice.

Payne was furious that Gil was getting off so lightly. By not assigning blame, Payne complained, the committee was avoiding the issue and "spinning our wheels. . . . To put this in perspective, in all the Society's history, this situation has not occurred before. . . . My concern is that there is a tendency to go in this direction and it seemed to me important to stop it and get back on track. That is the function of this committee—to make it plain that some articles— especially South Africa—did not conform with the editorial policy established many years ago."[56] Payne had overplayed his hand. The committee balked at is- suing an indictment. "The board cannot know in advance without reading the text whether an article will conform to policy," said Crawford Greenewalt, whose sympathies had shifted from Payne to Gil Grosvenor during the com- mittee's work. "Leave it to the editor to carry out a policy with full faith and trust."[57]

To save face for Payne, the committee finally agreed to issue a statement of ed- itorial policy, confirming *National Geographic*'s principles of editorial balance and fair-mindedness. The statement—to be drafted by Gil and published in a future issue of the magazine, signed by Gil as editor, Payne as chairman, and Bob Doyle as president—was designed to present a united front and to smooth over recent news of the schism between the organization's board and its editor. This brief statement, described as a "Reaffirmation of Editorial Policy," was published in the January 1978 *National Geographic*. The mission of National Geographic, it said,

> is to increase and diffuse geographic knowledge. Geography is defined in a
> broad sense: the description of land, sea, and universe; the interrelationship of
> man with the flora and fauna of earth; and the historical, cultural, scientific,
> governmental, and social backgrounds of people.
>
> The magazine strives to present timely, accurate, factual, objective mate-
> rial in an unbiased presentation. As times and tastes change, the magazine

slowly evolves its style, format, and subject matter to reflect that change without altering the fundamental policies above.

Excellence of presentation—accuracy, technical superiority in printing and photo reproduction, and clarity of meaning—remain traditional goals against which each article is measured.[58]

After the stresses of that long summer, Gil was happy to emerge from this crisis with such an amiable resolution. His steady demeanor, his grasp of editorial detail, his quiet toughness, and his restraint under fire had favorably impressed trustees who may have been previously uncertain of Gil's character. Had he been more strident and less respectful, he might have given the magazine's detractors an excuse for skewering him. Instead, he had won friends on the board, which was soon to consider him as a candidate for the president's job. By throwing this challenge Gil's way, Payne had inadvertently enhanced the editor's standing. Gil heaved a sigh of relief and declared victory. "Mel Payne was convinced that this committee was going to bounce me as editor," Gil said, "or at the very least he would become the supereditor of the magazine. He wanted the whole loaf—all he got was a slice. I saw this as a tremendous victory," he said of the policy statement, "because it was what we were doing already."[59]

But Payne's slice came at a cost. By challenging the editor's authority, Payne had managed to chill the magazine's fervor. As a result of the board's investigations, Gil grew more cautious in his approach to articles, easing back on his previous schedule of stories.[60] He scrapped a piece on war-torn Beirut, which had been in the works for months;[61] and he made sure that forthcoming articles on Moscow and Syria, while deeply researched and solidly reported, were scrupulously drained of controversy. In 1978, the year following Gil's trial by committee, the magazine reverted to its accustomed lineup on subjects such as Arkansas, Ladakh, Georgia, the Greek Bronze Age, Pennsylvania, balloon races, the Continental Shelf, the Gulf Intracoastal Waterway, four-eyed fish, tulips, and archaeology. Hardly anyone noticed that year when Bill Ellis, author of the maligned South Africa story, won an Overseas Press Club Award for the piece.

"The pressure was tremendous," Gil recalled.[62] Unbeknownst to Payne and others, Gil had weathered a parallel crisis that summer. As the power struggle unfolded in Washington, he had also been faced with the dissolution of his sixteen-year marriage to Donna Grosvenor. During their vacation in Baddeck she had begun a romance with Casey Baldwin, Jr., the gardener at Beinn Bhreagh and the grandson of Alexander Graham Bell's most loyal collaborator. When that summer ended, Gil was left alone to manage not only his office troubles but also this latest personal calamity.[63]

CHAPTER FIFTEEN
BROKEN FRIENDSHIP

ROBERT L. BREEDEN (*LEFT*) JOINS GILBERT M. GROSVENOR
AND EDITOR BILL GARRETT UNDER A DOME DECORATED WITH STARS
IN NATIONAL GEOGRAPHIC'S HEADQUARTERS, 1990.

There was something in Gilbert M. Grosvenor that recoiled at the messiness of daily life and reveled in the things that could be measured and quantified. He had, after all, been a statistical psychology major. Like his grandfather and namesake, Gil Grosvenor was a methodical man who yearned for order over chaos, thrift over profligacy, restraint over impulse. To their mutual friends, Donna and Gil Grosvenor had seemed an unlikely match from the start, with her flamboyance threatening to smother his instinctive diffidence. DIVERS DO IT DEEPER! screamed the bumper sticker on her car, which occasionally appeared in the executive lot at National Geographic. Few were surprised by their divorce, or that it was handled quietly, in large part because the memory of Melville's untidy breakup from Helen Grosvenor remained painful for the family.

"I didn't want to end up in an adversarial position," Gil said. "There was no attempt between my father and my mother to try and come up with any kind of relationship beneficial to the kids. It was so bitter that it was a burden to us—a burden to us," he said very softly. "It had a profound influence on my divorce. My kids are old enough now that they appreciate that effort by Donna and me. They didn't have to go through that turmoil." Shortly after his divorce from Donna, Gil married Wiley Jarman, a childhood friend from Washington whose family had summered in Baddeck, just over the mountain from the Grosvenors' Beinn Bhreagh estate.[1]

At work as at home, Gil tried to impose regularity on the disarray he found around him. One of his proudest achievements at the National Geographic Society was to begin testing book titles by direct mail in the 1970s; about the same time he inaugurated extensive surveys to determine how readers reacted to magazine articles. "I believe in statistics and I believe in surveys," he said, while acknowledging that this stance won little favor among his editorial staff. "Basically editorial people don't want to do that. They want to publish what they want to publish. . . . I believe in finding out what people like or don't like so that you can have a balanced magazine."[2]

As a young editor starting at National Geographic, Gil had been distressed to find that under Melville Grosvenor's exuberant influence, the editorial schedule resembled a constant work in progress, with most of the pages shuffled or hopelessly missing. Articles shifted in and out of the schedule without warning, text had to be rushed through when a batch of photographs appeared unexpectedly, and stories sometimes dragged in to the printer's a month behind schedule. "Nobody knew what anybody else was doing," Gil said.[3] To address this problem, he established the Control Center, where the magazine's entire editorial schedule and assignment list could be seen at a glance. Here, on a wall full of folding panels that Gil built in his basement workshop and installed at headquarters, one could follow the progress of each article, see how it meshed with other stories, and make orderly plans for the future. If his father was going to toss out assignments like so many Mardi Gras trinkets, at least now people could see where they were landing. The schedule board, with its crowded type and color-coded labels, looked like an oversized version of the periodic table familiar to any chemistry student—but it served a purpose, acting as a stabilizing influence on Melville's editorial extravagance.[4] In his plodding way, Gil had imposed some semblance of discipline on the family business, where he was destined for greater responsibility, as Melville Grosvenor grew frail with age and began to fade. (Almost thirty years later, the Control Center was still in operation, having become a central feature of National Geographic's day-to-day operations.)

With the crucial transition of 1980 approaching, Gil Grosvenor, now nearing fifty after a decade as editor, faced a difficult choice: He could remain as editor of National Geographic, a position he often called "the best job in journalism," or he could cross the hall to replace the retiring Robert E. Doyle as president, in effect the publisher and chief executive officer of the organization. But Gil could not take both jobs, as his father and grandfather had done; the board of trustees had decreed upon Melville Grosvenor's retirement in 1967 and twice reaffirmed since, in 1976 and 1980, that no person could serve in the dual role of editor and president. Gil deeply resented their decision, feeling that it had been aimed at him personally, which was very likely. His old adversary, Mel Payne, had had a hand in this decision. Payne still served as chairman of the board, still had his doubts about Gil, and still threw up roadblocks to his progress at every opportunity.[5]

Nonetheless, in a decision that would cause him no end of grief in the years ahead, Gil agreed in February 1980 to give up the editor's job and to take over as president, effective August 1980. This set off a behind-the-scenes scramble for the editor's chair—passing out of Grosvenor hands for only the third time in the twentieth century—Gil backing his friend and right-hand man, associate editor Bill Garrett, forty-nine, then responsible for all of the magazine's visual content, while Mel Payne pressed for his own candidate, Joseph R. Judge, an associate editor who was Garrett's counterpart on the text side.[6]

Cocky and outspoken, Garrett had come up the hard way, the son of a Missouri carpenter during the Depression. "He had nothing in Missouri," said a friend who admired Garrett's questing style. "He was going to get something here."[7] Others described Garrett as mercurial, stubborn, daring, demanding, and impulsive. "Ben Bradlee reinvented as Don Quixote," said another friend.[8] In a quarter century of working at the magazine, he established a reputation as a solid writer, a brilliant photographer, an inspired picture editor, and a gifted layout man. Even his detractors had to admit that Garrett was something of a genius when it came to the style of intelligent, illustrated journalism that he helped introduce at *National Geographic* magazine in the 1970s. This, of course, was Mel Payne's primary objection to Garrett's promotion. In the end, however, the board deferred to the wishes of Gil, their new president-elect, and Payne lost another battle with the Grosvenors—but it was a close call.

"It might have gone either way," said Ted Vosburgh, one of the board members who chose Garrett over Joe Judge. "We felt that Bill Garrett's breadth of experience and length of service gave him the edge. Bill knew photography; he knew the production side; and he was a natural leader. We were concerned that if he didn't get the job, he would leave. So he got it."[9]

Garrett first heard the news from Mel Payne. Just after the board meeting, the pale, sepulchral chairman appeared in Garrett's office, extended a bony hand, and said: "Congratulations, Mr. Editor." At the same time, true to form, Payne looked Garrett in the eye and told him that he had opposed the promotion, but did not tell him why. "I'm on my way to the airport now," he said, "but when I come back next week, we'll get together and talk about it."[10]

"Fine," said Garrett. As he watched the chairman disappear down the hall, Garrett had a good idea of what was troubling Payne, and when a reporter from the *Washington Post* called about his promotion, the new editor was primed.[11] "There absolutely will be no radical or visible changes in the magazine," Garrett announced. "It's like being given control of apple pie and motherhood; you don't fool around with it."[12] Next Garrett raced to Gil Grosvenor's corner office on the ninth floor. When he got there, he was surprised to find his friend fidgeting and grim-faced, seemingly on the verge of tears. Gil kept a tight lid on his emotional displays, but he was in fact a man with strong feelings, which bubbled to the surface in times of stress. Instead of offering congratulations to Garrett, Gil seemed brusque and businesslike when they met that morning. Who knows why? Perhaps Gil's demeanor had nothing to do with the promotion that day, but Garrett always supposed that his friend was still seething over the board's action, which stripped him of the editor's job. Whatever the cause, it was an awkward moment. Gil tried to take charge.

"I'll call the staff together to announce it at two o'clock," Gil said. "Don't tell anybody."

"It's too late," said Garrett. "The word is already out. I just heard from Mel Payne, who came by to congratulate me. Then Charlie Reichmuth called," Garrett added, naming the garrulous director of the National Geographic Society's dining rooms, who had very likely spread the word of Garrett's promotion halfway to Capitol Hill by then.

Gil looked annoyed, muttered something, and made plans for the announcement that afternoon.

Having arrived at the twin summits of power at the National Geographic Society, the new president and the new editor were already grappling for control. From that day in 1980, the ingredients for a showdown were in place.

When Payne returned to work the following week and Garrett phoned to schedule the agreed-upon appointment, Payne demurred. "We don't need to talk now," Payne told him. "I saw what you said in the *Washington Post* and we have no problems." By Garrett's testimony, he had few run-ins with Payne after that; in fact, the two came to respect one another, and Garrett even developed a grudging affection for the old man, as did others who worked closely with him on the Research and Exploration Committee.[13]

Garrett's association with Gil Grosvenor ran deeper. They were more than friends, according to Garrett's son Ken: "They were married for thirty-five years."[14] During the 1970s, when Grosvenor was editor, Garrett was one of his most effective lieutenants. Garrett had quickly befriended the troubled, uncertain son of National Geographic's dynasty, earning not only Gil's confidence but his affection. When Gil and Donna were building a new house on the Potomac River, Garrett called the architect, got the blueprints, produced a scale-model birdhouse in his woodworking shop, and proudly presented it to the Grosvenors; like the house itself, the model was made of Burmese teak.[15] The two men bought a motor home together and rented it to their colleagues. When Garrett injured his shoulder in a weekend football game and could not drive, Gil loaned him a Lincoln Town Car and a driver from the office pool—with Gil's blessing, Garrett continued the arrangement long after his shoulder had healed.

While Gil sometimes shied from asserting his authority on editorial matters, Garrett did not. Having earned the respect of the magazine's contributors, Garrett could drag in photographers and urge them to push a little harder on the next trip to Namibia or Peru, and he did not flinch from tearing apart and reassembling layouts at the last minute, with little concern for hurt feelings or fragile egos. Over their long collaboration, the complex bond between Gil Grosvenor and Bill Garrett began to look less like friendship than it did a marriage of mutual convenience, not unlike the successful alliance Alexander Graham Bell forged with the young Gilbert H. Grosvenor, or the one Bert, in turn, made with John Oliver LaGorce. Young Gil, introspective and tentative, gained strength from the gregarious, confident Garrett; and through Gil, the poor boy from Missouri gained entrée to Washington society, and instant influence at the highest levels of National Geographic. Mary Griswold Smith, a longtime *National Geographic* editor and a friend of both men, said the basis of the Grosvenor-Garrett attachment could be summed up in a word—opportunity. "Through Gil," she said, "Bill saw a way to be part of the Grosvenor clan. In Bill, Gil saw someone who could do things for him that he wouldn't or couldn't do for himself."[16]

The tension humming just beneath the surface of the Grosvenor-Garrett friendship might have been nothing more than the normal competition of contemporaries striving to outshine each other in the workplace. But perhaps other factors were at play. Gil's relationship with his father, while never adversarial, had been a tricky one. Garrett complicated it by establishing a closer relationship with Melville than Gil had ever enjoyed. Almost from the start, the old editor had bonded with the young, gap-toothed Missourian—like Melville, he was an instinctive operator who made no small plans. They delighted in one another's company. "We were on the same wavelength," Garrett recalled. "I think Melville liked having me there to push Gil beyond what he was willing to do."[17] It must have been embarrassing for Gil, who competed for his father's approval. "You don't have to be a psychiatrist to understand the effect this would have," said John M. Keshishian, a prominent Washington surgeon friendly with Garrett, Gil, and Melville.[18] Gil's way was to brood upon the slights, to nurse them along, and to avoid confrontation until it became unavoidable. "Then he goes off like a fucking roman candle!" said his longtime friend Bill Graves.[19]

It is unlikely that Melville Grosvenor saw the trouble brewing in the upper reaches of the National Geographic Society; he was happy to have placed his son at the tiller, with his friend Garrett close at hand. Melville withdrew from day-to-day operations, threw himself into sailing and gardening, and helped inaugurate the publishing career of yet another Grosvenor. With private backing from Laurance Rockefeller and other family friends, Eddie Grosvenor, twenty-eight, a Yale graduate who had earned advanced degrees in journalism and business administration at Columbia University, launched a new art magazine, *Portfolio,* in 1979. A handsomely illustrated bimonthly that looked like a cross between *Smithsonian* and *Connoisseur,* Eddie's journal earned glowing reviews, built a strong circulation in its niche, and gave every sign that the old family magic still worked.[20] Meanwhile, Melville gradually grew weaker, nursing the heart condition that finally felled him in April 1982. At eighty, he died quietly at a family compound near Miami. "The end was as unexpected as it was peaceful," his wife, Anne, reported that year, "and concluded a happy day of walking around to see our fruit trees, a nap, and a swim in the pool."[21] They laid him out at the National Presbyterian Church in Washington, with his favorite sailing cap close at hand, along with a chart for his beloved Bras d'Or Lakes.[22]

He took his leave at the best of times. *National Geographic*'s circulation topped out at 10.8 million in 1981.[23] From then until the end of the decade, membership ranged between 10.2 million and 10.6 million. But it never again climbed past the peak of 1981, despite a brilliant performance by Bill Garrett, who prodded the magazine to new levels of achievement. Garrett's magazine shone with clean design, thoughtful charts and artwork, detailed articles, timely maps, and documentary photography of surpassing quality, and *National Geographic* won recognition from the American Society of Magazine Editors, the White House Press Photographers' Association, and the Overseas Press Club. Like Alexander Graham Bell, Bill Garrett worried more about boring readers than he did about stretching them a bit. He loved to surprise them. "You have to make sure nobody gets bored," he said.[24]

In this endeavor, Garrett hit members with stories on rattlesnakes, lasers, cocaine, opium, China's economic reforms, Great Lakes flooding, sleep, elephant poaching, Eskimo hunters, women in Saudi Arabia, the world food supply, AIDS in Uganda, silk, wool, ants, fleas, pearls, jade, emeralds, soybeans, desertification, the Kremlin, the Vatican, the Houses of Parliament, the eruption of Mount St. Helens, the threatened temples of Angkor, the birth of democracy in Guatemala, El Niño, Antarctica, the microchip, smell, and the Yellowstone fires of 1988. He published the magazine's first holographic covers—in 1984, 1985, and 1988. He boldly commissioned Wally Herbert, a distinguished polar explorer, to examine the previously sealed notebooks and records of Rear Adm. Robert E. Peary, which led Herbert to cast doubts on Peary's North Pole claim. Garrett finally gave Matthew Henson, Peary's black comrade, due recognition for his role in the expedition. He launched explorer Bob Ballard on a search for the *Titanic* and published the first extensive photographs of the famous wreck. He provided readers with meticulous maps of the Persian Gulf, using satellite imagery to show that region's oil fields, pipeline routes, and shipping lanes—a study in the geography of future trouble.[25]

Because Garrett had sent him there, Steve McCurry was on photographic assignment in the Philippines when Imelda Marcos fled Malacanang Palace in February 1986. One of the first into the building, McCurry raced a mob of looters to the upstairs bedroom of Mrs. Marcos, where he sat at her desk, picked up the telephone, and dialed Bill Garrett in Washington to announce the abdication. "It will probably be the last phone call on the Marcos's bill," Garrett said.[26] For almost seven months McCurry had been traveling around the island nation, documenting the decline and fall of the regime.

Under Garrett, top editors at *National Geographic* were strongly encouraged to keep an eye on such hot spots and to get the magazine's writers and photographers in place when the time was right for a story. Garrett invented the photographic mantra, "f.8 and be there," which described his philosophy of blazing away in the thick of things. He had little interest in competing with newspapers or weeklies, but developed a class of interpretative reportage that added depth to the stories of the moment. Because they had weeks or months for a single assignment, *National Geographic*'s writers had time to interview the usual politicians, experts, and ministers in a country's capital, but they were also expected to range through the countryside and live with the farmers, fishermen, and tribal leaders who lent humanity to an article. Photographers might sit in a village for days, seemingly doing nothing until local citizens became so inured to their presence that the visitors seemed like furniture; then the work could begin. Photographers had been known to return from assignment with 35,000 frames for a thirty-four-page magazine piece, an unprecedented use of resources which raised eyebrows in the business office but helped the magazine achieve a three-dimensional quality of coverage that other publications had neither the time nor the money to produce. Contributors wanted to do their best for Garrett, not because they feared him but because, respecting him, they pushed themselves to measure up to his rigorous standards.

Like Melville Grosvenor, Garrett was inclined to say yes to most proposals. "I'd send Bill a little newspaper clipping on some obscure Maya discovery on Friday," said George E. Stuart, who served as the magazine's staff archaeologist, "and he would come right back to me: 'Let's go Monday!' He was wonderfully encouraging and enthusiastic."[27] When David Lamb, a well-traveled correspondent for the *Los Angeles Times,* proposed a story on minor-league baseball, Garrett seized upon the idea as a fresh avenue into small town America. "Crazy but good," he scrawled on Lamb's proposal. "Let's do it!"[28]

The magazine was able to attract respected professionals like David Lamb in the 1980s largely because Garrett had the revolutionary idea that *National Geographic's* writing, long the weakest link in an otherwise strong journal, should be as good as its photography. To make this happen, he hired Charles McCarry in 1983 and placed him in charge of freelance writers, who would become increasingly important to the journal. McCarry, a novelist, a steady contributor to *National Geographic,* and a former CIA agent, also happened to be a gifted editor with an impressive network of friends in the world of writing and publishing.* Garrett gave McCarry a blank check, an office on the ninth floor, and instructions to recruit new talent. "I also told Mac I would kill him if he used the magazine for CIA operations," Garrett said.[29] This threat, aimed at a man thoroughly trained in the use of small arms and other deadly arts, most likely provoked one of the loonlike laughs for which McCarry was famous.

He went to work. He dispatched Małgorzata Niezabitowska, a Neiman fellow visiting from Poland, on a yearlong journey across the United States for a new perspective on America. He sent Paul Theroux on assignments to Malawi, China, and India; Shana Alexander to the Serengeti; Ross Terrill to China's Sichuan Province; and James Fallows to the Vatican. He hired André Brink to write on the Afrikaners; Barry Lopez on California's deserts; and Tad Szulc on Poland's break with the Soviet Union. He sent Peter Benchley on diving expeditions to the Cayman Islands and the South Pacific. He drafted Larry L. King for a story on Anchorage, but stressed that *National Geographic* wanted the piece written in the Texan's distinctive voice: "You can say anything you like in *National Geographic,*" McCarry promised, "as long as it isn't a lie, a gratuitous in-

*With Bill Garrett's blessing, McCarry hired this book's author as his principal deputy in January 1989.

sult, or an obscenity." King answered: "You have just removed three-fourths of my arsenal!"[30]

Even with such writers to ennoble *National Geographic,* it was still easy to overlook the progress Garrett and McCarry had made. McCarry often joked that text was tolerated because the gray matter made the magazine's pictures look brighter. Approaching its centennial, *National Geographic* remained as Alexander Graham Bell and Gilbert H. Grosvenor intended, a journal driven by photographs and illustrations. These visual elements occupied 65 to 70 percent of all available space, crowding the text and occasionally overwhelming it. And, while the quality of freelance contributions improved markedly in the 1980s, much of the magazine was still being written by people Melville Grosvenor had hired twenty years before; a few did outstanding work, but many others had grown complacent and unproductive, grinding out just over one story per year, on average. In addition, the magazine continued to rely on contributions from amateur writers—the scientists and adventurers whose first-person pieces were most often salvaged and patched together by staff ghosts, which did little to lift *National Geographic*'s tired literary reputation.

After seven years on the job, McCarry acknowledged that much remained to be done. The dazzling novelty of photography, he told Garrett in 1990, could no longer carry the magazine alone. "People have long since learned to take our brilliant illustrations . . . for granted," he said.

Even now, the pictures cannot obscure the weakness of the text. . . . In the future, the imbalance can only become more evident. A magazine is like a suspension bridge: all of its components must be in an equal state of tension or it falls into the drink and disappears. The Geographic's text is weak because, unlike Geographic photography, it is not rooted in the tradition of professionalism. GHG [Gilbert H. Grosvenor] lost any real interest in printed matter after he discovered the power of the photograph to attract members, and left the writing to amateurs. MBG [Melville Bell Grosvenor] may well have been a genius in visual terms, but he did not understand the dignity of reporting or the beauty of words. . . .[31]

Nor, it seemed, did Gilbert M. Grosvenor, who shared his father's indifference toward the written side of *National Geographic*. Easily bored by reading, Gil assumed that members would be too. Long before it became a fad in the magazine business, he campaigned to keep articles short and simple, and to have more of them. And he sniffed at Garrett's effort to improve the magazine's writing, as if that somehow conflicted with *National Geographic*'s mass appeal. "I'd rather have a taxi driver than a Ph.D.," Gil was fond of saying.[32] He may have underestimated the intelligence and stamina of his audience: Years of his own surveys showed that subscribers to *National Geographic* were well-educated, avid readers who routinely ranked the longest articles as their favorites.

When Gil became president in 1980, he rarely commented on manuscripts or gave any sign that he read them before publication. Not only that, the cadre of executives around Gil soon discovered that he seldom dipped into the torrent of memos that flowed across his desk each day. When Garrett fretted that his boss was ignoring his notes, a colleague flashed a smile of recognition. "That's because he never reads anything!" said Owen R. Anderson, an executive vice president who worked closely with Gil. "Write the memo to Gil," Anderson advised. "Go in and talk about it, summarize what it says, and don't leave until he signs off on it." Garrett followed this advice and discovered, to his delight, that it worked.

Less reassuring was Gil's growing detachment from *National Geographic* magazine, an aloofness that seemed to widen as Garrett's confidence and influence grew in the 1980s. Rightly or wrongly, Garrett believed that his old friend had lost all interest in the magazine from the day Gil ceased to be editor and became president. "Everything stopped then," Garrett told friends.[33] To some degree, Garrett's impression was bolstered by Gil's own comments; more than five years after he had been compelled to give up the editor's job, he was still stewing over it. "You are inevitably going to have conflicts when you have two people basically running an organization," Gil said in a 1986 interview with C. D. B. Bryan.

I clearly don't agree with a lot of what Bill does, and he doesn't agree with a lot of what I do. But basically you have an organization now which has two heads. You have one person sending out a magazine to 10.5 million people . . . that the president basically has nothing to do with, and you have a president over here who is basically involved in running the Society, responsible for the books, responsible for educational products, but basically, I mean, no responsibility for the magazine whatsoever. . . . I feel it's very awkward and I'm stuck with it. It's here to stay.[34]

Unable or unwilling to assert himself over Garrett and the magazine, Gil began to branch out in other directions, continuing the diversification campaign his father had begun in the 1960s. "Because Melville had done so much, there wasn't much left for Gil," said Howard E. Paine, a senior editor who served under both Grosvenors.[35] Nonetheless in the 1980s, Gil moved to build a legacy of his own. He expanded the organization's book-publishing program, which he placed under the direction of Robert L. Breeden, one of the original Missouri Mafiosi who had known Garrett in journalism school. Breeden's division was soon contributing a cool $125 million in annual revenues, or 40 percent of the organization's total sales.[36] Gil launched two new magazines, *National Geographic Traveler* and *National Geographic Research,* in part to fill gaps in coverage left by the flagship magazine.* He enlarged the organization's presence on television with *Explorer,* a two-hour show that ran weekly on the Turner Broadcasting System. And long before the Discovery Channel became a household brand (and a threat to National Geographic's broadcasting ambitions) Gil sought without success to form a cable television consortium of like-minded nonprofit organizations.[37]

His favorite initiative focused on the woeful condition of geographic literacy in the United States. Gil's crusade—so far removed from the limelight or the news—returned the National Geographic Society to the core values that Gardiner Greene Hubbard and other founders had espoused before the organi-

*After losing millions of dollars, *Traveler* finally broke into the black in 1994. *National Geographic Research* was not so lucky; launched as a glossy alternative to the venerable peer-reviewed journals *Science* and *Nature,* the National Geographic start-up sputtered and died before its sixth birthday.

zation broadened its popular appeal. Taking his message on the road, Gil visited schools, civic groups, and governors to campaign for better geographic education, reminding them that geography was more important than ever in a complex and interdependent world. "The world is too competitive and dangerous to be a vague blur of memorized names and places," he said. "We face a critical need to understand foreign consumers, markets, customs, strengths, and weaknesses. Without a thorough grasp of geography, we see the world from our own narrow perspective."[38] On a trip to Seattle, Gil gamely went toe-to-toe with a twelve-year-old whiz named Alex, who had challenged the National Geographic leader to a geographic bout ("Which state capital is surrounded by desert?" "What is the term for the political units of Japan?" "A ship canal connects which city with the Gulf of Mexico?")* When the smoke cleared, the score was Alex Kerchner, 6,876; Gilbert M. Grosvenor, minus 316. "I would do better shooting hoops with Michael Jordan," Gil confessed.[39]

Despite occasional humiliations, Gil remained committed to his campaign for geographic education. Instead of just talking about it, he established geographic literacy as an integral part of the National Geographic Society's mission. He arranged for $40 million in direct National Geographic grants and matching funds to establish the Geography Education Foundation, a subdivision of the National Geographic Society. To encourage the teaching of geography in public schools, the organization also began a National Geography Bee, which drew millions of applicants from across the country. In addition to its altruistic aims, Gil hoped that his campaign would recruit a new generation of members for the National Geographic Society.[40]

Where will the new members come from? Gilbert M. Grosvenor brooded constantly on this question as National Geographic swept into its centennial year on a wave of sumptuous parties, special events, and public goodwill. To all outward appearances, the sprawling family concern had grown into a prosperous empire of many parts. Magazine and book publisher, exhibition sponsor and

---

*Phoenix is the only state capital surrounded by desert. The political units of Japan are prefectures. A canal connects Houston to the Gulf of Mexico.

television producer, patron of science and benefactor of education, advocate of environmental causes, it had respect, and it had money. Gil had recently over-seen the construction of the organization's fourth office complex in Washington. Completed in 1984, the seven-story pyramid backed onto Hubbard Hall and the Sixteenth Street complex of 1932. The new M Street building, which would house the National Geographic Society's growing television department and other divisions, along with the four-hundred-seat Gilbert H. Grosvenor Auditorium, had cost some $34 million to build. The organization had paid for the project in cash, but this still left reserves of $100 million, prime real-estate holdings of at least $100 million, and a blue-chip portfolio generating $30 million in interest and dividends.[41]

Under federal tax regulations, nonprofits such as National Geographic were free to rack up hefty surpluses, to occupy handsome offices, and to spend liber-ally for articles and television programming—as long as these activities were related to the organization's core mission. However, at the peak of Melville Grosvenor's tenure, the enterprise had grown so quickly that many executives worried whether National Geographic's success—and its conspicuous display of wealth—might draw complaints from other publishers and unwanted atten-tion from the Internal Revenue Service. For this reason, Melvin Payne fought a constant battle against Melville's free-spending habits, refused raises for himself for many years, and encouraged colleagues to prepare for the end of the boom.

By the time of the centennial, even as a mighty tide of celebration and ap-plause swirled around him, Gil Grosvenor had to admit that his old adversary Payne had seen the future clearly—financial trouble, Gil believed, lurked just around the bend.

Since the late 1970s, he had been alarmed about an almost imperceptible but persistent slide in annual membership renewals. Once in the 90 percent range, renewals now seemed stalled at about 85 percent, where they had hov-ered for most of Gil's decade as president. Even at 85 percent, that renewal rate was quite high for the magazine industry, but in the calculus of National Geo-graphic's financial well-being, one also had to consider the difficulties inherent in the organization's elephantine size: To remain 10.5 million strong while losing 15 percent of its base each year, the National Geographic Society had to scramble for 1.5 million new members annually. This was ruinously expensive, as the cost of promotion mounted steadily, just as other costs—for paper, printing, postage, and payroll—had been rising with inflation. To cover these expenses, National Geographic had a choice—it could trim spending, or it could raise annual dues. For years, it had taken the latter course: In Melville Grosvenor's day, dues stood at $6.50; by the time of the centennial, they had almost tripled to $18, and they would jump again in 1989, to $21. Each time dues increased, circulation plummeted, which required more spending to build it back. It was a vicious circle.[42]

To make matters worse, the organization's mailing list, once golden, seemed to have reached a saturation point. In its drive to diversify, the National Geographic Society had brought out many new products in a very short period—atlases, maps, calendars, slipcases, *Traveler, World,* and at least six book titles per year. All of these, along with the flagship magazine, were largely sold by direct mail to the membership list. This closed circle of membership restricted the organization's marketing options, but it had some advantages: It made available to members many products that could not be bought elsewhere, enhancing the notion that there was something exclusive about belonging to the National Geographic Society; and it kept the organization's books and magazines off the newsstands, which guarded against charges from publishers envious of National Geographic's tax-exempt status. The snag was that each product, as it competed for slots in an increasingly crowded mailing schedule, also competed for the attention of members, who were now inundated and annoyed by the flood of National Geographic solicitations. On average, a member received a piece of mail from the organization every three weeks. As the volume increased and each mailing became less effective, National Geographic was forced to send out more mail to achieve the same return.

With the benefit of hindsight, one might conclude that old Gilbert H. Grosvenor was on to something years before, when he had warned Melville about expanding too far beyond the National Geographic Society's core business. Whatever the reasons, National Geographic seemed fractured and spread dangerously thin as its centennial excitement began to fade. Income for 1988 would exceed revenues by only $2 million, a new low for the organization in recent times; the previous year, by comparison, had produced a surplus of $29 million. Alarmed about the health of his inheritance, Gil Grosvenor kicked off National Geographic's second century not with a call to greatness, but with an austerity campaign.[43]

Citing "changing demographics of our country and the saturation of the Society's membership with direct mail promotion," Gil announced the cancellation of two book series for children and the prospect of staff reductions in 1989. Departments would be consolidated or eliminated. And more than a hundred positions out of the National Geographic Society's workforce of 2,885 would be targeted for cutting, the first wave of a multiyear retrenchment campaign. Where possible, Gil stressed, staff reductions would be accomplished by reassignment or attrition. All divisions were expected to sacrifice.[44]

Bill Garrett resisted. If reductions were necessary, he and his allies felt, perhaps they should come from areas of the organization that brought in less money than the magazine, which then produced $240 million in dues each year, some 60 percent of National Geographic's annual revenues. "Why would you have the magazine get smaller as the population increases?" Garrett asked. "The smaller it gets, the higher the unit costs become, the less you have a mass audience for other things—like books, atlases, globes, maps, television. The magazine brings people to all of those areas."[45]

He managed to stave off Gil's staff reductions for the time being, but he was compelled to slice editorial pages in 1989 to pay for one of the centennial's most noticeable excesses, a holographic cover, published in December 1988, which had cost some $2 million, at least seven times the normal rate.[46] Garrett had commissioned the three-dimensional image, of a bullet shattering a crystal globe, to symbolize threats to the environment, the subject of a special anniversary series, "Can We Save This Fragile Planet?" Although ingeniously conceived, the hologram was difficult to read; it worked only if one tilted it this way and that until the right angle was found and light penetrated the image just so. Garrett freely admitted that the hologram had been an expensive experiment, but he argued that such innovations helped call attention to the maga-zine at a time when direct-mail campaigns and traditional marketing methods were increasingly ineffective. "That was why I did funny things like holo-grams, to make sure that nobody got bored. A magazine should be able to sell itself."[47]

This was just the sort of grandstanding that drove Gil to distraction. The bigger Garrett got, the more he overshadowed Gil, or so it seemed. Gil began complaining to friends and associates about Garrett. Why was the magazine bringing foreigners from Poland to photograph and write about America?[48] What was Garrett's deal with NASA, for which he arranged $250,000 for satel-lite monitoring of the rain forest in Mesoamerica?* Did readers really want a special issue devoted to France on its two-hundredth anniversary? And why was *National Geographic* doing a story on Jimmy Swaggart and the Bible Belt, of all things? In a rare show of authority, Gil spiked that story, an action that barely dented Garrett's armor-plated self-assurance. One day in the heat of argument, Garrett asked if Gil wanted his resignation. Gil remembered answering: "No, Bill, I just want you to join the team."[49]

As relations deteriorated between the old friends, neither was willing to give the other the benefit of the doubt. When Garrett drifted off in a meeting one afternoon, Gil took it as a sign of disrespect. "He fell ostentatiously asleep," recalled Gil, who had been addressing an editorial group on the need for belt-tightening. "What he was trying to do was to send a message: 'Don't worry about this, guys; it isn't going to happen.' "[50] Yet those who worked with Gar-rett during this period grew accustomed to his afternoon catnaps. He was working harder than ever, sleeping little at home, and finding no release from the escalating tensions. He stiffened his resolve to resist budget cuts, and he be-gan to grumble about Gil quite freely—to friends, to board members, even to

---

*According to Garrett's allies, Gil signed off on this initiative, forgot about it, challenged Garrett's authority to implement it, and had to back down when the editor produced Gil's memo of approval.

Joyce Graves, Gil's executive assistant.[51] He criticized Gil's lack of vision and his cowardice. If times were tight at National Geographic, perhaps it was because affairs had been so poorly managed, with resources drained away for new magazines such as *Traveler* and *National Geographic Research,* with $2 million worth of new atlases given away to schools, and with Gil's $40 million commitment to geography education. Those things contributed *nothing* to the sacred bottom line. How could Gil expect Garrett to cut corners when the magazine was still the major source of revenue? One evening, as photographer Sam Abell was walking along Seventeenth Street near the National Geographic headquarters, Garrett's car pulled over, the door swung open, and the editor offered a ride. "Of course I got in," Abell recalled, "and without any prompting, Garrett launched into a tirade against Gil. He was red in the face, really mad. God knows what set him off, but I was thinking *whoa!*"[52] Friends warned Garrett to be careful. The criticisms were drifting back to Gil.[53]

The strain began to show. Gil checked into the Washington Hospital Center for a coronary bypass operation, which was successful.[54] When he returned to work, he discovered that the atmosphere remained as toxic as ever. Someone was circulating a *New York Times* piece on the aftereffects of bypass surgery, which caused irrational behavior in some patients. Gil immediately suspected Garrett of disseminating the story, which to this day Garrett denies having done. At this point, however, it hardly mattered what Garrett said or did. During Gil's enforced absence, he had determined to forestall the inevitable no longer: Garrett had to go. Gil dispatched a trusted ally, Alfred J. "Winky" Hayre, a vice president and treasurer of the National Geographic Society, to offer Garrett a face-saving way out. "You are on thin ice, Bill," Hayre reportedly told him over lunch early in 1990, "and you are going to lose the magazine." Then Hayre dangled a golden parachute before the editor. This provoked an indignant response from Garrett: "The board would never let me go!" he is supposed to have said.[55]

Brilliant though he was, Garrett had woefully misjudged his support on the board, just as he had Gil's resolve. When the trustees gathered for a regularly scheduled meeting on April 12, 1990, Gil went through the routine business of committee reports and financial presentations as if nothing extraordinary was afoot. When the meeting adjourned, he asked fellow members of the nine-person executive committee to stay behind. In the hour-long discussion that followed, Gil made it clear that his problems with Garrett were not primarily editorial—although he had groused plenty about recent articles—but managerial: Gil was convinced that the organization had to cut staff and think smaller if the National Geographic Society was to survive. Garrett was constitutionally unable to accept this diminished vision of the future. That was the crux of their disagreement. Gil got the executive committee's unanimous consent to fire Garrett the following Monday, April 16. With this sanction in hand, Gil consulted Thomas E. Bolger, then chairman of Bell Atlantic and a board member, who provided particular advice: Give Garrett no warnings. Just tell him he's fired. You don't have to tell him why. Don't get into a discussion. Just do it quickly. Bolger arranged for a "facilitator," an outside expert in such terminations, to be present at the firing, along with Winky Hayre, who would witness the procedure.[56]

For Garrett, that Monday began like any other. His driver, George White, picked him up and they cruised through the spring morning, where wood ducks worked the Potomac backwaters and the last cherry blossoms fell like snow. After a round of morning meetings, Garrett joined Jon T. Schneeberger, a senior illustrations editor, for a visit to Capitol Hill, where they called on a congressman to discuss the Ruta Maya, one of Garrett's pet projects for preserving cultural sites in Guatemala, Belize, and Mexico. The meeting ran late, so Schneeberger and Garrett piled into the car and raced back to National Geographic headquarters, where Gil had scheduled a meeting for 3:00 P.M. To save time, Garrett jumped out when they reached the office and, on the trot, shouted over his shoulder to Schneeberger: "I'll call you when it's over." He bounded up the stairs and disappeared into the building.[57]

Arriving on the ninth floor, Garrett rushed to Gil's office for what he expected to be a routine meeting. There he found Gil and Winky Hayre waiting for him. Neither looked particularly happy to see him. Precisely what was said at this moment remains unclear, largely because the participants have been bound by confidentiality agreements. But Gil made it plain that Garrett had to go. He did not tell him why. He produced a piece of paper and asked Garrett to sign it. Garrett refused. Then, true to form, he tried to talk Gil out of it. "Gil, this isn't the way to do things. It's not the way *we* do things. Let's sleep on it." Gil was adamant—it was over. He directed the stunned Garrett to an adjoining conference room, where the facilitator awaited. The facilitator, who almost instantly came to be known as the "bouncer" or the "shrink" in office lore, led Garrett out of Gil's suite, across the lobby, and into the editor's office for the last time. From start to finish, the exercise had taken seven minutes. Prevented from gathering files, Garrett looked around, talked briefly with the facilitator, and finally walked away from the place that had been his professional home for thirty-five years. He was on the street by 3:30 P.M.[58]

By this time, *National Geographic* had a new editor. Gil had chosen Bill Graves, sixty-three, a valued friend who had joined the magazine's staff in 1956, just after Gil and Garrett. Graves had most recently been the journal's expeditions editor. Even before Garrett was off the premises, Gil and Graves swooped down on Joe Judge, the magazine's deputy editor, and fired him too. Graves did not trust Judge's loyalty. As Judge was getting his walking papers, the magazine's senior staff was summoned for a special meeting at 3:30 in the Control Center.[59]

The journal's art director, Howard E. Paine, arrived on time, took a seat among his colleagues, and scanned the room. "Something's up," he whispered to a friend. "Where's Garrett? Where's Joe?"[60]

Just then Gil walked in with Graves, both looking determined. Graves eased into the front row, clearing his throat repetitively as he did when nervous. Gil began to speak, noticed that the doors were open to the hall outside, and crossed the room to close them.

"I asked you to gather so I could make an important announcement," Gil said. "Bill Garrett is no longer editor, no longer with the Society . . ."

"Oh, shit!" whispered an editor sitting next to Graves, a little too loud.

"And I've asked Bill Graves to replace him. . . ."

"Everything's going to be all right," said the man beside Graves, patting him on the knee.[61]

Gil explained that he and Garrett had irreconcilable differences and that the board of trustees supported his decision to let Garrett go.[62]

The room sat silent for a long, awkward moment. Howard Paine raised his hand: "Gil, was Bill Garrett fired?"

Gil flushed. "He no longer works here," he said. The discussion was over.[63]

CHAPTER SIXTEEN
THE THIRD FLOWERING

NATIONAL GEOGRAPHIC SOCIETY PRESIDENT AND
CEO JOHN M. FAHEY, JR., CUBA, 1998.

At most media companies, the firing of a well-regarded editor might generate a brief sensation, some talk in the halls, and perhaps a few days of
news coverage before the ripples subsided. At the National Geographic Society,
where the attitude was familial, the pace of change glacial, and firings virtually
unheard of, Bill Garrett's departure constituted an epochal event, like a great
flood or a volcanic eruption that altered the landscape forever. His leaving ushered in a wave of leadership changes and upheaval still being felt today.

Gil Grosvenor has trouble talking about it. "I thought it was going to be like
a bad haircut—after three days everything would be OK," he said over lunch in

Washington. "I was wrong. Before the Garrett thing, I was one of the good guys. After it, I was one of the bad guys."[1]

Friends said that ousting Garrett had been the hardest thing Gil had ever done.[2] The aftershocks were harder yet, as Gil discovered when reporters began quizzing him about Garrett's dismissal. Used to soft treatment from the hometown press, he bristled when a writer from the *Washington Post* reminded him that one of Garrett's last issues of *National Geographic,* a single-topic number on France, had won a National Magazine Award.

"I don't believe our 40 million readers are really enchanted with single-subject issues," Gil said.* "The only critics of interest to me are the readers of *National Geographic.* I have had my share of honors and they're all stored in the drawer."[3] A few months later, Gil was vexed to find himself depicted as the villain in a *Washingtonian* magazine account of Garrett's firing. "The National Geographic had it all," the *Washingtonian* reported, "loyal employees, devoted members, a proud history. Then Gil Grosvenor made his move, and all hell broke loose."[4] Nothing would be the same for Gil or for the National Geographic Society afterward.

"That was the end of the Old Geographic and the beginning of the New," said an editor who lived through the changes of the 1990s,[5] which transformed an organization known for its paternalism, insularity, and endearing idiosyncrasies into something new—leaner, colder, more market oriented, more, in short, like any other business and less like the quirky place it had always been, part newsroom, part faculty club, part benevolent patriarchy, where employees had turned out on spring evenings to glimpse Bert Grosvenor leaving work in his black Cadillac,[6] where the staff was treated to an annual Thanksgiving lunch for twenty-five cents a head, where workers found birthday greetings on their desks once a year, where a handwritten note from the editor honored one's arrival on the masthead, and where the sons and daughters of National Geographic families often found work as legacy hires. For them as for many others

---

*For years, Gil and others at National Geographic used the forty million figure, which assumed that each of the magazine's ten-million members shared the journal with at least three other readers.

the place offered sanctuary, security, and opportunity in an uncertain world. Here Maynard Owen Williams, Luis Marden, Jane Goodall, Bob Ballard, and other unorthodox talents found patrons who understood them and encouraged them to shine, which accounted for National Geographic's occasional flashes of brilliance.

"Being there," said Mary Griswold Smith, "was like assisting a powerful, eccentric Victorian era collector. Everyone on the staff was encouraged to find and return to headquarters with the biggest lizard or the brightest butterfly or the oldest treasure from the farthest away place."[7]

But nothing lasts forever, and Gil began to worry that he might be the Grosvenor to preside over the demise of the National Geographic Society. After reaching a peak of 10.9 million circulation began dropping in 1990. Expenses continued rising, and for several years the organization had to take out short-term loans of up to $35 million in June to meet expenses until magazine renewals arrived between July and October. The alternative was to dip into reserves, which Gil refused to do. "It was a tough time for the economy," Gil recalled. "It was a high inflationary period. We were having trouble with promotion. We were finding it very difficult to replace our circulation. . . . We had to downsize. We had to downsize fairly rapidly." It also became clear, Gil said, that if National Geographic was to thrive "in a zero growth environment in membership, we were going to have to develop alternative sources of income. It was axiomatic."[8] At age sixty, having just recovered from a heart operation and contemplating his own retirement, Gil realized that time was running out.

His goal, largely defensive, was to place the organization on a sound financial footing before he stepped down as president in five years. He sketched out a retrenchment campaign that would occupy the remainder of his tenure: Bil' Garrett's firing represented the first in a program of forced retirements; would be cut by more than a third, from a peak of almost 2,900 to some today. Certain services previously performed by National Geographic en ees would be farmed out to contractors. The organization's unwieldy ment operation would be modernized. As an alternative to its tra flagship magazine, National Geographic would extend its reach into te launch a line of new products, and chase the emerging chimera of electronic publishing. With this lengthy agenda in hand, Gil had little time for second-guessing Bill Graves, whom he had entrusted to run *National Geographic* magazine.

Like Gil, Bill Graves came from a National Geographic family. His father, Ralph, had been a respected editor in the 1920s under Gilbert H. Grosvenor. When Ralph Graves died young, his widow married a prominent Washingtonian named Francis Sayre, who became the U.S. high commissioner to the Philippines before World War II. Bill Graves, then fourteen, was there with the family when the Japanese invaded in 1941. After dodging bombs and machine gun fire for several days, the Sayre family escaped Corregidor aboard the submarine *Swordfish*. They were seen off by Gen. Douglas MacArthur. Bill Graves eventually made his way to Harvard and to the Foreign Service, which posted him to Germany and Japan after the war. He returned to Washington as a newspaper reporter before joining *National Geographic*'s writing staff in 1956. (Journalism ran in the family—Bill's brother Ralph Graves rose through the ranks at Time-Life to become the last managing editor of the weekly *Life* magazine.)[9]

Thrust into *National Geographic*'s top job with little warning, Bill Graves found himself in the unenviable position of following the legendary Bill Garrett, and at a time when tensions over job security and crimped budgets had shaken the once placid organization to its core. Moreover, at age sixty-three, Graves was seen by many of his colleagues as a transitional figure who—like Jack LaGorce or Ted Vosburgh before him—would merely warm the editor's chair until a permanent occupant took it. Cynics even suggested that, with Graves as his puppet, Gil would finally regain control of the magazine he had lost to Garrett in 1980. Graves quickly dashed these perceptions. Within days of Garrett's firing, he called the staff together and announced that he intended to be editor for the foreseeable future. His appointment was open-ended. And with Gil nodding behind him, Graves made it clear that he—and not Gil Grosvenor—would be running the magazine. Then Graves tracked down one f the editors who had been spreading rumors about his transitional status and couraged him to leave *National Geographic* for another department.

Bright, decisive, and volatile, Graves was not to be easily crossed or dismissed. He could be frightening in a fit of anger, when he flushed deep purple and the veins in his neck stood out, a display that once prompted poker playing friends to lock him out of his own house until he calmed down. But Graves also had a generous side and a deep capacity for friendship. He was always the first to visit colleagues in the hospital and he signed memos with red hearts. He greeted visitors to his office with a paralyzing bear hug, and when the occasion demanded, he often found the right words to soothe frayed feelings.

While the memory of Garrett's departure was still fresh, Graves disarmed the staff by admitting that he was no Bill Garrett, and that he needed their help to keep the magazine going. "You are the people who make it work," he told his associates. "I have a lot to learn from you." He emphasized that, while some necessary budget-cutting was in store, it would be accomplished humanely and with little diminution of *National Geographic*'s emphasis on quality. Finally he promised that the magazine would not retreat from the real-world journalism Garrett had emphasized.[10]

"We're not going back to the rose-colored glasses," he announced in a meeting with hundreds of staff members in the week after Garrett's departure. "We should take that word out of our dictionaries. We're going ahead, to good places, fascinating places—and we're going to some terrible places too."[11]

For most *National Geographic* readers, the magazine under Graves seemed to change little from Garrett's day. In part, this was because so many of Garrett's articles were in the pipeline that it would take more than a year to publish them. Bylines remained familiar, as did subject matter and the high quality of *National Geographic*'s production values. Graves's tenure also coincided with a time of unprecedented geographical change, as the Soviet Union broke apart, the Berlin Wall fell, Eastern Europe took on a new character, war rocked tʰ Persian Gulf, China reabsorbed Hong Kong, and the place-names and bou aries on maps shifted from one month to the next. Graves dispatched the writers and photographers to document these changes and prepared new to keep readers current.

When the Persian Gulf erupted in 1991, the cartographic department had a new map, "Middle East/States in Turmoil," ready for publication that February. In addition to the copies provided with the magazine, an extra fifty thousand were printed and given to Allied forces in the region. Graves rushed a science writer and a photographer to report on the environmental aftermath of the war. He also commissioned timely coverage on the Kurds, on the Palestinians, and on the politics of water in the region. This fare was supplemented with the usual history, adventure, science, and natural history readers expected from the magazine.

"I had a lot of leeway with Gil because he trusted me," Graves recalled. "And I knew he wasn't going to fire another editor, so it gave me the confidence to do the stories I thought were right. I kept to the schedule, though—I didn't have the imagination Garrett did to change things at the last minute."[12] Indeed, Graves displayed little of Garrett's flair for the dramatic or the surprising, but he put new energy into the magazine, which soon adjusted to the editor's own frenetic style. He began each day the same way: At 4:00 or 4:30 A.M., he would sit bolt upright in bed and announce, "I'm late!" before racing off to work in the dark. Arriving at the office while the stars shone, he would proceed to his ninth-floor roost, zoom through the piles of memos and manuscripts on his desk and render his comments ("Jesus!" "Marvelous!" "Sob!" "Jesus Christ!!") with a slashing number-two pencil.* Then, gathering up his papers, he would race through the lower floors, distributing work for his editors, turning on their lights as he went. He dashed through the day this way, jotting notes on his hand with a ballpoint pen and showing up early for meetings and lunches. Colleagues soon learned to follow suit. In the Control Center, where a collection of clocks announced local times in Washington, London, Moscow, and Tokyo, a fifth clock was added to the wall: Labeled "Graves Time," it ran ten minutes ahead of normal.

*Editors and writers grew wary if Graves broke his pencil more than three times on a manuscript—sure sign the article was in trouble.

Under Graves's spur, staff writers began to meet deadlines. Those who did not were weeded out. Greater scrutiny was given to story budgets, which began to shrink. Layouts got finished without the sort of last-minute tweaking for which Garrett had been famous, and the magazine reached the printer on time. "The magazine expenses just plummeted under Bill Graves," said Gil Grosvenor. "We had been spending a tremendous amount of money reworking engravings at the printer, being late to the printer, deciding at the last moment to throw half an issue out. . . . Through his entire editorship Graves was a bear about deadlines and doing it once rather than twice on the presses."[13] Graves encouraged many veterans of the writing and editing staff to accept early retirement packages, just as those in other departments of the National Geographic Society did in the early 1990s. As the staff shrank, the magazine farmed out an increasing share of work to freelance writers and photographers. This made room for new contributors, such as Geoffrey C. Ward, Bill Bryson, T. R. Reid, Donovan Webster, Jennifer Ackerman, Richard Conniff, and others, whose writing brought new style and depth to the magazine. Well-known contributors such as Peter Benchley, Paul Theroux, Tad Szulc, and Ross Terrill continued writing for *National Geographic,* which often collected National Magazine Awards, including the much-coveted one for general excellence, on Graves's watch.*

Despite the magazine's solid performance under Graves, circulation continued to drop, plunging the organization into a long period of uncertainty and trouble. Memberships fell below the 10 million mark in 1992, hitting 9.7 million. Maintaining circulation proved increasingly difficult, and for the first time, the costs of promoting the magazine surpassed its editorial budget—a hint of the shifting values that would dominate the future. Under pressure from the board and determined to arrest the financial slide, Gil Grosvenor kept a hiring freeze in place and continued to cut staff to reduce costs through the 1990s.

---

*Shortly after Garrett's firing, Charles McCarry resigned in sympathy. On McCarry's recommendation, Bill Graves placed the author in charge of all text for the magazine.

With his own retirement fast approaching, he also began a search for his successor.[14]

For the first time in its history, the National Geographic Society had no member of the Bell-Grosvenor family in position to take over the organization. This shift arose for a combination of reasons. The family's influence, while still strong, had been diluted by the rapid growth of the organization, by the splitting of responsibility for editorial and executive jobs in 1967, and by the increasing influence of trustees sensitive to nepotism. None of Gil Grosvenor's oldest children had expressed interest in the family business.[15] His son Hovey, who had worked as an intern for National Geographic television, had moved to Oregon, where he played in a band called Orange Splat and struggled to make a living as an independent television producer. His daughter Alexi was finishing medical school in North Carolina and would soon become a pediatrician. His son Graham, still in high school, was too young.

Then there were Ed and Sara Grosvenor, the children of Melville Grosvenor's second marriage, whose hopes for National Geographic careers seemed to die with their father in 1982. Ed Grosvenor had considerable editorial experience; he had photographed at least five articles for *National Geographic* and he had launched two well-received, but ultimately unsuccessful, magazines on his own. But he had taken himself out of the National Geographic orbit to pursue these ventures at a critical time, and he was never able to gain reentry. In part this was because Ed had offended some colleagues at the magazine, where he had demonstrated little of the modesty other family members had displayed when they were in training there. "He was too bumptious when he was young," said his Aunt Mabel Grosvenor. "I don't think it would have ever worked."[16] Neither, apparently, did Gil. Having survived the trial of dealing with Mel Payne and Bill Garrett so recently, Gil had little interest in inviting another management challenge into the tent—even if that person was his half brother.

Ed's sister, Sara Grosvenor, a Stanford graduate who had studied at the

Missouri School of Journalism, had been an intern for *World,* National Geographic's children's magazine, and had taken contract assignments in other departments before joining *U.S. News & World Report* as a picture editor. She was smart, capable, and very much committed to the family's legacy, but she never seemed to achieve traction at the National Geographic Society. According to friends, she suspected that Gil had barred her progress, perhaps out of resentment over Melville's second marriage. There is no evidence of this. But it is also clear that Gil felt none of the protective parental feeling for his half siblings that had propelled previous generations of Grosvenors up the ladder at the National Geographic Society. And Gil was sensitive about nepotism, the old ghost that had haunted most of his career.[17]

The arguments for nepotism—a tradition frowned upon in a culture that esteems individual merit—are that it allows families to test future leaders under close observation, provides for continuity of values, and usually assures orderly transitions.[18] None of these was in the cards for National Geographic, which now went through a confused period of executive succession. Looking outside the family, Gil considered and rejected several National Geographic candidates for the president's job. John T. Howard, who joined National Geographic from Time Customer Service, Incorporated (TCS), the Tampa-based fulfillment part of the Time-Life empire, was to update operations at the Membership Center Building (MCB), the National Geographic's fulfillment operation in Gaithersburg, Maryland. But after barely a year on duty, Howard accepted a handsome settlement package and disappeared from the upper reaches of *National Geographic*'s masthead in 1992. Timothy T. Kelly, a rising star who headed the organization's television division, took the spotlight after Howard's departure. Kelly soon fell out of favor with Gil, who then put his hopes on Michela English, a highly intelligent and personable executive in charge of National Geographic's marketing and book-publishing program. After a few months, Gil lost interest in her as well. The pattern was becoming all too familiar: Like a teenager in love, Gil would seize upon an heir apparent, idealize that executive, only to cool on him or her at the first signs of imperfection. Toward the end of 1992, as it became evident that Gil's successor would not emerge from the ranks of National Geographic, board members prevailed upon him to consider candidates from the outside, using an executive recruiter. The successful applicant would have two assignments—to sort out the chaos at MCB and to prepare for National Geographic's presidency when Gil stepped down in 1995.[19]

For reasons still mysterious to many of his colleagues, Gil settled on Reg Murphy, then vice president of the United States Golf Association. Murphy, fifty-nine, a low-key Georgian with a bloodhound's furrowed visage and a disarming aw-shucks style, had a reputation for toughness in the newspaper world, where he had spent much of his career trying to shore up struggling dailies. As chairman and publisher of the *Baltimore Sun* in the 1980s, he had introduced new technologies, emphasized a move to zoned editions, shut down many of the paper's foreign bureaus, slashed costs, and engineered the sale of the family-owned *Sunpapers* to the Times-Mirror Company, earning himself an estimated $13 million to $16 million in the transaction.[20]

He also earned the enmity of many in the Baltimore newsroom, where he was known for his hard-nosed negotiating tactics, his self-promotion, and his single-minded attention to the bottom line. During the last major newspaper strike in that city, reporters on picket lines had chanted "Reg must go!" One carried a sign reading PUT REG BACK IN THE TRUNK![21] a rather insensitive reference to the time that Murphy, while editor of the *Atlanta Constitution,* was kidnapped, locked in the trunk of a car, and held hostage by right-wing terrorists in 1974. After forty-nine hours, he was released unharmed. His captors were convicted and sent to prison, while Murphy wrote about the ordeal for his newspaper and won a Pulitzer Prize that year.

Murphy reported for work as executive vice president at the National Geographic Society in May 1993. His assignment was to put the organization on a sound financial footing, to develop a strategic plan for its future, and to impose for-profit sensibilities on an institution proud of its nonprofit traditions. Shortly after his arrival at National Geographic, Murphy accelerated Gil's austerity program, urging other executives to join him in scrounging to curb expenses. He ridiculed the magazine's fact-checking process, which subjected manuscripts to intense scrutiny in the name of absolute accuracy, the first of the guiding principles laid down by Gilbert H. Grosvenor. "I don't want them to waste their time any more calling the Library of Congress to find out how high is an elephant's eye," Murphy explained to a reporter, repeating a line he often used in meetings with the editorial staff.[22]

Much of Murphy's economizing was long overdue, but a great many employees who suffered from it were the least able to afford it. Most of the organization's modestly paid service workers—custodians, gardeners, security officers, and cafeteria staff—were laid off and replaced by contract employees. Such moves, coming from a multimillionaire who seemed to value few of National Geographic's paternal practices, did little to endear Murphy to his new associates. Nor did an expensive salary study he commissioned by Towers Perrin in 1996. The consulting firm, examining how the National Geographic Society's salaries compared with other publishing and broadcasting companies, concluded that most employees at the Society were paid fairly. The exceptions were long-serving administrative assistants (paid too much, according to Towers Perrin) and top executives like Murphy (paid too little). Adjustments were made accordingly. Reg Murphy's salary jumped from $550,000 to $600,000; under recommendations from Towers Perrin, a new executive bonus structure was adopted, which sent compensation soaring for National Geographic officers in future years. The president's compensation jumped from $453,500 in 1996 to $813,740 in 2002; in the same period, the pay for National Geographic's editor zoomed from $275,406 to $476,925.[23] A new executive culture, at once mercenary and more devoted to the bottom line, was taking hold.

Meanwhile, the troubles continued at National Geographic's MCB fulfillment center, a one-hundred-acre facility in Gaithersburg, Maryland, where 350 Society employees kept track of subscriptions and membership records. Faced with modernizing the operation or closing it, Murphy decided to sell the Gaithersburg property. Its workforce was laid off and its functions assigned to outside contractors in 1997; one of these, ironically, was Time Customer Service (TCS), the Tampa establishment that had provided National Geographic with the star-crossed John T. Howard a few years before. During the transition, National

Geographic headquarters was flooded with calls from frustrated members complaining about delayed subscriptions and indifferent service at TCS.[24] This may have been one of many factors contributing to *National Geographic* magazine's continuing slide in circulation, which had dropped to 8.5 million by 1998.[25]

On the positive side, the Gaithersburg sale produced $22.2 million—far less than the property was worth, but a welcome infusion of cash for National Geographic.* The transaction also reduced a $15 million drain on the organization's annual budget. Murphy's penny-pinching, combined with the timely sale of land and securities, put National Geographic comfortably in the black for a few years, from 1997 to 1999.[26] "The bottom line was restored to where it was in the late eighties, which was very comforting," Gil Grosvenor recalled.[27]

While the belt tightening helped to slow the losses, austerity measures could not guarantee the organization's future prosperity. The magazine continued to falter, which roused Murphy to explore new sources of income. He expanded National Geographic's book-publishing program, which would produce forty or fifty titles a year, including trade books; these would be sold in stores and taxed at the normal rate. The organization also waded into the crowded world of outdoor magazines, hiring John Rasmus, the respected founding editor of *Outside,* to launch *National Geographic Adventure* in 1999. And to stretch beyond its traditional audience, the organization brought out its first foreign editions of *National Geographic,* which eventually reached 2.4 million subscribers in twenty-six languages, supplementing the erosion of readers at home.

In addition to this expansion, both Reg Murphy and Gil Grosvenor saw greater opportunity for growth in television, film, and electronic media, perhaps a logical progression for an organization which, after all, had made its reputation disseminating visual images to a mass audience. Yet this new terrain, in

---

*One year after paying for the MCB property, Manor Care resold it for $55 million in 1999; less than three years later, it was sold again for $62.6 million.

addition to being notoriously boggy, was also very expensive real estate to acquire. Even with its healthy reserves, the National Geographic Society could ill afford the substantial start-up costs of film, television, and electronic publishing without raising massive capital or forming partnerships with those who already had it. At the same time, the organization had to preserve its tax-exempt status, which might be threatened if the Internal Revenue Service determined that it was straying too far from its core educational mission. To break free of these strictures, Reg Murphy devised a restructuring of the National Geographic Society that would change it forever: From 1995, the organization would bifurcate, with the National Geographic Society remaining a tax-exempt publisher of books and magazines and a dispenser of grants, while its wholly owned subsidiary, National Geographic Ventures, would operate for profit, create new businesses, form partnerships, and pay taxes.

To run this new enterprise, Murphy and Grosvenor chose John M. Fahey, Jr., a forty-four-year old M.B.A. who had most recently been president of Time-Life, Incorporated, the direct-marketing arm of Time Warner in Alexandria, Virginia. Fahey, a prematurely gray man with a smooth style and a forthright manner, proved to be a born deal maker. Under his guidance, and with an infusion of more than $150 million from the parent National Geographic Society, Ventures was soon launched in a frenzy of new directions—making feature films, building a Web site, and compiling a CD-ROM of more than one hundred years of *National Geographic* magazine. It formed a pact with Rupert Murdoch to ease National Geographic television into cable markets at home and abroad. In the same entrepreneurial spirit, Venture's parent National Geographic Society branched out into the marketplace by licensing its brand for martini glasses, microfiber barn coats, lamps, and toys.[28]

For a fundamentally conservative organization that had always been slow to change, this rush of commercial activity produced tectonic consequences, with National Geographic veterans and Ventures newcomers staring at one another across a widening fault line, new media on one side, old media on the other. Competing for the same internal resources with little common ground between them, members of the divided National Geographic family proved to be reluctant collaborators. Murphy tried to bully the traditionalists, whom he accused of arrogance and elitism. They viewed Murphy, Fahey, and their followers as quick-buck artists who demonstrated little appreciation for the value of tradition.

But the old order was fading. Bill Graves had been out for a year, reluctantly retiring as editor at the end of 1995. His replacement as editor was William L. Allen, a mild-mannered Texan who had spent most of his National Geographic career as a picture editor. Although conscientious and bright, Allen demonstrated neither Bill Garrett's editorial panache nor Bill Graves's strength of character. Averse to confrontation and afraid of Reg Murphy, Allen watched the magazine shrink, both in literal terms and in terms of the standing it had held within the organization. He cut pages, lightened the magazine's paper weight, reduced the number of map supplements, slashed field time, and gave over an increasing share of editorial pages to advertising—all to meet the straitened circumstances imposed upon him by a new administration that viewed *National Geographic* magazine as a relic of the glory days but still needed its revenues. Allen yielded them, while producing a magazine that remained quite good, but a bony, lightweight version of its former self, at higher cost to members.*

After seething in private for months, the cultural clashes at the National Geographic Society finally broke into the open in August 1997, when the *New York Times* published a lengthy article, "Seeing Green in a Yellow Border," by Constance L. Hays. Documenting the magazine's decline and the rise of a brash new regime at the National Geographic Society, she detailed how conflicting styles had split old-timers and newcomers, threatening harmony at the $500 million-a-year enterprise. "With change there is nearly always protest," she wrote, but at National Geographic, "the protest is so sustained that it suggests the society may be abandoning what has made it unique all these years—and in

---

*Circulation plummeted while Allen was editor. When he retired in 2005, Chris Johns, a veteran *National Geographic* photographer, was named to replace him.

the process, trading in its rather classy image for a more commonplace devotion to the bottom line."[29]

Almost unnoticed amidst the changes he had unleashed, Gil Grosvenor had gradually withdrawn to the sidelines, retiring as president in 1996 but retaining the largely ceremonial title of chairman, just as his father and grandfather had done. At Gil's insistence there was no ceremony to mark this passage; worried that he might "puddle up" in front of colleagues, he stepped aside as quietly as he had arrived on the scene four decades earlier.[30] But he had not stepped out. He began to get complaints about Reg Murphy, who had with power grown contentious and impatient toward those who questioned the pace of change at the National Geographic Society. Gil breathed a sigh of relief when Murphy abruptly resigned as president in December 1997 after only eighteen months on the job. Murphy took a final salary and accrued benefits of $904,000 with him.[31] He announced that he wanted to write books and play more golf.[32] He was named vice chairman of the board, Gil said, "so it didn't look like he was getting canned."[33]

John Fahey left Ventures to accept appointment as Murphy's replacement. He would continue to press most of the initiatives Murphy had launched, but do a better job of selling these programs within National Geographic. He kept the door of his office open to all comers, answered a flood of e-mails from the staff each day, and met weekly with small groups of employees, working through the alphabet again and again. He shared information willingly, encouraged others to do so, and sweet-talked reluctant colleagues into his new vision of the National Geographic Society. "Just because we're aggressively going after these new initiatives doesn't mean that we're abandoning the magazine," he said recently. "We're not. But we have faced the question: Are we all about being a magazine, or should we be something beyond that?[34] . . . We want to make sure that we accomplish the mission in every way possible. . . . We have to adapt the delivery to the changes—the sometimes revolutionary changes—in technology and behavior on a worldwide basis."[35]

It could be argued that this recent expansion of National Geographic— into cable television, electronic media, foreign languages, and feature films— constitutes the third great flowering of the organization. The first had come in

1899, when Alexander Graham Bell took over a barely viable society, saw its promise, and hired the young Gilbert H. Grosvenor to transform *National Geographic* magazine into an illustrated popular journal for a national audience. The second renaissance began when Melville Grosvenor got control of the organization in 1957 and put new life into it, boosting color photography to new levels, branching into television for the first time, and hiring new talent. A similar democratic impulse inspires today's National Geographic Society, which distributes *National Geographic* to 9 million subscribers in translation and English; proselytizes to 1.1 million children through *NG Kids,* successor to *World* magazine; and delivers stories by cable television to hundreds of millions of subscribers in the United States and abroad. Millions of viewers regularly visit its Web site, *nationalgeographic.com,* each month. In addition to these editorial programs, the organization devotes an increasing share of its annual budget to scientific research, conservation, and geography education.[36]

Keeping track of this sprawling enterprise is no easy task, so the upper echelon of the organization has grown thick with executive talent in recent years: Where there was one president, one vice president, and one editor at the National Geographic Society, now the organization and its subsidiaries employ six chairmen, twenty-two presidents, six executive vice presidents, twenty-five senior vice presidents, and fifty vice presidents. There are four staff photographers. The old culture at National Geographic, which had been driven and directed by editorial talent, has been replaced by a new culture dominated by marketing experts, lawyers, advertising executives, and accountants. They are paid very well.

John Fahey's most recently reported compensation, for the 2002 tax year, was $813,740 in base salary and bonuses. This is almost double the $420,000 paid that year to Steven J. McCormick, president of the Nature Conservancy, an Arlington, Virginia, nonprofit with assets of $3 billion (almost triple National Geographic's), 3,200 employees (double National Geographic's workforce), and annual expenses of $661 million ($200 million more than National Geographic's). Other prominent nonprofits pay their chief executives considerably less: The president of the American Red Cross earns $377,000; the president of the Rockefeller Foundation, $598,479; the president of the Ford Foundation, $651,713. Fahey defends his pay at National Geographic, where executive salaries are set by the board's compensation committee based on comparable positions in the job market. No comparable positions for his job exist in the nonprofit world, according to Fahey. "The comparables for running a high-end magazine operation are going to be in for-profit industry," he said.[37]

It would be nice to report that the new National Geographic Society is thriving in its tertiary efflorescence. But even John Fahey says that it is too soon to declare victory. "The jury is still out to see how successful we will be up against all the other guys out there," he said. "Our history and our reputation are at once our greatest assets. But they put us in a position to take fewer chances."[38]

Some chances have already paid off, such as the organization's foray into foreign-language editions for books and magazines, which produced $12 million in revenues for 2003.[39] Like other publishers, National Geographic also placed great hopes—and sizable investment—in electronic publishing in the 1990s, but has not, as yet, received much of a return. After the organization's online operation lost $30 million over several years, its staff was cut back sharply in 2003.[40] Its CD-ROM compilation of National Geographic magazine, "The Complete Geographic," sold briskly, racking up net revenues of $2 million until it was pulled off the shelves in 2003 because of a copyright dispute with writers and photographers.[41] To date, several federal courts have ruled in cases brought by the artists, who charged that the National Geographic Society infringed on their copyrights by publishing the CD-ROM without their permission; while the artists have won in a few instances, the major court rulings have gone National Geographic's way—in effect holding that the CD collection is an archival reprint of the magazine itself. With further appeals pending, however, a protracted—and costly—legal wrangle appears likely. National Geographic has already accumulated some $10 million in legal fees, $9 million of which will be paid by insurance.[42]

In March of 2000 the lobby of the National Geographic headquarters became a $10 million state-of-the-art television studio, the flashy focus of Explorers Hall. Here National Geographic Today, the live daily news show and the flagship of the organization's new television channel, was launched in January 2001. It premiered to good reviews but failed to attract many viewers. It went from being live to being taped, from running an hour to a half hour, finally disappearing from the screen in 2003. Its staff was dismissed and the studio closed. The National Geographic Channel is in the red now, but Fahey expects it to be making money in two or three years. Until then it will scramble for advertisers and viewers, many of whom get the channel but do not realize it amidst the clutter of so much other programming.[43]

Since many of the organization's new businesses are still in the formative stages, few of them are producing much revenue. Most money still comes from old reliable *National Geographic,* which provides, even in its reduced state, about half of the organization's total revenues, the rest coming from book sales, rents, royalties, and dividend income. With advertising taking over an increasing share of *National Geographic,* ad sales have increased in recent years. So has the price of the magazine, which has climbed from $21 in 1990 to $34 today. The magazine still attracts the best photographers and writers; although it has skimped on story budgets, it pays well by comparison to other periodicals, and it covers the considerable expenses required for preparing a *National Geographic* article. The organization's other major magazines, *Traveler* and *Adventure,* have contributed little to the bottom line. *Traveler,* set back by the aftershocks from September 11, has broken even by cutting costs in some years, and has lost money in others; *Adventure,* now in its sixth year, is still considered a start-up and is yet to turn a profit.[44]

The cost of its many start-ups, the wobbly performance of the stock market, and increasing promotion costs have conspired to drive National Geographic's total revenues down, from $542 million in 1999 to $505 million in 2000, to $486 million in 2001, to $434 million in 2002. In that same period, surplus revenues—National Geographic does not have profits—have proven erratic, from an excess of $59 million in 1999 to $17 million in 2000, to a deficit of $23,424 in 2001, to a razor-thin surplus of $2.9 million in 2002.[45] Had National Geographic not sold securities in 1999 and 2000, its surplus would have been much less—or, in 2000, nonexistent. The narrow margins are a recent worry for an institution that paid little attention to the bottom line for most of its life and piled up substantial reserves in the process. Ironically, in the years since National Geographic Society has put more emphasis on making money, its surplus has dwindled.

John Fahey predicts a few more lean years, followed by better times. "We'll have a few years ahead of us that will be tough," Fahey told the staff recently, "but I'm feeling confident. I see a big light at the end of the tunnel. If we keep being diligent, we'll be in good shape."[46]

For his part, Gilbert M. Grosvenor liked and trusted John Fahey, whom he considered an able custodian of the family legacy. "John is absolutely committed to the core mission," Gil said recently. "It's music to my ears."[47] To that musical accompaniment, Gil retreated from day-to-day involvement at National Geographic, but he remained restless and busy, growing azaleas and breeding spotted horses in the Virginia countryside, presiding at board meetings, and traveling to make speeches for the National Geographic Society. He almost single-handedly filled the coffers for his pet project, the Geography Education Fund, which had amassed an endowment of some $100 million by the end of 2003.[48] The fund, a subdivision of the National Geographic Society, dispenses money for teacher training and other programs to raise geographic literacy throughout the country.*

On the far side of the Potomac, Bill Garrett settled into a fitful retirement, tending his vineyard, bottling his own wine, giving advice to the Nature Conservancy, and keeping quiet about the fate of *National Geographic* magazine. Under the terms of a generous separation agreement, Garrett was forbidden to criticize his old employer, to compete with the magazine, or to enter the headquarters building. Friends said that he survived the first years of exile by staying in perpetual motion and by wrapping himself in a protective carapace of denial.

---

*President George W. Bush, citing Gil's contributions to geographic education, awarded him the nation's highest civilian honor, the Presidential Medal of Freedom, in the summer of 2004.

"He could not accept that it was over," said his friend John M. Keshishian.[49] To this day, Garrett does not know why he was fired. "Nobody ever told me," he said. But almost fourteen years after the event, Garrett finally began to mellow. "I should probably call Gil and buy him a drink," Garrett said recently. "If I had stayed at National Geographic I probably would have had a heart attack."[50]

Like Gil Grosvenor, Garrett had entered his seventies. Both were obliged to attend more funerals than they cared to. They often saw each other at such gatherings but they never spoke, a situation that Ken Garrett, Bill's son and a National Geographic photographer, found absurd. He proposed a reunion for the old comrades. Gil readily agreed. Bill Garrett was less receptive. "No way!" he said. But he did not mean it. The families gathered at Ken's house in rural Virginia, where they shared a Sunday lunch, watched a Redskins game, and celebrated Ken's fiftieth birthday—this was the present he wanted. "We talked as old friends," Bill Garrett recalled. "He never brought up the firing."[51]

This time they parted without bitterness, two battle-scarred old men who had set out together as Young Turks not so long ago, eager to conquer the world and all that was in it.

ACKNOWLEDGMENTS

This book began to take shape long before I realized it. The germination probably dates to April 1995 when I agreed to conduct an oral history interview with my friend and colleague Luis Marden, the legendary character who appears elsewhere in these pages. That single meeting, taped for the National Geographic Society's archives division, began a conversation that stretched over the course of many months and many reels of tape, eventually encompassing half a century and most of the world. Hearing Luis relive the old days reminded me what a remarkable home the National Geographic Society had been for both of us. I could also see how the place was changing as the twentieth century faded, the founding family withdrew, and the institution struggled to define itself in new terms. The rapid nature of that change convinced me that it was time for a book—to document what the National Geographic Society had been and to suggest what it was becoming.

Having determined to tell that story, I realized that I could not remain at the National Geographic Society if I wanted to write candidly about it. So, after describing my plans to John M. Fahey, Jr., and to Gilbert M. Grosvenor and winning a pledge of cooperation from each, I retired as executive editor of *National Geographic* magazine and began the book in July 2001.

This project could not have been completed without the enthusiastic support of Fahey and Grosvenor. With no expectation of reviewing my work before publication, they provided me with unrestricted access to staff members, to the institution's archives, and to many hours of taped and unpublished oral history interviews. Fahey and Grosvenor were more than generous with their own time, sitting for several interviews, fielding e-mail queries, and steering me to helpful sources. Both executives helped me to make this book as honest as I could.

The first time I knocked on Dr. Mabel H. Grosvenor's door and she answered with a resonant "Hoy! Hoy!" I knew I was in luck; it was the same interjection her Grandfather Alexander Graham Bell had used to welcome visitors and the one he advocated for answering the telephone. It never caught on, but more than a century later the family tradition lives on, courtesy of Dr. Mabel. A retired pediatrician who closely observed her kin for most of a century, she became my best source for family lore. In a series of eight interviews begun in 2002, she vividly recalled her Grampy Bell, her father, Gilbert H. Grosvenor, her brother Melville Bell Grosvenor, and other figures who wander through this volume. I consider myself fortunate indeed to have had the benefit of her sharp memory and frank commentary. Her nephew Edwin Augustus Grosvenor Blair supplemented Dr. Mabel's recollections with other anecdotes and insights, as did his cousins Hugh Muller and Helene Pancoast. Other members of the Grosvenor clan—Aleck Myers, Bert Coville, Jim Watson, and Bob Watson—helped me sort out the fine points of family history. I am especially indebted to Anne Revis Grosvenor, who unearthed and generously shared a wealth of unpublished correspondence that shed light on her husband Melville Grosvenor's life at the National Geographic Society. For helping to organize that correspondence, thanks to Sara Revis, Anne Grosvenor's sister.

Ansley MacFarlane, manager of the Alexander Graham Bell National Historical Site in Baddeck, Nova Scotia, kindly made her museum's archives and staff available to me, fielded questions, and provided suggestions for interviews. Thanks to her, and to Gil and Wiley Grosvenor, who invited me to the family sanctuary in Baddeck, introduced me to relatives and friends there, and provided a tour of the old Bell estate at Beinn Bhreagh. To Bill Stephens and Jimmy McKillop of Baddeck, my gratitude for their hospitality and guidance.

I owe a special debt to the unfailingly professional staff of the Manuscript Reading Room at the Library of Congress, where the Hubbard, Bell, and Grosvenor family papers are housed. Not only did the library staff keep the boxes of letters coming my way for months on end, they also helped to decipher passages of correspondence that seemed positively hieroglyphic. One day I was surprised that Professor Edwin A. Grosvenor had described Elsie Bell, his prospective daughter-in-law, as "a tramp." I passed the letter to the reference desk for a second opinion—a moment of scrutiny, a wrinkled brow, a check of the unabridged dictionary, and the code breaker rendered his verdict: "A trump!" he said triumphantly. "The professor called her a *trump*." For setting me straight on that and other mysteries, my sincere thanks to the aces of the reading room: Fred Bauman, Ernie Emrich, Jeff Flannery, Ahmed Johnson, Patrick Kerwin, Bruce Kirby, and Laura Gottesman.

Allan J. Teichroew, who helped organize the Grosvenor Family Papers for the Library of Congress, also provided me with invaluable guidance.

For information on the National Geographic Society's 1963 Everest Expedition, I am grateful for the insights of Dr. Thomas F. Hornbein, one of the survivors of that historic climb. Also thanks to Dr. Rob Schaller, who provided details of his participation in several CIA expeditions to the Himalayas in the mid-1960s; my appreciation also to mountaineer Jim Wickwire and to researcher Carroll Dunham for perspective on those clandestine missions. Washington journalist Ronald Kessler kindly provided background on the FBI's secret work at the National Geographic Society.

For their help in other areas of research I express sincere thanks to Yolanda Maddux; to the arctic explorer Sir Wally Herbert; to William S. Dudley, director of the Naval Historical Center; to George M. Elsey, a trustee emeritus of the National Geographic Society; to John M. Keshishian; and to the writer Paul Dickson, who generously shared his research files on National Geographic.

From the beginning of this project, my former colleagues at the National Geographic Society expressed keen interest and offered their help. Three former editors in chief of *National Geographic* magazine—Bill Garrett, Bill Graves, and Ted Vosburgh—graciously took time for interviews and offered suggestions; Vosburgh also gave me permission to draw upon his unpublished autobiography, which added valuable material on the Chandler case and the crucial transitions of the 1970s and 1980s.*

Other current or former National Geographic veterans who shared their recollections and offered encouragement were Sam Abell, Tom Allen, Andy Brown, Bob Cline, Wade Davis, Lilian Davidson, Bill Ellis, Ken Garrett, Bob Gilka, Wendy Glassmire, Delores Granberg, Joyce Graves, David Alan Harvey, Gifford Hampshire, Chris Johns, Emory Kristof, Chris Liedel, Ed Linehan, Charles McCarry, Mary McPeak, Ethel Marden, Karen Marsh, Maura Mulvihill, George Newstedt, Howard Paine, Oliver Payne, Dick Pearson, Greg Platts, Jennifer Reek, Jon Schneeberger, Dan Shaffer, Mary Griswold Smith, Joe Steptoe, George Stuart, Joel Swerd-

*Several of my most faithful informants died while this book was in preparation: Luis Marden, Bill Graves, Ted Vosburgh, Andy Brown, Gifford Hampshire, and Jon Schneeberger. All will be sorely missed.

low, Charlene Valeri, Kurt Wentzel, and Peter White. Other friends from the National Geographic Society assisted me in numerous ways: Karen Sligh transcribed hours of taped interviews and helped with research; Katie Wogec provided picture research; Marilza Iriarte copied mountains of memos and letters from the archives. The keepers of those archives—Renee Braden, Cathy Hunter, and especially Mark Jenkins—helped guide me through a rich and seemingly inexhaustible source of material. Roger Nathan, a friend and an avid collector of all things relating to the National Geographic Society, opened his collections in Woodstown, New Jersey, to me. Thank you all.

Terrence B. Adamson, an executive vice president of the National Geographic Society, went out of his way to back me from the inception of this project. He argued my case at the highest levels, offered invaluable advice, and gave his unstinting friendship—all gratefully appreciated.

The camera never lies, but in the hands of photographer Marion Ettlinger, it can transform the most unpromising subject into a worthy one. Bless you, Marion.

My agent, Melanie Jackson, deserves special accolades, as does her associate, Andrea Schaefer. Melanie spotted this book when its form was vague and my path uncertain. She set me on the way, encouraged me to keep at it, and delivered the project into the gifted hands of Ann Godoff, my editor at The Penguin Press. Ann's steadfastness, wisdom, and conscience made my work better than I ever imagined it could be, for which I thank her in the Spanish style, *con el corazón en la mano,* "with my heart in hand." Thanks also to Jason Brody, a splendid copy editor, and to Sharon Gonzalez, Meredith Blum, Sophie Fels, and Liza Darnton, the Penguin Press professionals who kept the book moving toward completion. Darren Haggar provided the handsome jacket design. Tracy Locke and Ashmimi Ramaswamy brought their inventiveness and boundless enthusiasm to bear on publicity for the book. My gratitude to all.

Finally, thanks to my beloved wife and chief researcher, Suzanne K. Poole. She dug through the archives with me, hurtled down a hundred blind alleys at my urging, provided timely and thoughtful readings of the manuscript, and absorbed the stresses of the creative process with unflinching good humor and style. She did so because she is a trump.

NOTES

The primary sources for this book consist of interviews by the author; the extensive collection of Hubbard, Bell, and Grosvenor family papers in the Library of Congress and in the Alexander Graham Bell National Historic Site in Baddeck, Nova Scotia; a private collection of papers in the possession of Anne Revis Grosvenor; and the official archives of the National Geographic Society.

I am also indebted to the excellent work of C. D. B. Bryan and Charles McCarry, who have written previously about the National Geographic Society and its founding family; to Robert V. Bruce, whose biography of Alexander Graham Bell stands as the definitive work on the inventor; to Sir Wally Herbert, for his exhaustive study of Robert E. Peary's quest for the North Pole; and to James Ramsey Ullman and Thomas F. Hornbein for their accounts of the American Everest Expedition of 1963.

The following are abbreviations for archival sources appearing below:

AGBNHS   Alexander Graham Bell National Historic Site, Baddeck, Nova Scotia.
AGP   Anne Grosvenor Papers, privately held, Bethesda, Md.
BFP   Bell Family Papers, Library of Congress, Washington, D.C.
GFP   Grosvenor Family Papers, Library of Congress, Washington, D.C.
HFP   Hubbard Family Papers, Library of Congress, Washington, D.C.
NGA   National Geographic Society Archives, Washington, D.C.

PROLOGUE

1. Flannery O'Connor to anonymous correspondent, June 28, 1956, *Flannery O'Connor: The Habit of Being,* ed. Sally Fitzgerald (New York: Farrar, Straus & Giroux, 1979), 164.
2. Ralph Gray, *That We May All Know More of the World: The Founding of the National Geographic Society And Its Century-Long Association with the Cosmos Club* (Washington, D.C.: The Cosmos Club, 1987).
3. Gardiner Greene Hubbard, "Introductory Address by the President, Mr. Gardiner Greene Hubbard," *National Geographic,* Oct. 1988, 3.
4. Alexander Graham Bell to Gilbert H. Grosvenor, Mar. 5, 1909, Aug. 7, 1899, Sept. 21, 1899, Jan. 21, 1901, Mar. 7, 1901, Mar. 8, 1901, Oct. 29, 1903, Oct. 17, 1907, Oct. 21, 1907, Nov. 1, 1907, GFP.
5. Alexander Graham Bell to Gilbert H. Grosvenor, Mar. 5, 1900, GFP.

CHAPTER ONE: ALEC AND MABEL

1. Robert V. Bruce, *Alexander Graham Bell and the Conquest of Solitude* (Boston: Little, Brown and Company, 1973), 91–92.

2. Mabel H. Grosvenor, interviewed by author, Dec. 19, 2002.

3. Elsie Bell Grosvenor, address at the dedication of Alexander Graham Bell National Historical Site, Baddeck, N.S., Aug. 16, 1956, GFP.

4. Mabel Hubbard Bell, unpublished autobiography, HFP; Helen E. Waite, *Make a Joyful Sound* (Philadelphia: MacRae Smith Company, 1961), 53.

5. Waite, *Joyful Sound,* 87.

6. Bruce, *Conquest of Solitude,* 101; Mabel Hubbard Bell autobiography.

7. Lilias M. Toward, *Mabel Bell: Alexander's Silent Partner* (Wreck Cove, N.S.: Breton Books, 1996), 22–23.

8. Mabel Hubbard Bell, "Gardiner Greene Hubbard," unpublished biography transmitted to Library of Congress with Hubbard's collection of engravings, 1898, HFP; "Gardiner Greene Hubbard," *Report of the Council of the Proceedings of the American Antiquarian Society,* Apr. 1898.

9. Letter of introduction from Secretary of State Hamilton Fish, July 23, 1870, HFP.

10. William G. Heill to Alexander Graham Bell, Sept. 6, 1901, HFP.

11. J. M. Caznem to Gardiner Greene Hubbard, Mar. 14, 1866, HFP.

12. Alexander Graham Bell to Alexander Melville and Eliza Symonds Bell, Feb. 12, 1876, BFP.

13. Bruce, *Conquest of Solitude,* 238.

14. Ibid., 30–34.

15. Ibid., 87.

16. Toward, *Silent Partner,* 23; Waite, *Joyful Sound,* 54–55; Mabel Hubbard Bell autobiography, HFP.

17. Waite, *Joyful Sound,* 55.

18. Bruce, *Conquest of Solitude,* 331.

19. Gardiner Greene Hubbard, deposition, *American Bell Telephone Co. vs. People's Telephone Co.,* June 25, 1879, 955, BFP; Waite, *Joyful Sound,* 89.

20. Bruce, *Conquest of Solitude,* 129.

21. Ibid., 134.

22. Ibid., 144.

23. Mabel Hubbard Bell to Mrs. George F. Kennan, undated, BFP.

24. Bruce, *Conquest of Solitude,* 144.

25. Alexander Graham Bell to Gertrude McCurdy Hubbard, Aug. 1, 1875, BFP.

26. Alexander Graham Bell to Mabel Hubbard, Aug. 8, 1875, BFP.

27. Alexander Graham Bell, Journal, Aug. 8, 1875, BFP.

28. Alexander Graham Bell, Journal, Aug. 26, 1875, BFP.

29. Bruce, *Conquest of Solitude,* 157.

30. Gardiner Greene Hubbard to Alexander Graham Bell, Oct. 29, 1875, BFP.

31. Bruce, *Conquest of Solitude,* 160; Alexander Graham Bell to Gardiner Greene Hubbard, Nov. 23, 1875, BFP.

32. Alexander Graham Bell to Gardiner Greene Hubbard, Aug. 23, 1875, BFP.

33. Bruce, *Conquest of Solitude,* 161.

34. Ibid.

35. Alexander Graham Bell to Mabel Hubbard, Nov. 25, 1875, BFP.

36. Mabel Hubbard to Mary True, Dec. 2, 1875, BFP.

37. Bruce, *Conquest of Solitude*, 165.

38. The century-old challenge of Antonio Meucci resurfaced in 2002, when the U.S. House of Representatives, acting out of motives more political than scientifically correct, passed a nonbinding resolution recognizing him as the real inventor of the telephone. In fact, Meucci's claim was without foundation. It had been dismissed by courts in the nineteenth century and no new evidence has emerged to support him.

39. Mabel Hubbard to Caroline McCurdy, Feb. 16, 1877, BFP.

40. Alexander Graham Bell to Alexander Melville and Eliza Symonds Bell, Dec. 7, 1875, BFP.

41. Bruce, *Conquest of Solitude*, 233–234.

42. Gardiner Greene Hubbard to Gertrude McCurdy Hubbard, Aug. 26, 1879; Sept. 12, 1829, HFP; Bruce, *Conquest of Solitude*, 292.

43. "Suburban Homes," *Washington Evening Star*, undated, GFP.

44. John G. Barker to Mabel Hubbard Bell, Oct. 22, 1898, HFP.

45. Bruce, *Conquest of Solitude*, 336–43.

46. John A. Garraty and Mark C. Carnes, eds., *American National Biography*, vol. 2 (Oxford and New York: Oxford University Press, 1999), 499.

47. Bruce, *Conquest of Solitude*, 394.

48. Alexander Graham Bell to Mabel Hubbard Bell, July 1881, HFP.

49. Bruce, *Conquest of Solitude*, 373–76.

50. Frank Luther Mott, *A History of American Magazines 1885–1905* (Cambridge, Mass.: The Belknap Press, 1957), 306–8; Bruce, *Conquest of Solitude*, 376–78.

51. Gardiner Greene Hubbard to Alexander Graham Bell, Jul. 11, 1882, Aug. 4, 1882, Nov. 1883, Dec. 4, 1884, BFP.

52. Bruce, *Conquest of Solitude*, 378.

53. Gardner Greene Hubbard to Alexander Graham Bell, Apr. 6, 1892, BFP.

CHAPTER TWO: A QUIET BIRTH

1. Mark Jenkins, "Clarence Edward Dutton," NGA.

2. Invitation, Gardiner Greene Hubbard and others, Jan. 10, 1888, NGA.

3. Mark Jenkins, "John Wesley Powell," NGA.

4. NGA; Ralph Gray, *That We May All Know More of the World: The Founding of the National Geographic Society And Its Century-Long Association with the Cosmos Club* (Washington, DC.: The Cosmos Club, 1987), 6.

5. Frank Luther Mott, *A History of American Magazines 1885–1905* (Cambridge, Mass.: The Belknap Press, 1957), 624.

6. Gray, *Founding*, 7.

7. Gardiner Greene Hubbard, "Introductory Address," *National Geographic*, October 1888.

8. *National Geographic*, 1888–1898; Gray, *Founding*, 6–7.

9. Mott, *American Magazines*, 621.

10. William Morris Davis, "The Rivers and Valleys of Pennsylvania," *National Geographic*, July 1889.

11. Charles Willard Hayes and Marius R. Campbell, "Geomorphology of the Southern Appalachians," *National Geographic,* May 1894.

12. Henry Gannett, "The Annexation Fever," *National Geographic,* Dec. 1897.

13. Henry Gannett, "The Movements of Our Population," *National Geographic,* Mar. 1893.

14. Gardiner Greene Hubbard, "Geographic Progress in Civilization," *National Geographic,* Feb. 1894.

15. Gardiner Greene Hubbard, "Africa, Its Past and Future," *National Geographic,* Apr. 1889.

16. J. B. Hatcher, "Patagonia," *National Geographic,* Nov. 1897.

17. John Hyde, "Some Recent Geographic Events," *National Geographic,* Dec. 1897.

18. Israel C. Russell, "An Expedition to Mount St. Elias, Alaska," *National Geographic,* May 1891; "NGS Place Names," NGA.

19. Edward Everett Hayden, "The Great Storm of March 11–14, 1888," *National Geographic,* Oct. 1888.

20. Eliza Ruhamah Scidmore, "The Recent Earthquake Wave on the Coast of Japan," *National Geographic,* Sept. 1896.

21. Wally Herbert, "Commander Robert E. Peary: Did He Reach the Pole?" *National Geographic,* Sept. 1988.

22. Wally Herbert, *The Noose of Laurels,* (New York: Atheneum, 1989).

23. C. D. B. Bryan, *The National Geographic Society* (New York: Abrams, 1987), 53; Herbert, "Did He Reach the Pole?"

24. Robert E. Peary, "Across Nicaragua with Transit and Machete," *National Geographic,* May 1889.

25. "Proceedings of the National Geographic Society," *National Geographic,* Jan. 1892.

26. Gilbert H. Grosvenor, interviewed by Allan C. Fisher, Jr., Aug. 25, 1962, NGA.

27. "Proceedings of the National Geographic Society," 1892.

28. *National Geographic,* April 1889; Robert V. Bruce, *Alexander Graham Bell and the Conquest of Solitude* (Boston: Little, Brown and Company, 1973), 423.

29. Mabel Hubbard Bell to Alexander Graham Bell, May 12, 1903. AGBNHS; Alexander Graham Bell, Home Notes, Nov. 2, 1915, BFP.

CHAPTER THREE: THE FIRST GROSVENOR

1. Gilbert H. Grosvenor, "The Romance of the Geographic," *National Geographic,* Oct. 1963; Charles McCarry, "Three Men Who Made the Magazine," *National Geographic,* Oct. 1988; Robert V. Bruce, *Alexander Graham Bell and the Conquest of Solitude* (Boston: Little, Brown and Company, 1973), 423.

2. Alexander Graham Bell, "The National Geographic Society," *National Geographic,* Feb. 1912.

3. Alexander Graham Bell, "Address of the President to the Board of Managers, June 1, 1900," *National Geographic,* Oct. 1900.

4. John Hyde to Alexander Graham Bell, Jan. 22, 1899, NGA.

5. Anne Rothe, ed. *Current Biography* (New York: H. W. Wilson, 1946), 228–30.

6. Gilbert H. Grosvenor, "The National Geographic Society and Its Magazine," National Geographic Society, 1957.

7. Gilbert H. Grosvenor to Lilian Waters Grosvenor, Mar. 5, 1899, GFP.

8. Gilbert H. Grosvenor to Edwin A. Grosvenor, Mar. 5, 1899, GFP.

9. Gilbert H. Grosvenor, undated video interview by Dennis Kane, NGA.

10. Gilbert H. Grosvenor to Lilian Waters Grosvenor, Mar. 26, 1899, GFP.

11. Gilbert H. Grosvenor to Edwin A. Grosvenor, Mar. 21, 1899, GFP.

12. Gilbert H. Grosvenor, interviewed by Allan C. Fisher, Jr., Aug. 25, 1962, NGA.

13. Mabel H. Grosvenor, interviewed by author, Apr. 18, 2002.

14. Lilian Waters Grosvenor to Gilbert H. Grosvenor, Aug. 9, 1900, GFP.

15. Gilbert H. Grosvenor to Lilian Waters Grosvenor, Apr. 2, 1899, GFP.

16. *National Geographic,* June–Oct. 1899.

17. Gilbert H. Grosvenor to Edwin A. Grosvenor, Apr. 13, 1899, GFP.

18. Gilbert H. Grosvenor to Edwin A. Grosvenor, June 7, 1899, GFP.

19. Alexander Graham Bell to Gilbert H. Grosvenor, July 13, 1899, GFP.

20. Gilbert H. Grosvenor to Lilian Waters Grosvenor, Apr. 24, 1899, GFP.

21. "Miscellanea," *National Geographic,* July 1899.

22. Alexander Graham Bell, address to the board of managers, *National Geographic,* Oct. 1900.

23. Alexander Graham Bell to Gilbert H. Grosvenor, July 12, 1899, GFP.

24. Mabel Hubbard Bell to Alexander Graham Bell, May 24, 1895, BFP.

25. Gilbert H. Grosvenor to Edwin A. Grosvenor, Apr. 5, 1899, GFP.

26. Mabel Hubbard Bell to Alexander Graham Bell, May 31, 1899, BFP.

27. Mabel Hubbard Bell to Alexander Graham Bell, June 4, 1899, BFP.

28. Mabel Hubbard Bell to Alexander Graham Bell, May 20, 1899, BFP.

29. Mabel Hubbard Bell to Alexander Graham Bell, May 12, 1899, BFP.

30. Edwin A. Grosvenor to Gilbert H. Grosvenor, June 21, 1899, GFP.

31. Gilbert H. Grosvenor to Edwin A. Grosvenor, June 20, 1899, GFP.

32. Alexander Graham Bell to Gilbert H. Grosvenor, Sept. 24, Sept. 19, Sept. 24, Sept. 6, Sept. 28, Aug. 15, Aug. 16, Sept. 28, Aug. 15, July 13, Sept. 19, July 13, Aug. 3, Sept. 24, July 14, 1899, BFP.

33. Gilbert H. Grosvenor, private memorandum, Feb. 14, 1963, GFP.

34. Gilbert H. Grosvenor to Edwin A. Grosvenor, Apr. 11, 1899, and to Lilian Waters Grosvenor, Apr. 24, 1899, GFP.

35. Gilbert H. Grosvenor to Edwin A. Grosvenor, Dec. 13, 1899, GFP; Grosvenor, "The National Geographic Society and Its Magazine."

36. Gilbert H. Grosvenor to Lilian Waters Grosvenor, May 7, 1899, GFP.

37. John Hyde to Gilbert H. Grosvenor, June 14, 1899, NGA.

38. Gilbert H. Grosvenor, undated notation on John Hyde letter, NGA.

39. Gilbert H. Grosvenor to Alexander Graham Bell, May 30, 1899; Gilbert H. Grosvenor to Elsie May Bell, Aug. 10, 1900, GFP.

40. Grosvenor, "The National Geographic Society and Its Magazine."

41. Mabel Hubbard Bell to Alexander Graham Bell, May 4, 1900, BFP.

42. Mabel Hubbard Bell to Alexander Graham Bell, May 30, 1899, BFP.

43. Gilbert H. Grosvenor to Lilian Waters Grosvenor, June 5, 1900, GFP.

44. John Hyde to Alexander Graham Bell, Nov. 18, 1899, NGA.

45. Edwin A. Grosvenor to Alexander Graham Bell, July 31, 1900, GFP.

46. Edwin Augustus Grosvenor Blair, interviewed by author, June 14, 2002.

47. Walter A. Dyer, "Grosvie," *Amherst Graduates Quarterly,* Feb. 1936.

48. Edwin A. Grosvenor to Gilbert H. Grosvenor, June 6, 1900, GFP.

49. Ibid.

50. Edwin A. Grosvenor to Gilbert H. Grosvenor, July 1, 1900, GFP.

51. Ibid.

52. Gilbert H. Grosvenor to Elsie May Bell, Aug. 2, 1900, GFP.

53. Gilbert H. Grosvenor to Elsie May Bell, Aug. 10, 1900, GFP.

54. Edwin A. Grosvenor to Alexander Graham Bell, July 31, 1900, GFP.

55. Edwin A. Grosvenor to Gilbert H. Grosvenor, Aug. 20, 1900, GFP.

56. Mabel Hubbard Bell to Lilian Waters Grosvenor, Aug. 22, 1900, GFP.

57. Gilbert H. Grosvenor, "The Romance of the Geographic," *National Geographic,* Oct. 1963.

58. Gilbert H. Grosvenor to Mabel Hubbard Bell, Sept. 17, 1900, GFP.

59. Mabel H. Grosvenor, interviewed by author, April 18, 2002.

60. Grosvenor, "The Romance of the Geographic."

61. Ibid.

62. Gilbert H. Grosvenor to Edwin A. Grosvenor, Jan. 11, 1902, GFP.

CHAPTER FOUR: A NEW GENERATION

1. Gilbert H. Grosvenor to Edwin A. Grosvenor, Sept. 17, 1901, GFP.

2. Melville Bell Grosvenor, "Life with Grandfather," address to the Literary Society of Washington, Feb. 10, 1968, NGA.

3. Gilbert H. Grosvenor to Edwin A. Grosvenor, Nov. 30, 1902, GFP.

4. Gilbert H. Grosvenor, "National Geographic Society," *National Geographic,* Dec. 1904.

5. Alexander Graham Bell, "The National Geographic Society," *National Geographic,* Feb. 1912.

6. Alexander Graham Bell to Gilbert H. Grosvenor, Dec. 7, 1905, GFP.

7. Ishbel Ross, "Geography, Inc.," *Scribner's,* June 1938.

8. Gilbert H. Grosvenor, interviewed by Allan C. Fisher, Jr., Aug. 25, 1962, NGA.

9. Gilbert H. Grosvenor, letter written for Secretary O. P. Austin's signature, Dec. 11, 1907, NGA.

10. Ross, "Geography, Inc."

11. Mark Jenkins, "Format Six: 1st Oak and Laurel Cover (February 1910)," *National Geographic Timeline—1910,* NGA.

12. Eliza R. Scidmore, "Koyasan, the Japanese Valhalla," *National Geographic,* Oct. 1907.

13. Melville E. Stone, "Race Prejudice in the Far East," *National Geographic,* Dec. 1910.

14. J. R. Hildebrand, "Revolution in Eating," *National Geographic,* Mar. 1942.

15. Robert DeC. Ward, "Our Immigration Laws from the Viewpoint of National Eugenics," *National Geographic,* Jan. 1912.

16. Robert V. Bruce, *Alexander Graham Bell and the Conquest of Solitude* (Boston: Little, Brown and Company, 1973), 417–20.

17. Gilbert H. Grosvenor, "Report of the Editor to the Board of Trustees for the Year 1921," NGA.

18. William Howard Taft, "Some Recent Instances of National Altruism," *National Geographic,* July 1907.

19. John W. Foster, "China," *National Geographic,* Dec. 1904.

20. Gilbert H. Grosvenor, Proceedings, American Antiquarian Association, vol. 5, Oct. 1942.

21. Alexander Graham Bell to Gilbert H. Grosvenor, May 2, 1901, GFP.

22. Charles F. Thompson, quoted by Melville Bell Grosvenor, "Life with Grandfather II," address to the Literary Society of Washington, Dec. 8, 1973, NGA.

23. Gilbert H. Grosvenor, "National Geographic and Its Magazine," National Geographic Society, 1957.

24. Ibid.

25. Alexander Graham Bell, Home Notes, Oct. 23, 1907, BFP.

26. Alexander Graham Bell to Gilbert H. Grosvenor, Oct. 7, 1905, GFP.

27. Gilbert H. Grosvenor, interviewed by Allan C. Fisher, Jr., Aug. 25, 1962, NGA.

28. Mabel Hubbard Bell to Gilbert H. Grosvenor, Oct. 19, 1904, BFP.

29. Grosvenor, "The National Geographic Society and Its Magazine."

CHAPTER FIVE: EDUCATING MELVILLE

1. Robert V. Bruce, *Alexander Graham Bell and the Conquest of Solitude* (Boston: Little, Brown and Company, 1973), 445–47.

2. Ibid., 446–49.

3. Alexander Graham Bell to Gilbert H. Grosvenor, July 4, 1917, GFP.

4. Mabel Hubbard Bell to Alexander Graham Bell, Oct. 21, 1901, BFP.

5. Bruce, *Conquest of Solitude,* 300–1.

6. Mabel Hubbard Bell to Gertrude McCurdy Hubbard, undated, BFP.

7. Lilian Grosvenor, "My Grandfather Bell," *The New Yorker,* Nov. 11, 1950.

8. Dorothy Harley Eber, *Genius at Work* (Halifax: Nimbus, 1991), 34.

9. David Fairchild, *The World Was My Garden* (New York: Charles Scribner's Sons, 1938), 330.

10. Mabel H. Grosvenor, interviewed by author, May 28, 2002.

11. "Bell at Baddeck," oral history, Canadian Broadcasting Corp., AGBNHS.

12. Mabel Hubbard Bell to a daughter (unidentified), Feb. 23, 1909, BFP.

13. "Bell at Baddeck."

14. Mabel H. Grosvenor, interviewed by author, April 4, 2002.

15. Mabel Hubbard Bell to Elsie May Bell, May 12, 1891, BFP.

16. Fairchild, *Garden,* 335.

17. Melville Bell Grosvenor, "Life with Grandfather," address to the Literary Society of Washington, Feb. 10, 1968, NGA; Lilian Grosvenor, "My Grandfather Bell."

18. Alexander Graham Bell to Gilbert H. Grosvenor, Feb. 3, 1909, BFP.

19. Gilbert H. Grosvenor to Mabel Hubbard Bell, Feb. 13, 1909, GFP.

20. Alexander Graham Bell to Gilbert H. Grosvenor, Sept. 19, 1899, GFP.

21. Special Meeting of the Board of Managers, National Geographic Society, Nov. 11, 1908, NGA.

22. Adolphus W. Greely to O. P. Austin, Nov. 11, 1908, NGA.

23. Alexander Graham Bell, telegram to Gilbert H. Grosvenor, Nov. 13, 1908, NGA.

24. Mabel H. Grosvenor, interviewed by author, May 28, 2002.

25. Elsie Bell Grosvenor, address at the dedication of Alexander Graham Bell National Historical Site, Baddeck, N.S., Aug. 16, 1956.

26. Mabel H. Grosvenor, interviewed by author, May 28, 2002.

27. Ibid.

28. Grosvenor, "Life with Grandfather."

29. Mabel Hubbard Bell to Gilbert H. Grosvenor, Oct. 17, 1914, AGBNHS.

30. Grosvenor, "Life with Grandfather."

31. Mabel Hubbard Bell to Elsie Bell Grosvenor, Oct. 8, 1914, AGBNHS.

32. Alexander Graham Bell to Gilbert H. Grosvenor, Dec. 1914, BFP.

33. Grosvenor, "Life with Grandfather."

34. Alexander Graham Bell to Gilbert H. Grosvenor, Dec. 1914, BFP.

35. Alexander Graham Bell, Home Notes, Sept. 16, 1914, BFP.

36. Mabel H. Grosvenor, interviewed by author, April 22, 2002.

37. Melville Bell Grosvenor to Gilbert H. Grosvenor, May 5, 1914, GFP; Gilbert H. Grosvenor to Mabel Hubbard Bell, Sept. 19, 1914, GFP; Gilbert H. Grosvenor to Alexander Graham Bell, Sept. 14, 1914, GFP.

38. Melville Bell Grosvenor to Gilbert H. Grosvenor, Nov. 5, 1914, GFP.

39. Gilbert H. Grosvenor to Alexander Graham Bell, Sept. 14, 1914, GFP.

40. Gilbert H. Grosvenor to Alexander Graham Bell, June 28, 1915, GFP.

41. Mabel Hubbard Bell to Elsie Bell Grosvenor, undated, but probably 1909, GFP; Alexander Graham Bell to Gilbert H. Grosvenor, Dec. 1914, BFP.

42. Alexander Graham Bell to Gilbert H. Grosvenor, December 1914, BFP.

43. Mabel H. Grosvenor interviewed by author, April 4, 2002.

44. Gilbert H. Grosvenor to Edwin A. Grosvenor, May 15, 1919; to Edwin P. Grosvenor, May 23, 1919, GFP.

45. Luis Marden, interviewed by author, May 18, 1995, NGA.

CHAPTER SIX: THE FIRST HERO

1. "Peary's Work in 1900 and 1901," *National Geographic*, Oct. 1901.

2. "Farthest North," *National Geographic*, Nov. 1906.

3. Wally Herbert, *The Noose of Laurels* (New York: Atheneum, 1989), 22–23.

4. "Honors to Peary," *National Geographic*, Jan. 1907.

5. Edmund Morris, *Theodore Rex* (New York: Random House, 2001), 193.

6. C. D. B. Bryan, *The National Geographic Society* (New York: Abrams, 1987), 94–95.

7. Gilbert H. Grosvenor, interviewed by Allan C. Fisher, Jr., Aug. 25, 1962, NGA.

8. Gilbert H. Grosvenor to Ida M. Tarbell, July 25, 1900, as quoted in *Proceedings of the American Antiquarian Society,* Oct. 1942.

9. Herbert, *Laurels,* 22.

10. "Honors to Peary."

11. "President Gives Medal to Peary," *New York Herald,* Dec. 16, 1906.

12. "Honors to Peary"; "From the Hand of the President," *Washington Sunday Star,* Dec. 16, 1906; "Must Find the Pole," *Washington Post,* Dec. 16, 1906; "President Praises Peary's Hardy Virtue," *New York Times,* Dec. 16, 1906.

13. Gilbert H. Grosvenor, interviewed by Allan C. Fisher, Jr., Aug. 25, 1962, NGA.

14. Program, "Annual Banquet of the National Geographic Society," Dec. 15, 1906, NGA.

15. Mabel H. Grosvenor, interviewed by author, Apr. 18, 2002.

16. Luis Marden, interviewed by author, May 18, 1995, NGA.

17. Herbert, *Laurels,* 21–33.

18. "Honors to Peary"; Herbert, *Laurels,* 21–33.

19. "From the Hand of the President."

20. "Honors to Peary."

21. Herbert, *Laurels,* 32; "Honors to Peary."

22. Robert E. Peary, undated audiotape, NGA.

23. "Honors to Peary."

24. "Recognition of Robert E. Peary, the Arctic Explorer," Committee on Naval Affairs, U.S. House of Representatives, 61st Congress, 3rd Session, Jan. 21, 1911.

25. Howard S. Abramson, *Hero in Disgrace* (San Jose: toExcel Press, 1991), 61.

26. Herbert, *Laurels,* 33.

27. Robert E. Peary to members of the Peary Arctic Club, May 1908, NGA.

28. Herbert, *Laurels,* 212–13.

29. Ibid., 217–18.

30. Robert E. Peary to Gilbert H. Grosvenor, May 3, 1908, NGA.

31. Herbert, *Laurels,* 216.

32. Committee on Naval Affairs, 3.

33. Herbert, *Laurels,* 216–17.

34. Gilbert H. Grosvenor to Willis L. Moore, July 26, 1908, NGA.

35. Gilbert H. Grosvenor to Edwin A. Grosvenor, July 12, 1909, GFP.

36. Willis L. Moore to Gilbert H. Grosvenor, July 30, 1908, GFP.

37. Gilbert H. Grosvenor, undated memorandum appended to letter from Alexander Graham Bell, Feb. 3, 1909, GFP.

38. Gilbert H. Grosvenor to Board of Managers, Apr. 19, 1911, NGA.

39. Edwin A. Grosvenor to Gilbert H. Grosvenor, May 22, 1907, GFP.

40. Herbert, *Laurels,* 281.

41. Robert E. Peary, telegram to Gilbert H. Grosvenor, Sept. 6, 1909, NGA.

42. Herbert, *Laurels,* 282.

43. Ibid., 282.

44. Bryan, *National Geographic,* 65.

45. Ibid., 65–66.

46. Frederick A. Cook, "The Discovery of the Pole," *National Geographic,* Oct. 1909.

47. Robert E. Peary, "The Discovery of the Pole," *National Geographic,* Oct. 1909.

48. Herbert, *Laurels,* 291.

49. Ibid., 288.

50. Adolphus W. Greely to Gilbert H. Grosvenor, Oct. 5, 1909, NGA.

51. Alexander Graham Bell to Gilbert H. Grosvenor, Oct. 19, 1909, NGA.

52. Gilbert H. Grosvenor to Adolphus W. Greely, Oct. 6, 1909, NGA.

53. Gilbert H. Grosvenor to Alexander Graham Bell, Nov. 5, 1909, GFP.

54. "Peary Satisfies Board of Experts," *New York Times,* Nov. 2, 1909; Committee on Naval Affairs.

55. Wally Herbert, "Did He Reach the Pole?" *National Geographic,* Sept. 1988; Committee on Naval Affairs.

56. Matthew A. Henson, *A Negro Explorer at the North Pole* (New York: Stokes, 1912), 135; "Henson's Story," *Boston American,* July 17, 1910.

57. John W. Godsell, *On Polar Trails* (Austin: Eakin Press, 1983), 183.

58. Herbert, *Laurels,* 239.

59. "Peary Brings Proofs," *Washington Post,* Nov. 3, 1909.

60. "Peary Satisfies Board of Experts," *New York Times,* Nov. 2, 1909.

61. Henry Gannett, C. M. Chester, O. H. Tittman, Report to the Board of Managers, National Geographic Society, Nov. 3, 1910, NGA.

62. "Accepts Peary Proof," *Washington Post,* Nov. 4, 1909.

63. Herbert, *Laurels,* 305.

64. Ibid., 305.

65. Ibid., 320–21.

66. Committee on Naval Affairs.

67. Adolphus W. Greely, journals, NGA.

68. Adolphus W. Greely, "The North Pole Question," undated, unpublished memorandum, NGA.

69. When Robert Peary died in 1920, the cause was pernicious anemia, a condition Dr. Frederick A. Cook had diagnosed many years earlier when the two friends were traveling in Greenland. (Herbert, *Laurels,* 25.)

70. Mark Muro, "A Geographic Retreat," *The Boston Globe,* Oct. 2, 1988.

CHAPTER SEVEN: OFF TO WAR

1. Alexander Graham Bell, telegram to Willis L. Moore, Oct. 20, 1909, NGA.

2. Gilbert H. Grosvenor, interviewed by Allan C. Fisher, Jr., Feb. 1963, NGA.

3. Mabel Hubbard Bell to Alexander Graham Bell, Nov. 16, 1909, BFP.

4. Mabel Hubbard Bell to Alexander Graham Bell, Nov. 17, 1909, BFP.

5. William W. Chapin, "Glimpses of Korea and China," *National Geographic,* Nov. 1910.

6. Gilbert H. Grosvenor, interviewed by Allan C. Fisher, Jr., Feb. 1963, Aug. 25, 1962, NGA.

7. Eliza R. Scidmore to Gilbert H. Grosvenor, 1912, NGA.

8. Adolphus W. Greely to O. P. Austin, April 16, 1912, NGA.

9. John Noble Wilford, "'Lost City' Yielding Its Secrets," *New York Times,* Mar. 18, 2003, D-1.

10. Hiram Bingham, "In the Wonderland of Peru," *National Geographic,* Apr. 1913.

11. Adolphus W. Greely to Gilbert H. Grosvenor, July 8, 1916, NGA.

12. David Fairchild, *The World Was My Garden* (New York: Charles Scribner's Sons, 1938), 451; O. F. Cook, "Staircase Farms of the Ancients," *National Geographic,* May 1916.

13. Mabel H. Grosvenor, interviewed by author, Apr. 18, May 21, 2002.

14. Alexander Graham Bell, Home Notes, Mar. 5–6, 1914, BFP.

15. Gilbert H. Grosvenor, interviewed by Allan C. Fisher, Jr., Feb. 1963, NGA.

16. Ibid.

17. Gilbert H. Grosvenor, "Young Russia—The Land of Unlimited Possibilities," *National Geographic,* Nov. 1914.

18. Sir Edward Grey, *Fly Fishing* (London: The Flyfisher's Classic Library, 1992), v.

19. Gilbert H. Grosvenor, interviewed by Allan C. Fisher, Jr., Feb. 1963, NGA.

20. Alexander Graham Bell to Gilbert H. Grosvenor, Aug. 13, 1914, BFP.

21. Gilbert H. Grosvenor, "Report of the Director and Editor of the National Geographic Society for the Year 1914," *National Geographic,* Mar. 1915.

22. C. D. B. Bryan, *The National Geographic Society* (New York: Abrams, 1987), 90.

23. William Howard Taft, "The League of Nations, What It Means and Why It Must Be," *National Geographic,* Jan. 1929.

24. Margaret MacMillan, *Paris, 1919* (New York: Random House, 2001), 10–14.

25. Gilbert H. Grosvenor to Edwin A. Grosvenor, Mar. 13, 1919, GFP.

26. Gilbert H. Grosvenor to Alexander Graham Bell, May 13, 1921, GFP.

27. Gilbert H. Grosvenor to Edwin A. Grosvenor, Dec. 24, 1918, NGA.

28. Thomas Whittemore, "The Rebirth of Religion in Russia," *National Geographic,* Nov. 1918.

29. Edwin A. Grosvenor to Gilbert H. Grosvenor, Oct. 6, 1909, Mar. 20, 1919, GFP.

30. Grosvenor, "Young Russia."

31. William Eleroy Curtis, "The Revolution in Russia," *National Geographic,* May 1907.

CHAPTER EIGHT: THE CHANGING ORDER

1. Gilbert H. Grosvenor, "The National Geographic Society," *Proceedings of the American Antiquarian Society,* Oct. 1942, 13; "Colleague of the Golden Years: John Oliver LaGorce," *National Geographic,* Mar. 1960.

2. Grosvenor, "Colleague of the Golden Years."

3. Mabel H. Grosvenor, interviewed by author, Apr. 22, 2002.

4. Wells W. Cooke, "Saving the Ducks and Geese," *National Geographic,* Mar. 1913.

5. Mark Jenkins, "National Conservation Congress," *National Geographic Timeline—1913,* NGA.

6. Horace M. Albright and Marian Albright Schenck, *Creating the National Park Service* (Norman: University of Oklahoma Press, 1999), 69.

7. Robert Shankland, *Steve Mather of the National Parks* (New York: Alfred A. Knopf, 1970), 68–75; Mark Jenkins, "The Mather Mountain Party," *National Geographic Timeline—1913*, NGA.

8. Shankland, *Steve Mather*, 71.

9. Albright, *Creating National Park Service*, 69.

10. Shankland, *Steve Mather*, 101.

11. Gilbert H. Grosvenor, "Our Big Trees Saved," *National Geographic,* Jan. 1917.

12. Mark Jenkins, "The Society and Sequoia National Park," *National Geographic Timeline—1916*, NGA.

13. Alexander Graham Bell to Gilbert H. Grosvenor, Oct. 31, 1916, GFP.

14. Robert V. Bruce, *Alexander Graham Bell and the Conquest of Solitude* (Boston: Little, Brown and Company, 1973), 474.

15. Attributed to Catherine MacKenzie, Bell's longtime secretary, in Bruce, *Conquest of Solitude,* 487.

16. Alexander Graham Bell to Gilbert H. Grosvenor, Feb. 3, 1909, GFP.

17. Mabel H. Grosvenor, interviewed by author, Dec. 19, 2002; Gilbert H. Grosvenor to Edwin A. Grosvenor, Apr. 12, 1909, GFP.

18. Gilbert H. Grosvenor, unpublished notes for an interview with Allan C. Fisher, Jr., 1963, GFP.

19. Mabel Hubbard Bell to Gilbert H. Grosvenor, Jan. 18, 1920, GFP.

20. Mabel H. Grosvenor, interviewed by author, Apr. 22, 2002.

21. Gilbert H. Grosvenor, telegram to Alexander Graham Bell, Jan. 21, 1920, GFP.

22. Gilbert H. Grosvenor to Edwin P. Grosvenor, Mar. 10, 1920; Edwin P. Grosvenor to Gilbert H. Grosvenor, Apr. 1, 1920, GFP.

23. Gilbert H. Grosvenor, interviewed by Allan C. Fisher, Jr., Aug. 25, 1962, NGA.

24. Alexander Graham Bell to Gilbert H. Grosvenor, Apr. 27, 1920, GFP.

25. Gilbert H. Grosvenor to Alexander Graham Bell, Apr. 23, 1920, GFP.

26. Gilbert H. Grosvenor, interviewed by Allan C. Fisher, Jr., Aug. 25, 1962, NGA.

27. Gilbert H. Grosvenor to Mabel Hubbard Bell, Jan. 18, 1921, GFP.

28. Mark Jenkins, "First Chaco Canyon–Pueblo Bonito Expedition," *National Geographic Timeline—1921*, NGA.

29. Gilbert H. Grosvenor, "Report of the Editor of the National Geographic Society, Gilbert Grosvenor for the Year 1921," NGA.

30. C. D. B. Bryan, *The National Geographic Society* (New York: Abrams, 1987), 183.

31. Luis Marden, interviewed by author, Apr. 11, 1995, NGA.

32. Ibid.

33. Luis Marden, interviewed by author, May 18, 1995, NGA.

34. Ibid.; the note is from a book inscribed from Gilbert H. Grosvenor to Luis Marden in 1939.

35. Gilbert H. Grosvenor to Mabel Hubbard Bell, Jan. 18, 1921, GFP.

36. Mabel H. Grosvenor, interviewed by author, Apr. 18, 2002.

37. Memorandum of payment to Dr. Edwin A. Grosvenor, Jan. 20, 1919, NGA.

38. Alexander Graham Bell to Gilbert H. Grosvenor, July 8, 1919, GFP.

39. Edwin P. Grosvenor to Gilbert H. Grosvenor, Jan. 10, 1921, GFP.

40. Grosvenor, "Report of the Editor."

41. Gilbert H. Grosvenor to Edwin A. Grosvenor, June 6, 1919, GFP.

42. Lilias M. Toward, *Alexander Graham Bell's Silent Partner* (Wreck Cove, N.S.: Breton Books, 1996), 244.

43. Bruce, *Conquest of Solitude,* 486.

44. Walter Pinaud, quoted in "Bell at Baddeck," oral history, Canadian Broadcasting Corp., AGBNHS.

45. David Fairchild to Gilbert H. Grosvenor, Aug. 6, 1922, GFP.

46. Catherine MacKenzie, *Alexander Graham Bell: The Man Who Contracted Space* (Boston: Houghton Mifflin Company, 1928), 363.

47. David Fairchild to Gilbert H. Grosvenor, Aug. 6, 1922; Daisy Fairchild to Elsie Bell Grosvenor, undated; Daisy Fairchild to Gilbert H. and Elsie Bell Grosvenor, Aug. 6, 1922; Mabel Hubbard Bell to Elsie Bell Grosvenor, Aug. 2, 1922, GFP.

CHAPTER NINE: THE CRASH

1. Mabel Hubbard Bell to Elsie Bell Grosvenor, Aug. 2, 1922, BFP.

2. Mabel Hubbard Bell, Home Notes, Aug. 31, 1922, BFP.

3. Gilbert H. Grosvenor to Elsie Bell Grosvenor, Oct. 5, 1922, GFP.

4. Gilbert H. Grosvenor to Elsie Bell Grosvenor, Oct. 5, Oct. 4, 1922, GFP.

5. Elsie Bell Grosvenor, undated to Gilbert H. Grosvenor, GFP.

6. Gilbert H. Grosvenor to Elsie Bell Grosvenor, Oct. 4, 1922, GFP.

7. Elsie Bell Grosvenor, undated to Gilbert H. Grosvenor, GFP

8. Gilbert M. Grosvenor, memorandum of conversation with Mabel Hubbard Bell, Jan. 1, 1923, GFP.

9. Mabel Hubbard Bell to Gilbert H. Grosvenor, May 4, 1922, GFP.

10. Gilbert H. Grosvenor to Elsie Bell Grosvenor, Oct. 4, 1922, GFP.

11. Mabel H. Grosvenor, interviewed by author, July 11, 2003.

12. Ibid.

13. Lilias M. Toward, *Alexander Graham Bell's Silent Partner* (Wreck Cove, N.S.: Breton Books, 1996), 256.

14. Grosvenor Blair, interviewed by author, June 14, 2002.

15. Ibid.

16. Gilbert H. Grosvenor to Edwin P. Grosvenor, Jan. 30, 1926, GFP.

17. Edwin P. Grosvenor to Gilbert H. Grosvenor, Jan. 26, 1927, GFP.

18. "NGS Place Names," NGA; Geoffrey T. Hellman, "Geography Unshackled," *The New Yorker,* Sept. 25, Oct. 2, Oct. 9, 1943.

19. Hellman, "Geography Unshackled," Oct. 9, 1943, 27.

20. Mark Jenkins, "The Society versus the U.S. Board of Geographic Names," *National Geographic Timeline—1925,* NGA.

21. Gilbert H. Grosvenor, memorandum, May 26, 1958, NGA.

22. Mark Jenkins, "The Society versus the U.S. Board of Geographic Names."

23. Gilbert H. Grosvenor to Edwin A. Grosvenor, May 3, 1929, GFP.

24. Gilbert H. Grosvenor to Edwin P. Grosvenor, Oct. 6, 1921, GFP.

25. John Oliver LaGorce to Gilbert H. Grosvenor and Frederick V. Coville, July 2, 1927, NGA.

26. Charles McCarry, "Three Men Who Made the Magazine," *National Geographic*, Sept. 1988. Ms. Patterson never slowed down. Eventually promoted to the magazine's legends, or caption-writing, department, she became the first woman on staff to appear as a senior editor on the masthead.

27. John Oliver LaGorce, memorandum to Mrs. Evans, Sept. 12, 1924, NGA.

28. "Misquoted on Peary Amundsen Now Says," *New York Times,* Jan. 25, 1926, 4.

29. Howard S. Abramson, *Hero in Disgrace* (San Jose: toExcel Press, 1991), 208.

30. Gilbert H. Grosvenor to Edwin P. Grosvenor, Mar. 5, 1926, GFP.

31. John Oliver LaGorce to Gilbert H. Grosvenor, Jan. 24, 1926, NGA; "Misquoted on Peary Amundsen Now Says"; "Explains Cancelling of Invitation," *New York Times,* Mar. 4, 1926, 23.

32. Edwin P. Grosvenor to Gilbert H. Grosvenor, Mar. 8, 1926, GFP.

33. Gen. Umberto Nobile, "Navigating the *Norge* from Rome to the North Pole and Beyond," *National Geographic,* Aug. 1927.

34. Anne Chamberlain, "Two Cheers for the National Geographic," *Esquire,* Dec. 1964.

35. Mabel H. Grosvenor, interviewed by author, Apr. 4, 2002.

36. Melville Bell Grosvenor to Gilbert H. Grosvenor, June 1923, GFP.

37. Melville Bell Grosvenor to Elsie Bell Grosvenor, Feb. 26, 1923, GFP.

38. Lilian Grosvenor, unpublished diary, Jan. 3, 1924, private collection.

39. Edwin P. Grosvenor to Gilbert H. Grosvenor, June 8, 1923, GFP.

40. Lilian Grosvenor, unpublished diary, Jan. 3, 1924.

41. Helen Rowland Grosvenor to Elsie Bell Grosvenor, Mar. 1924, GFP.

42. Gilbert H. Grosvenor to Elsie Bell Grosvenor, Oct. 15, 1924, GFP.

43. Gilbert H. Grosvenor to Elsie Bell Grosvenor, undated, GFP.

44. Grosvenor Blair, interviewed by author, June 14, 2002; Mabel H. Grosvenor, interviewed by author, Apr. 22, 2002.

45. Gilbert H. Grosvenor to Mabel Hubbard Bell, Oct. 16, 1921, GFP.

46. Gilbert H. Grosvenor to Elsie Bell Grosvenor, Apr. 28, 1914, Sept. 3, 1914, Sept. 5, 1914, GFP.

47. Grosvenor Blair, interviewed by author, June 14, 2002.

48. Mabel Hubbard Bell, unfinished and undated, to Melville Bell Grosvenor, BFP.

49. *Melville B. Grosvenor v. Helen Rowland Grosvenor,* Equity No. 14227, Circuit Court, Montgomery County, Maryland, 1949–1950.

50. Edwin P. Grosvenor to Gilbert H. Grosvenor, Apr. 16, 1924, Sept. 22, 1928, GFP.

51. T. H. Watkins, *The Hungry Years* (New York: Henry Holt, 1999), 32–33.

52. Edwin A. Grosvenor to Gilbert H. Grosvenor, Nov. 3, 1929, GFP.

53. Mabel H. Grosvenor, interviewed by author, Apr. 22, 2002; Gilbert H. Grosvenor to Edwin A. Grosvenor, Nov. 14 and 17, 1929, GFP.

54. Gilbert H. Grosvenor to Edwin A. Grosvenor, Nov. 14, 1929, GFP.

55. Edwin A. Grosvenor to Gilbert H. Grosvenor, Nov. 7, 1929, GFP.

56. Edwin P. Grosvenor to Gilbert H. Grosvenor, Nov. 15, 1929, GFP.

57. Gilbert H. Grosvenor to Edwin A. Grosvenor, Nov. 14, 1929, GFP.

58. Edwin A. Grosvenor to Gilbert H. Grosvenor, Nov. 4, 1929, GFP.

59. Gilbert H. Grosvenor to Edwin A. Grosvenor, Nov. 14, 1929; Edwin P. Grosvenor to Gilbert H. Grosvenor, Dec. 21, 1929, GFP.

CHAPTER TEN: A GENTLEMANLY STANDARD

1. Maynard Owen Williams to Gilbert H. Grosvenor, Apr. 6, 1931, NGA.

2. Eric Sevareid, quoted in T. H. Watkins, *The Hungry Years* (New York: Henry Holt and Company, 1999), 70.

3. John Carver Edwards, *Berlin Calling: American Broadcasters in Service to the Third Reich* (New York: Praeger, 1991), 116–17, 120.

4. John Oliver LaGorce to Douglas Chandler, Aug. 5, 1936, NGA.

5. Watkins, *The Hungry Years*, 131–36.

6. Gilbert H. Grosvenor to Edwin A. Grosvenor, June 9, 1932, GFP.

7. Watkins, *The Hungry Years*, 131–41.

8. "The Society's Special Medal Awarded to Amelia Earhart: First Woman to Receive Geographic Distinction at Brilliant Ceremony in the National Capital." Address by Amelia Earhart, *National Geographic*, Sept. 1932; Mark Jenkins, "Amelia Earhart awarded Special Gold Medal," *National Geographic Timeline—1932*. NGA.

9. Gilbert H. Grosvenor to Edwin A. Grosvenor, June 22, 1932, GFP.

10. Watkins, *The Hungry Years*, 131–41.

11. Gilbert H. Grosvenor to Edwin A. Grosvenor, Oct. 10, 1932, GFP.

12. IRS Form 1099 for Gilbert H. Grosvenor and John Oliver LaGorce, Calendar Year 1928.

13. Mark Jenkins, "New Headquarters Building Finished," *National Geographic Timeline—1932*, NGA.

14. Luis Marden, interviewed by author, May 18, 1995, NGA.

15. Cathy Hunter, "A Half Mile Down: William Beebe and the Bathysphere," *National Geographic Timeline—1934*, NGA.

16. Mark Jenkins, "NGS–U.S. Army Air Corps Stratosphere Flight: Explorer I," *National Geographic Timeline—1934*, NGA.

17. Mark Jenkins, "Explorer II: the National Geographic–U.S. Army Air Corps Stratosphere Flight," *National Geographic Timeline—1935*, NGA.

18. Maynard Owen Williams to Gilbert H. Grosvenor, May 30-31, 1931, NGA.

19. Maynard Owen Williams, "The Citroën-Haardt Trans-Asiatic Expedition Reaches Kashmir," *National Geographic*, Oct. 1931; Mark Jenkins, "Maynard Owen Williams and Bamian," *National Geographic Timeline—1931*, NGA.

20. Maynard Owen Williams to Gilbert H. Grosvenor, May 30, 1931, NGA.

21. Maynard Owen Williams to Gilbert H. Grosvenor, May 30, Aug. 20-23, 1931, NGA.

22. Maynard Owen Williams to Gilbert H. Grosvenor, Nov. 7, 1931, NGA.

23. Maynard Owen Williams to Gilbert H. Grosvenor, Apr. 6, 1931, NGA.

24. Maynard Owen Williams to Gilbert H. Grosvenor, Aug. 20-23, 1931, NGA.

25. Maynard Owen Williams to John Oliver LaGorce, Sept. 8, 1931, NGA.

26. Maynard Owen Williams to John Oliver LaGorce, Feb. 22, 1932, NGA.

27. Gilbert H. Grosvenor and John Oliver LaGorce, copy of undated cable to Maynard Owen Williams, NGA.

28. Maynard Owen Williams to Gilbert H. Grosvenor, Mar. 3, 1948, NGA.

29. Joseph Rock to David Fairchild, July 8, 1923, NGA.

30. Joseph Rock to Ralph Graves, Nov. 26, 1931, NGA.

31. Gilbert H. Grosvenor, memorandum to Franklin L. Fisher, Feb. 2, 1934, NGA.

32. George W. Hutchison, memorandum to John Oliver LaGorce, June 5, 1931, NGA.

33. Mabel H. Grosvenor, interviewed by author, May 21, 2002.

34. Luis Marden, interviewed by author, May 18, 1995, NGA.

35. Melville Bell Grosvenor, related by Frederick G. Vosburgh, *The Century As I Saw It*, unpublished memoir, 2000; 116–117, NGA.

36. Gilbert H. Grosvenor, memorandum to John Oliver LaGorce, Jan. 4, 1934, NGA.

37. Gilbert H. Grosvenor to Edwin A. Grosvenor, Oct. 10, 1932, GFP.

38. *Melville B. Grosvenor v. Helen Rowland Grosvenor,* Equity No. 14227, Circuit Court, Montgomery County, Maryland, 1950.

39. John Oliver LaGorce to Gilbert H. Grosvenor, Sept. 19, 1935, NGA.

40. "Grosvenor Talks to College Group," *Washington Evening Star,* June 9, 1930.

41. Mabel H. Grosvenor, interviewed by author, Apr. 18, 2002.

42. Gilbert H. Grosvenor, memorandum to Jesse R. Hildebrand, Mar. 19, 1936, NGA.

43. Maynard Owen Williams, memorandum to Gilbert H. Grosvenor, Dec. 28, 1933, NGA.

44. Gilbert H. Grosvenor, memorandum to Maynard Owen Williams, Dec. 28, 1933, NGA.

45. Mabel H. Grosvenor, interviewed by author, Dec. 19, 2002.

46. Gilbert H. Grosvenor, memorandum to John Oliver LaGorce, Oct. 31, 1933, NGA.

47. Gilbert H. Grosvenor, memorandum to John Oliver LaGorce, May 12, 1933, NGA.

48. Douglas Chandler, "Changing Berlin," *National Geographic,* Feb. 1937.

49. Ibid.

50. John Patric, "Imperial Rome Reborn," *National Geographic,* Mar. 1937.

51. John Oliver LaGorce to Gilbert H. Grosvenor, Sept. 1, 1939, NGA.

52. Vosburgh, *The Century As I Saw It,* 156, NGA; Thomas W. McKnew to George W. Hutchison, June 1 and June 4, 1938, NGA; John Oliver LaGorce to William H. Danforth, May 17, 1939, NGA.

53. William H. Danforth to John Oliver LaGorce, May 15, 1939, NGA.

54. Edwards, *Berlin Calling,* 120.

55. Douglas Chandler, quoted in *Time,* June 9, 1941, 47, 48.

56. John Oliver LaGorce to Virginius Dabney, May 21, 1940, NGA.

57. John Oliver LaGorce to Charles O. Collett, Apr. 17, 1942, NGA.

58. John Oliver LaGorce to Alan C. Collins, May 26, 1942, NGA.

59. Frederick G. Vosburgh, interviewed by author, Aug. 12, 2001; *The Century As I Saw It,* 149.

60. Gilbert H. Grosvenor, memorandum to John Oliver LaGorce, Nov. 20, 1936, NGA.
61. Edwards, *Berlin Calling*, 145–47.

CHAPTER ELEVEN: THE BIRTHRIGHT

1. "An Exhibition of Natural History Illustrations," Brandywine River Museum, Chadds Ford, Pennsylvania. Mar. 26–May 22, 1983.
2. Two former chief editors of *National Geographic*, Frederick G. "Ted" Vosburgh and William P. E. "Bill" Graves, confirmed that Murayama was fired; three of his former colleagues, Andrew H. Brown, Volkmar K. Wentzel, and Luis Marden corroborated this in interviews with the author.
3. Luis Marden, undated conversation with author. Marden gave the purloined watercolors, still wrapped in brown paper, to the author several years ago. The author returned the two pieces of art—one portraying sturgeon, another depicting herring—to the National Geographic Society in 2003.
4. This story, related to the author by William P. E. "Bill" Graves, could not be confirmed. In Graves's version of it, the cherry continued to sprout stubbornly for years. Interviewed by author, Aug. 23, 2002.
5. Andrew H. Brown, interviewed by author, Sept. 6, 2002.
6. Johana Fiedler, *Molto Agitato* (New York: Nan Talese, Doubleday, 2002), 34–35.
7. Mark Jenkins, "The National Geographic Society and World War II," *National Geographic Timeline—1945*, NGA.
8. Gilbert H. Grosvenor, memorandum to John Oliver LaGorce, Dec. 3, 1943, NGA.
9. Gilbert H. Grosvenor, memorandum to John Oliver LaGorce, Feb. 11, 1942, NGA.
10. George M. Elsey, "Blueprints for Victory," *National Geographic*, May 1995.
11. Ibid.
12. John Oliver LaGorce to Gilbert H. Grosvenor, Aug. 17, 1945, NGA.
13. John Oliver LaGorce, memorandum to Gilbert H. Grosvenor, Nov. 29, 1948, NGA.
14. Gilbert H. Grosvenor, memorandum to John Oliver LaGorce, Nov. 30, 1948, NGA; William H. Nichols, "Biggest Worm Farm Caters to Platypuses," *National Geographic*, Feb. 1949.
15. Mark Jenkins, "The 35mm./Kodachrome Revolution," *National Geographic Timelines—1935, 1936, 1937, 1938*, NGA.
16. Maynard Owen Williams to Gilbert H. Grosvenor, May 9, 1930, NGA.
17. Luis Marden, interviewed by author, May 18, 1995, NGA.
18. Mark Jenkins, "The 35mm./Kodachrome Revolution," *National Geographic Timeline—1935*, NGA.
19. Luis Marden, interviewed by author, Apr. 11, 1995, NGA.
20. Gilbert H. Grosvenor and Melville B. Grosvenor, quoted by Mark Jenkins, "The 35mm./Kodachrome Revolution," *National Geographic Timeline—1936*, NGA.
21. Wilbur E. "Bill" Garrett, interviewed by author, Apr. 17, 2002.
22. Mark Jenkins, "The 35mm./Kodachrome Revolution," *National Geographic Timeline—1937*, NGA.
23. Melville Bell Grosvenor to Gilbert H. Grosvenor, Mar. 23, 1945, GFP.

24. Melville Bell Grosvenor to Gilbert H. Grosvenor, Sept. 24, 1945, AGP.

25. Melville Bell Grosvenor to Gilbert H. and Elsie Bell Grosvenor, Oct. 2, 1944, GFP.

26. Gilbert H. Grosvenor, memorandum to Melville Bell Grosvenor, Sept. 12, 1953, GFP.

27. Gilbert H. Grosvenor, memorandum to Melville Bell Grosvenor, May 8, 1940, NGA.

28. John Oliver LaGorce to Gilbert H. Grosvenor, Aug. 6, 1937, NGA.

29. Gilbert M. Grosvenor, interviewed by Leah Bendavid-Val, *Stories on Paper and Glass* (Washington: National Geographic Society, 2001), 59.

30. Gilbert H. Grosvenor, memorandum to Melville Bell Grosvenor, Apr. 20, 1936, NGA; to Melville Bell Grosvenor and Jesse R. Hildebrand, July 16, 1949, GFP; to Melville Bell Grosvenor, Apr. 2, 1949, NGA.

31. Gilbert H. Grosvenor, memorandum to Melville Bell Grosvenor, Nov. 19, 1951; to Melville Bell Grosvenor, March 6, 1947, GFP; to Melville Bell Grosvenor and Thomas W. McKnew, Nov. 16, 1949, NGA.

32. Melville Bell Grosvenor, memorandum to Gilbert H. Grosvenor, Mar. 5, 1951, GFP.

33. Grosvenor Blair, interviewed by author, June 14, 2002.

34. *Melville B. Grosvenor v. Helen Rowland Grosvenor,* Equity No. 14227, Circuit Court, Montgomery County, Maryland, Aug. 29, 1949; 1–8. Hereafter *Grosvenor v. Grosvenor.*

35. *Grosvenor v. Grosvenor,* Examination of Helen Rowland Grosvenor, Dec. 16, 1949, 9.

36. Ibid.

37. Ibid.

38. Mabel H. Grosvenor, interviewed by author, Apr. 22, 2002.

39. Grosvenor Blair, interviewed by author, June 14, 2002.

40. Gilbert H. Grosvenor to Elsie Bell Grosvenor, Oct. 26, 1916, GFP.

41. Gilbert H. Grosvenor to Melville Bell Grosvenor, Feb. 22, 1948, AGP.

42. Elsie Bell Grosvenor to Gilbert H. Grosvenor, Feb. 23, 1948, GFP.

43. "Melville B. Grosvenor, Geographic Magazine Aide, Asks Divorce," *Washington Evening Star,* Aug. 30, 1949, A-3; "Son of Editor: M. B. Grosvenor Seeks Divorce; Calls Wife 'Cruel, Inhuman,'" *Washington Post,* Aug. 31, 1949, A-1; "Mrs. Grosvenor: Editor's Wife Is Granted Divorce," *Washington Post,* Jan. 18, 1950.

44. *Grosvenor v. Grosvenor,* Final Decree, Dec. 28, 1949.

45. Mabel H. Grosvenor, interviewed by author, May 21, 2002.

46. Melville Bell Grosvenor to Gilbert H. Grosvenor, July 22 and 26, 1948, AGP.

47. Gilbert H. Grosvenor, memorandum to Melville Bell Grosvenor, Oct. 27, 1948, GFP.

48. Melville Bell Grosvenor, memorandum to Gilbert H. Grosvenor, Oct. 27, 1948, GFP.

49. Marie McNair, "Weddings Take Center of Stage; Grosvenor Rite Held in Virginia," *Washington Post,* Aug. 14, 1950.

50. Anne Revis, "Mr. Jefferson's Charlottesville," *National Geographic,* May 1950.

51. Melville Bell Grosvenor, memorandum to Gilbert H. Grosvenor, Nov. 13, 1940, AGP; Nov. 21, 1949, Jan. 23, 1950, GFP.

52. Gilbert H. Grosvenor to the Board of Trustees, Apr. 1, 1954, NGA.

CHAPTER TWELVE: MELVILLE TAKES CHARGE

1. Maynard Owen Williams to Melville Bell Grosvenor, undated, probably from May 1954, when LaGorce was named editor and president, NGA.
2. Priit J. Vesilind, "National Geographic and Color Photography," Master of Arts thesis, Syracuse University, Dec. 1977; Frederick G. Vosburgh to C. D. B. Bryan, Nov. 14, 1987, NGA.
3. John Oliver LaGorce to Gilbert H. Grosvenor, Aug. 6, 1937, NGA.
4. Charles McCarry, "Three Men Who Made the Magazine," National Geographic, Sept. 1988.
5. Gilbert H. Grosvenor, interviewed by Allan C. Fisher, Jr., Aug. 25, 1962, NGA.
6. Melville Bell Grosvenor, memorandum to Gilbert H. Grosvenor, Nov. 19, 1948, NGA.
7. John Oliver LaGorce, memorandum to Gilbert H. Grosvenor, Nov. 22, 1948, NGA.
8. Frank Greve, "The Passion and the Power," Tropic, May 15, 1977, 24–30.
9. Gilbert H. Grosvenor to Melville Bell Grosvenor, Aug. 28, 1951, Jan. 24, 1955, Nov. 24, 1955, AGP.
10. Gilbert H. Grosvenor to Melville Bell Grosvenor, July 14, 1952, AGP.
11. Gilbert H. Grosvenor to Melville Bell Grosvenor, Dec. 3, 1952, AGP.
12. Gilbert H. Grosvenor to Melville Bell Grosvenor, Nov. 24, 1955, Jan. 18, 1956, AGP.
13. Gilbert H. Grosvenor to Melville Bell Grosvenor, Nov. 25 and 27, 1955, AGP.
14. Gilbert H. Grosvenor to Melville Bell Grosvenor, Nov. 25, 1955, AGP.
15. Gilbert H. Grosvenor to Melville Bell Grosvenor, Jan. 24, 1955, AGP.
16. Gilbert H. Grosvenor to Melville Bell Grosvenor, Oct. 19, 1955, AGP.
17. Mary Griswold Smith, e-mail to author, Jan. 30, 2004.
18. Wilbur E. "Bill" Garrett, interviewed by author, Apr. 17, 2002.
19. Melville B. Grosvenor to Gilbert H. Grosvenor, Feb. 20, 1952, AGP.
20. McCarry, "Three Men."
21. Gilbert M. Grosvenor, interviewed by author, July 18, 2001.
22. Mabel H. Grosvenor, interviewed by author, May 21, 2002.
23. Grosvenor Blair, interviewed by author, June 14, 2002.
24. Mabel H. Grosvenor, interviewed by author, May 21, 2002; Gilbert M. Grosvenor, interviewed by author, June 24, 2002.
25. Wilbur E. "Bill" Garrett, interviewed by author, Aug. 20, 2001.
26. Mabel H. Grosvenor, interviewed by author, May 21, 2002.
27. Gilbert M. Grosvenor and Charles Neave, "Helping Holland Rebuild Her Land," National Geographic, Sept. 1954.
28. Tom Kelly, "The Magic Mountain," Washingtonian, Nov. 1967, 36–64.
29. Melville Bell Grosvenor, interviewed by Dennis Kane, Jan. 1982, NGA; Kelly, "Magic Mountain," 62.
30. Wilbur E. "Bill" Garrett, interviewed by author, Aug. 17, 2002.
31. Jean White, "Dignified Geographic Society Bows to Coffee-Break (One) for Its Staff," Washington Post, Sept. 21, 1960.
32. A. Kent MacDougall, "National Geographic Society Members Build Big Sales for Magazine," Wall Street Journal, Apr. 24, 1966, 1.
33. Melville Bell Grosvenor, interviewed by Dennis Kane, Jan. 1982, NGA.

34. Mark Jenkins, "The MBG Era: The Board's Resolution Upon His Retirement," *National Geographic Timeline—1967,* NGA.

35. Mark Jenkins, "A Launching Pad for Growth: The Geographic, Printing Presses, and the Donnelley Years," *National Geographic Timeline—1957,* NGA.

36. Wilbur E. "Bill" Garrett, interviewed by author, Apr. 17, 2002.

37. Peter T. White, interviewed by author, May 9, 2002.

38. Luis Marden, interviewed by author, May 18, 1995, NGA.

39. Frederick G. Vosburgh, *The Century As I Saw It,* unpublished memoir, 2000, 264–65.

40. Martin Weil, "Melville Grosvenor, Former Editor of National Geographic, Dies," *Washington Post,* Apr. 24, 1982.

41. Mabel H. Grosvenor, interviewed by author, Apr. 22, 2002.

42. Greve, "The Passion and the Power," 28–30.

43. Luis Marden, interviewed by author, May 18, 1995, NGA.

44. McCarry, "Three Men."

45. Frederick G. Vosburgh, interviewed by author, Aug. 12, 2001.

46. Confidential communication.

47. Peter T. White, interviewed by author, May 9, 2002.

48. Richard E. Pearson, interviewed by author, Aug. 27, 2002.

49. William P. E. "Bill" Graves, interviewed by author, June 10, 2002; "Between the Lines," undated video produced for National Geographic editorial seminar, NGA.

50. Gilbert M. Grosvenor, interviewed by author, May 23, 2002.

51. Luis Marden, interviewed by author, May 18, 1995.

52. Vosburgh, *The Century As I Saw It,* 281–82.

53. MacDougall, "Build Big Sales."

54. Joseph E. Steptoe, interviewed by author, May 2, 2002.

55. Wilbur E. "Bill" Garrett, interviewed by author, May 16, 2002.

56. Chester Cooper, Sr., National Security Council, to Melville Bell Grosvenor, Dec. 10, 1965, confidential source; Melville Bell Grosvenor, memorandum of conversation with Richard M. Helms, Apr. 18, 1966, confidential source; Bromley K. Smith, National Security Council, to Melville Bell Grosvenor, Oct. 21, 1966, Lyndon B. Johnson Library and Museum, Austin, Texas; M. S. Kohli and Kenneth Conboy, *Spies in the Himalayas* (Lawrence: The University Press of Kansas, 2002), 1–11.

57. Wilbur E. "Bill" Garrett, interviewed by author, May 16, 2002.

58. George E. Stuart, undated interview, NGA.

59. Mabel Hubbard Bell to Alexander Graham Bell, May 12, 1899, BFP.

60. Gilbert H. Grosvenor to Melville Bell Grosvenor, Oct. 19 and Nov. 25, 1955, AGP.

61. Gilbert H. Grosvenor to Melville Bell Grosvenor, Nov. 18, 1962, AGP.

62. Cited by Grosvenor Blair, "My Grandmother," unpublished memoir, Mar. 12, 1998.

63. Grosvenor Blair, interviewed by author, June 11, 2002.

64. William P. E. "Bill" Graves, interviewed by author, Aug. 23, 2002; Mabel H. Grosvenor, interviewed by author, Dec. 19, 2002.

CHAPTER THIRTEEN: THE BEST OF TIMES

1. Luis Marden, "I Found the Bones of the *Bounty*," *National Geographic,* Dec. 1957.
2. Thomas J. Abercrombie, interviewed by Mark Jenkins, Feb. 11, 1995, NGA.
3. Gilbert M. Grosvenor, interviewed by Mark Jenkins, July 22, 1997, NGA; Virginia Morell, *Ancestral Passions: The Leakey Family and the Quest for Human Beginnings* (New York: Simon & Schuster, 1995), 195.
4. Luis Marden, interviewed by author, May 18, 1995, NGA.
5. Marden, "I Found the Bones of the *Bounty*."
6. Luis Marden, interviewed by author, Apr. 27, 1995, NGA.
7. Marden, "I Found the Bones of the *Bounty*."
8. James Conaway, "Inside the Geographic," *Washington Post,* Dec. 18, 1984, B-1.
9. Luis Marden, interviewed by author, Apr. 27, May 18, 1995, NGA; Mark Jenkins, "Origins of NGS Television," *National Geographic Timeline—1958*; Lawrence Laurent, Television column, *Washington Post,* Feb. 11, 1958.
10. Gilbert M. Grosvenor, interviewed by Mark Jenkins, Mar. 19, 1998, NGA.
11. Alexander Wetmore, *Song and Garden Birds of North America* (Washington: The National Geographic Society, 1964), 267.
12. W. Robert Moore, "Cities of Stone in Utah's Canyonland," *National Geographic,* May 1962, 664.
13. Malcolm Ross, "North Carolina, Dixie Dynamo," *National Geographic,* Feb. 1962, 167.
14. *Newsweek,* "Dear Aunt Sally," Nov. 11, 1963, 82.
15. Allan C. Fisher, Jr. to Anne R. Grosvenor, May 18, 1982, AGP.
16. Anne R. Grosvenor to Elsie Bell Grosvenor, Feb. 11, 1959, AGP; Joseph R. Judge, "The Greatest Job in the World," *National Geographic,* Sept. 1988; Mary Leakey to Anne R. Grosvenor, June 4, 1982, AGP.
17. Anne R. Grosvenor to Elsie Bell Grosvenor, Feb. 11, 1959, GFP.
18. Mabel H. Grosvenor, interviewed by author, Apr. 22, 2002; Bart McDowell to Anne R. Grosvenor, Apr. 23, 1982, AGP.
19. Melville Bell Grosvenor to Ralph Mooney, Oct. 1, 1951, AGP.
20. Anne R. Grosvenor to Gilbert H. and Elsie Bell Grosvenor, Feb. 6, 1958, GFP.
21. Melville Bell Grosvenor to Elsie Grosvenor, Mar. 31, 1955, GFP.
22. Melville Bell Grosvenor to Gilbert H. Grosvenor, May 22, 1958, AGP.
23. Anne R. Grosvenor to Gilbert H. and Elsie Bell Grosvenor, Apr. 22, 1958, GFP.
24. Ada Louise Huxtable, "National Geographic Society's Building Sets a Standard for Washington," *New York Times,* Dec. 10, 1963.
25. Tom Kelly, "The Magic Mountain," *Washingtonian,* Nov. 1967.
26. Kelly, "The Magic Mountain"; "The World of National Geographic," *Metropolitan,* Mar.–Apr. 1972; A. Kent MacDougall, "National Geographic Society Members Build Big Sales for Magazine," *Wall Street Journal,* Apr. 24, 1966, 1.
27. MacDougall, "Big Sales."
28. Norman G. Dyhrenfurth, "Scientific Program for the American Mount Everest Expedition." Application to National Geographic Committee for Research and Exploration, Feb. 25, 1962, NGA.

29. Andrew H. Brown, memorandum of meeting with Melville Grosvenor, Norman Dyhrenfurth, and others at National Geographic, Mar. 23, 1962; Norman Dyhrenfurth to Andrew H. Brown, Feb. 12, 1962; "Minutes of Meeting, Committee for Research and Exploration," May 7, 1962, NGA; James Ramsey Ullman, *Americans on Everest* (Philadelphia and New York: J. P. Lippincott, 1964).

30. Ullman, *Americans on Everest*.

31. Ibid., 107–10.

32. Ibid., 182–89.

33. Ibid., 372; Barry C. Bishop, "How We Climbed Everest," *National Geographic,* Oct. 1963; Luther G. Jerstad, "Reaching Beyond the Grasp," *National Observer,* Mar. 11, 1968, 26.

34. Bishop, "How We Climbed Everest"; Barry C. Bishop, interviewed by Matt C. McDade, June 5, 1963, NGA.

35. Barry C. Bishop, interviewed by Matt C. McDade, June 5, 1963, NGA; Barry C. Bishop, lecture on "Exploration," summer 1991, NGA; Thomas F. Hornbein, *Everest: The West Ridge* (Seattle: The Mountaineers, 1999), 172–81; Ullman, *Americans on Everest, 274–81.*

36. Walt Unsworth, *Everest: A Mountaineering History* (Boston: Houghton Mifflin, 1981).

37. Jon Krakauer, *Into Thin Air* (New York: Anchor Books, 1998).

38. Finn-Olaf Jones, "Into Finn Air," *Forbes FYI,* Winter 2001, 55.

39. Cathy Hunter, "Hubbard Medal to the Members of the American Mount Everest Expedition," *National Geographic Timeline—1963,* NGA.

40. Barry C. Bishop to Norman Hardie, Aug. 22, 1963, NGA.

41. *National Geographic,* Oct. 1963.

42. Barry C. Bishop to Sayre Rodman, June 20, 1963, NGA.

43. M. S. Kohli and Kenneth Conboy, *Spies in the Himalayas* (Lawrence: University Press of Kansas, 2002), 43.

44. Thomas F. Hornbein, interviewed by author, Oct. 1, 2002.

45. Rob Schaller, interviewed by author, Mar. 17, 2003; Kohli and Conboy, *Spies,* 1–11; Howard Kohn, "The Nanda Devi Caper," *Outside,* May 1978.

46. Melville Bell Grosvenor, memorandum of conversation with Richard M. Helms, Apr. 18, 1966, confidential source; Chester Cooper, Sr., National Security Council, to Melville Bell Grosvenor, Dec. 10, 1965, confidential source; Bromley K. Smith, National Security Council, to Melville Bell Grosvenor, Oct. 21, 1966, Lyndon B. Johnson Library and Museum, Austin, Texas; Barry C. Bishop, "Application to the National Science Foundation," Doctoral Dissertation Research Grant, Nov. 27, 1967, NGA.

47. Rob Schaller, interviewed by author, Mar. 17, 2003; Kohli and Conboy, *Spies,* 31–48.

48. Kohli and Conboy, *Spies,* 64.

49. Rob Schaller, interviewed by author, Mar. 17, 2003.

50. Kohli and Conboy, *Spies,* 99.

51. Kohli and Conboy, *Spies,* 212–13; Rob Schaller, interviewed by author, Mar. 17, 2003; Kohn, "Nanda Devi Caper."

52. "India Irked at Report U.S. Nuclear Device Was Left on a Peak," Associated Press report in *New York Times,* Apr. 14, 1978, A-9.

53. "N-device installed with India's approval," *Times of India,* Apr. 18, 1978, A-1.

54. Kohli and Conboy, *Spies,* 204.

55. Melville Bell Grosvenor, confidential memorandum to Personnel Department, Dec. 10, 1965, NGA.

56. Robert Doyle, interviewed by Renee Braden, Dec. 6, 1998, NGA.

57. Melville Bell Grosvenor, speech in Fiji on U.S. National Parks, 1969, AGP.

CHAPTER FOURTEEN: FACING LIFE

1. William P. E. "Bill" Graves, interviewed by author, June 10, 2002; Mary Griswold Smith, e-mail to author, Jan. 30, 2004; Gilbert M. Grosvenor, interviewed by author, June 4, 2002.

2. Gilbert M. Grosvenor, interviewed by author, June 4, 2002.

3. Melville Bell Grosvenor to William O. Douglas, Oct. 17, 1967, AGP.

4. Gilbert M. Grosvenor, interviewed by author, June 4, 2002.

5. C. D. B. Bryan, *The National Geographic Society* (New York: Abrams, 1987), 379.

6. Melvin M. Payne, transcript of undated interview by C. D. B. Bryan, NGA.

7. William P. E. "Bill" Graves, interviewed by author, June 10, 2002. The capillary quote, now in the realm of legend at National Geographic, has been attributed to several authors, but the most frequently mentioned are Howard LaFay and Franc Shor, both of whom wrote for the magazine.

8. Charles McCarry, "Three Men Who Made the Magazine," *National Geographic,* Sept. 1988.

9. Gilbert M. Grosvenor, interviewed by author, June 24, 2002.

10. Ethel Marden, interviewed by author, Jan. 4, 2002.

11. Gilbert M. Grosvenor, interviewed by author, June 24, 2002.

12. Gilbert M. Grosvenor, interviewed by author, June 24, 2002; William P. E. "Bill" Graves, interviewed by author, June 10, 2002.

13. Gilbert M. Grosvenor, interviewed by author, June 24, 2002.

14. Ibid.

15. William P. E. "Bill" Graves, interviewed by author, June 10, 2002.

16. Gilbert M. Grosvenor, interviewed by author, June 24, 2002.

17. Richard E. Pearson, interviewed by author, Aug. 27, 2002.

18. Edwards Park, quoted in "Between the Lines," transcript of an undated video, NGA.

19. Melvin M. Payne, transcript of undated interview by C. D. B. Bryan, NGA.

20. William P. E. "Bill" Graves, interviewed by author, Aug. 23, 2003.

21. Gilbert M. Grosvenor, interviewed by author, June 24, 2003.

22. Joyce Graves, interviewed by author, Aug. 23, 2002.

23. Elsie Bell Grosvenor to Melville and Anne Grosvenor, Sept. 1960; Mary Griswold Smith, e-mail to author, Jan. 30, 2004.

24. Gifford D. Hampshire, interviewed by author, July 26, 2002.

25. Ibid.; Wilbur E. "Bill" Garrett, interviewed by author, June 6, 2002.

26. Gilbert M. Grosvenor, interviewed by author, May 23, 2002.

27. "Donna Kerkam, Sweet Briar '60, to Wed June 16," *New York Times,* Apr. 16, 1961; "Miss Kerkam to Wed Mr. Gilbert Grosvenor," *Washington Sunday Star,* Apr. 16, 1961.

28. Boris Weintraub, "Searching for the Mystical Blue Whale," *Washington Star,* Oct. 28, 1976, D-3.

29. Sarah Booth Conroy, "Close to Outdoors, But Not Really," *Washington Post,* July 15, 1973, H-1.

30. Gifford D. Hampshire, interviewed by author, July 26, 2002.

31. Gilbert M. Grosvenor, interviewed by Mark Jenkins, Mar. 19, 1998, NGA.

32. Frederick G. Vosburgh, *The Century As I Saw It,* unpublished memoir, 2000, 314; Bryan, *The National Geographic Society,* 379; Ibid., 378.

33. Carolyn Bennett Patterson, *Of Lands, Laughter, and Legends* (Golden, Col.: Fulcrum Publishing, 1998), 47; Mark Jenkins, "Clearing Hyenas: A Story from the Washington Riots," *National Geographic Timeline—1968,* NGA; Vosburgh, *The Century As I Saw It,* 327–29.

34. Vosburgh, *The Century As I Saw It,* 334–35; Melville Bell Grosvenor, "North Through History Aboard *White Mist,*" *National Geographic,* July 1970.

35. Gilbert M. Grosvenor, interviewed by author, May 23, 2002.

36. Tom Buckley, "With the National Geographic on Its Endless, Cloudless Voyage," *New York Times Magazine,* Sept. 6, 1970.

37. Gilbert M. Grosvenor, interviewed by author, June 24, 2002.

38. Melvin M. Payne, memorandum to Gilbert M. Grosvenor, June 15, 1973, NGA.

39. Gordon Young, "Pollution: Threat to Man's Only Home," *National Geographic,* Dec. 1970.

40. Gilbert M. Grosvenor, editor's page, *National Geographic,* Nov. 1974; Tom Shales, "A 'Geographic' Correction," *Washington Post,* Sept. 20, 1974, B-6.

41. Conrad Wirth et al., "Report on the Continuity of Management," Apr. 8, 1976, NGA.

42. Robert C. Seamans, Jr., *Aiming at Targets* (Washington, D.C.: NASA, 1996), 225–27; Conrad Wirth et al., "Continuity Report," NGA.

43. A. Kent MacDougall, "Geographic: From Upbeat to Realism," *Los Angeles Times,* Aug. 5, 1977, A-1.

44. Edwin S. Grosvenor, interviewed by author, June 7, 2003.

45. Lloyd Elliot, quoted in "Minutes of Board of Trustees Meeting," June 9, 1977, NGA.

46. Bryan, *The National Geographic Society,* 392; Melville Bell Grosvenor, quoted in "Minutes of Board of Trustees Meeting," June 9, 1977, NGA.

47. Frederick G. Vosburgh, quoted in "Minutes of Board of Trustees Meeting," June 9, 1977, NGA.

48. Crawford H. Greenewalt, quoted in "Minutes of Board of Trustees Meeting," June 9, 1977, NGA.

49. Earl Warren, quoted in "Minutes of Board of Trustees Meeting," June 9, 1977, NGA.

50. Kathy Sawyer, "Change at the Geographic," *Washington Post,* July 17, 1977, A-1.

51. MacDougall, "From Upbeat to Realism"; Tony Schwartz with Jane Whitmore, "The Geographic Faces Life," *Newsweek,* Sept. 12, 1977, 111.

52. MacDougall, "From Upbeat to Realism."

53. Gilbert M. Grosvenor, interviewed by author, June 26, 2002.

54. Gilbert M. Grosvenor, quoted in "Minutes of Meeting, Ad Hoc Committee on Editorial Policy," June 27, 1977, NGA.

55. Lloyd Elliott, quoted in "Minutes of Meeting, Ad Hoc Committee on Editorial Policy," Aug. 11, 1977, NGA.

56. Melvin M. Payne, quoted in "Minutes of Meeting, Ad Hoc Committee on Editorial Policy," Aug. 11, 1977, NGA.

57. Crawford H. Greenewalt, quoted in "Minutes of Meeting, Ad Hoc Committee on Editorial Policy," Aug. 11, 1977, NGA.

58. Gilbert M. Grosvenor, Melvin M. Payne, and Robert E. Doyle, "Reaffirmation of Editorial Policy," National Geographic, Jan. 1978.

59. Gilbert M. Grosvenor, interviewed by author, June 24, 2002.

60. Gilbert M. Grosvenor, interviewed by Mark Jenkins, May 10, 2000, NGA.

61. MacDougall, "From Upbeat to Realism."

62. Gilbert M. Grosvenor, interviewed by author, June 24, 2002.

63. Joyce Graves, interviewed by author, Aug. 23, 2002.

CHAPTER FIFTEEN: BROKEN FRIENDSHIP

1. Gilbert M. Grosvenor, interviewed by author, June 24, 2002.

2. Gilbert M. Grosvenor, interviewed by C. D. B. Bryan, Oct. 7, 1986, NGA.

3. Ibid.

4. Gilbert M. Grosvenor, interviewed by Mark Jenkins, Mar. 1998.

5. Gilbert M. Grosvenor, interviewed by author, June 24, 2002.

6. Gilbert M. Grosvenor, interviewed by Mark Jenkins, Sept. 17, 1998, NGA.

7. John M. Keshishian, interviewed by author, Feb. 9, 2004.

8. Allen Weinstein, quoted in Louis Jacobson, "International Geographic," Washington City Paper, Jan. 9, 1998.

9. Frederick G. Vosburgh, interviewed by author, Aug. 12, 2001.

10. Wilbur E. "Bill" Garrett, interviewed by author, Aug. 20, 2001.

11. Ibid.

12. Jake Newman and Tom Zito, "Personalities," Washington Post, July 11, 1980.

13. Wilbur E. "Bill" Garrett, interviewed by author, Aug. 20, 2001.

14. Kenneth L. Garrett, interviewed by author, Mar. 14, 2004.

15. Wilbur E. "Bill" Garrett, interviewed by author, Feb. 4, 2004.

16. Mary Griswold Smith, e-mail to author, Jan. 30, 2004.

17. Wilbur E. "Bill" Garrett, interviewed by author, May 16, 2002.

18. John M. Keshishian, interviewed by author, Feb. 9, 2004.

19. William P. E. "Bill" Graves, interviewed by author, June 6, 2002.

20. Grace Glueck, "Art People," New York Times, Mar. 23, 1979; Josine Ianco-Starrels, "Portfolio Makes Debut," Los Angeles Times, Apr. 8, 1979.

21. Anne R. Grosvenor to Henry King Stanford, June 5, 1982, AGP.

22. Robert C. Seamans, Jr., "Melville Bell Grosvenor 1901–1982," Cruising Club News, May 1982, 23.

23. Mark Jenkins, "1981, Society Membership: 10,861,186," "1982, Society Membership: 10,610,000," National Geographic Timeline, NGA.

24. Wilbur E. "Bill" Garrett, interviewed by author, Apr. 17, 2002.

25. Wilbur E. "Bill" Garrett, interviewed by author, Aug. 20, 2001.

26. Margot Gibb-Clark, "The Features Page," *Toronto Globe and Mail,* Apr. 26, 1986, A-10; "On Assignment," *National Geographic,* July 1986.

27. George E. Stuart, interviewed by author, Feb. 16, 2004.

28. Charles McCarry, e-mail to author, Feb. 27, 2004.

29. Wilbur E. "Bill" Garrett, interviewed by author, Aug. 20, 2001.

30. Charles McCarry, e-mail to author, Feb. 27, 2004.

31. Charles McCarry, memorandum to Wilbur E. "Bill" Garrett, Feb. 5, 1990.

32. James Conaway, "Inside the Geographic," *Washington Post,* Dec. 18, 1984.

33. John M. Keshishian, interviewed by author, Feb. 9, 2004.

34. Gilbert M. Grosvenor, interviewed by C. D. B. Bryan, Oct. 7, 1986, NGA; C. D. B. Bryan, *The National Geographic Society* (New York: Abrams, 1987), 430–33.

35. Howard E. Paine, interviewed by author, Feb. 4, 2004.

36. William Hoffer, "All the World's a Page," *Regardie's,* Mar. 1985, 51–64.

37. Gilbert M. Grosvenor to Charles Hunt, Chancellor, University of California, Los Angeles, Dec. 16, 1982, NGA.

38. Gilbert M. Grosvenor, "Those Panamanian Pandas," *New York Times,* July 31, 1988, E-25.

39. Katherine Long, "On Top of the World," *Washington Post,* July 4, 1997.

40. Gilbert M. Grosvenor, interviewed by Mark Jenkins, Oct. 14, 1998, NGA.

41. Hoffer, "All the World's a Page"; Howard S. Abramson, "Geographic Travels Tax-Free: Nonprofit Status Gives Magazine Competitive Edge," *Washington Post,* Mar. 8, 1987, H-1.

42. Gilbert M. Grosvenor, interviewed by author, May 23, 2002; interviewed by Mark Jenkins, Sept. 17, 1998, NGA; Charles Trueheart, "Garrett, Grosvenor, and the Great Divide," *Washington Post,* May 7, 1990, C-1.

43. Gilbert M. Grosvenor, interviewed by author, May 23, 2002; Charles Trueheart, "Austerity Drive at National Geographic," *Washington Post,* Dec. 9, 1988, C-1.

44. Trueheart, "Austerity Drive."

45. Wilbur E. "Bill" Garrett, interviewed by author, May 16, 2002.

46. Howard Means, "Earthquake," *Washingtonian,* Dec. 1990.

47. Wilbur E. "Bill" Garrett, interviewed by author, Apr. 17, 2002.

48. Gilbert M. Grosvenor, interviewed by Mark Jenkins, Sept. 17, 1998, NGA.

49. Trueheart, "Great Divide."

50. Gilbert M. Grosvenor, interviewed by Mark Jenkins, Sept. 17, 1998, NGA.

51. John Keshishian, interviewed by author, Feb. 9, 2004.

52. Sam Abell, interviewed by author, Sept. 20, 2002.

53. John M. Keshishian, interviewed by author, Feb. 9, 2004.

54. Ibid.

55. William P. E. "Bill" Graves, interviewed by author, July 25, 2002; Robert E. Gilka, interviewed by author, May 13, 2002; Charles Trueheart, "Great Divide."

56. William P. E. "Bill" Graves, interviewed by author, Aug. 23, 2002; John M. Keshishian, interviewed by author, Feb. 9, 2004; Robert E. Gilka, interviewed by author, May 13, 2002; Trueheart, "Great Divide."

57. Jon T. Schneeberger, interviewed by author, Feb. 4, 2004.

58. John M. Keshishian, interviewed by author, Feb. 9, 2004; Trueheart, "Great Divide."

59. William P. E. "Bill" Graves, interviewed by author, July 25 and Sept. 20, 2002; Howard E. Paine, interviewed by author, Feb. 4, 2004.

60. Howard E. Paine, interviewed by author, Feb. 4, 2004.

61. William P. E. "Bill" Graves, interviewed by author, July 25 and Aug. 23, 2002; Howard E. Paine, interviewed by author, Feb. 4, 2004.

62. Thomas Y. Canby, *From Botswana to the Bering Sea* (Washington, D.C.: Island Press, 1998), 206–10.

63. Howard E. Paine, interviewed by author, Feb. 4, 2004.

CHAPTER SIXTEEN: THE THIRD FLOWERING

1. Gilbert M. Grosvenor, interviewed by author, July 19, 2001.

2. William P. E. "Bill" Graves, interviewed by author, June 10, 2002.

3. Charles Trueheart, "Garrett, Grosvenor, and the Great Divide," *Washington Post,* May 7, 1990, C-1.

4. Howard Means, "Earthquake," *Washingtonian,* Dec. 1990.

5. Mary Griswold Smith, e-mail to author, Jan. 30, 2004.

6. Richard E. Pearson, interviewed by author, Aug. 27, 2002.

7. Mary Griswold Smith, e-mail to author, Jan. 30, 2004.

8. Gilbert M. Grosvenor, interviewed by Mark Jenkins, Sept. 17, 1998, Apr. 6, 1999, Feb. 21, 1999, NGA.

9. William P. E. "Bill" Graves, "Corregidor Revisited," *National Geographic,* July 1986.

10. William P. E. "Bill" Graves, address to the staff, Apr. 18, 1990; Trueheart, "Great Divide."

11. Trueheart, "Great Divide."

12. William P. E. "Bill" Graves, interviewed by author, July 17, 2002.

13. Gilbert M. Grosvenor, interviewed by Mark Jenkins, Sept. 17, 1998, NGA.

14. Gilbert M. Grosvenor, interviewed by author, May 23, 2002; interviewed by Mark Jenkins, Feb. 2, 1999.

15. Constance L. Hays, "Seeing Green in a Yellow Border," *New York Times,* Aug. 3, 1997, F-1.

16. Mabel H. Grosvenor, interviewed by author, May 21, 2002.

17. Mary Griswold Smith, e-mail to author, Jan. 30, 2004.

18. Adam Bellow, *In Praise of Nepotism* (New York: Doubleday, 2003).

19. Gilbert M. Grosvenor, interviewed by author, May 23, 2002; interviewed by Mark Jenkins, Feb. 2, 1999.

20. Susan Baer, "Reg Murphy to be No. 2 at Geographic," *Baltimore Sun,* Apr. 9, 1993; "New Blood Line at Geographic?" *Washingtonian,* June 1993.

21. Tim Wheeler, interviewed by author, Mar. 5, 2004.

22. Hays, "Seeing Green."

23. IRS Form 990, "Return of Organization Exempt From Income Tax," National Geographic Society, 1996, 2002.

24. Gilbert M. Grosvenor, interviewed by author, May 23, 2002.

25. National Geographic Society, IRS Form 990 filing for 1998, Nov. 11, 1999.

26. Gilbert M. Grosvenor, interviewed by author, May 23, 2002.

27. Gilbert M. Grosvenor, interviewed by Mark Jenkins, Feb. 2, 1999.

28. John M. Fahey, Jr., interviewed by author, Mar. 15, 2004.

29. Hays, "Seeing Green."

30. Joyce Graves, interviewed by author, Aug. 23, 2002.

31. National Geographic Society, IRS Form 990 for 1998, Nov. 11, 1999.

32. Constance L. Hays, "National Geographic Society Chief Resigns," *New York Times,* Dec. 1997; "Reg Murphy Resigns as National Geographic Chief Executive," Associated Press, Dec. 11, 1997; Kate Tyndall, "Catching up with Reg," *Style,* Jan.–Feb. 1998.

33. Gilbert M. Grosvenor, interviewed by author, June 24, 2002.

34. Chuck Salter, "New Frontiers to Explore," *Fast Company,* Sept. 9, 2001, 76.

35. John M. Fahey, Jr., interviewed by author, Mar. 1, 2004.

36. Christopher A. Liedel, interviewed by author, Mar. 15, 2004; John M. Fahey, Jr., address to employees, Jan. 22, 2004; Terrence B. Adamson, interviewed by author, Mar. 1, 2004; John M. Fahey, Jr., interviewed by author, Mar. 1, 2004; National Geographic Society, IRS Form 990 for 2000, Nov. 13, 2003.

37. David B. Ottaway and Joe Stephens, "Nonprofit Land Bank Amasses Billions," *Washington Post,* May 4, 2003, A-1; Elizabeth Schwinn and Ian Wilhelm, "Nonprofit CEO's See Salaries Rise," *The Chronicle of Philanthropy,* Oct. 2, 2003; John M. Fahey, Jr., interviewed by author, Mar. 1, 2004.

38. John M. Fahey, Jr., interviewed by author, Mar. 1, 2004.

39. Terrence B. Adamson, e-mail to author, Mar. 14, 2004.

40. John M. Fahey, Jr., interviewed by author, Mar. 15, 2004.

41. Terrence B. Adamson, e-mail to author, Mar. 14, 2004.

42. Ibid.

43. John M. Fahey, Jr., interviewed by author, Mar. 15, 2004.

44. John M. Fahey, Jr., address to employees, Jan. 22, 2004, Feb. 14, 2003, Feb. 7, 2004.

45. National Geographic Society, IRS Form 990 filings, 1999, 2000, 2001, 2002.

46. John M. Fahey, Jr. address to employees, Jan. 22, 2004.

47. Gilbert M. Grosvenor, interviewed by Mark Jenkins, Feb. 2, 1999, NGA.

48. John M. Fahey, Jr., address to employees, Jan. 22, 2004.

49. John M. Keshishian, interviewed by author, Feb. 9, 2004; Robert E. Gilka, interviewed by author, May 13, 2004.

50. Wilbur E. "Bill" Garrett, interviewed by author, Feb. 4, 2004.

51. Wilbur E. "Bill" Garrett, interviewed by author, Feb. 4, 2004; Kenneth L. Garrett, interviewed by author, Mar. 14, 2004.

ILLUSTRATION CREDITS

INDEX

FOR THE BEST IN PAPERBACKS, LOOK FOR THE

In every corner of the world, on every subject under the sun, Penguin represents quality and variety—the very best in publishing today.

For complete information about books available from Penguin—including Penguin Classics, Penguin Compass, and Puffins—and how to order them, write to us at the appropriate address below. Please note that for copyright reasons the selection of books varies from country to country.

In the United States: Please write to *Penguin Group (USA), P.O. Box 12289 Dept. B, Newark, New Jersey 07101-5289* or call 1-800-788-6262.

In the United Kingdom: Please write to *Dept. EP, Penguin Books Ltd, Bath Road, Harmondsworth, West Drayton, Middlesex UB7 0DA.*

In Canada: Please write to *Penguin Books Canada Ltd, 90 Eglinton Avenue East, Suite 700, Toronto, Ontario M4P 2Y3.*

In Australia: Please write to *Penguin Books Australia Ltd, P.O. Box 257, Ringwood, Victoria 3134.*

In New Zealand: Please write to *Penguin Books (NZ) Ltd, Private Bag 102902, North Shore Mail Centre, Auckland 10.*

In India: Please write to *Penguin Books India Pvt Ltd, 11 Panchsheel Shopping Centre, Panchsheel Park, New Delhi 110 017.*

In the Netherlands: Please write to *Penguin Books Netherlands bv, Postbus 3507, NL-1001 AH Amsterdam.*

In Germany: Please write to *Penguin Books Deutschland GmbH, Metzlerstrasse 26, 60594 Frankfurt am Main.*

In Spain: Please write to *Penguin Books S. A., Bravo Murillo 19, 1° B, 28015 Madrid.*

In Italy: Please write to *Penguin Italia s.r.l., Via Benedetto Croce 2, 20094 Corsico, Milano.*

In France: Please write to *Penguin France, Le Carré Wilson, 62 rue Benjamin Baillaud, 31500 Toulouse.*

In Japan: Please write to *Penguin Books Japan Ltd, Kaneko Building, 2-3-25 Koraku, Bunkyo-Ku, Tokyo 112.*

In South Africa: Please write to *Penguin Books South Africa (Pty) Ltd, Private Bag X14, Parkview, 2122 Johannesburg.*